W9-DJM-116

HANDBOOK OF PRECISION ENGINEERING

HANDBOOK OF PRECISION ENGINEERING

Volume 1 Fundamentals
Volume 2 Materials
Volume 3 Fabrication of Non-metals
Volume 4 Physical and Chemical Fabrication Techniques
Volume 5 Joining Techniques
Volume 6 Mechanical Design Applications
Volume 7 Electrical Design Applications
Volume 8 Surface Treatment
Volume 9 Machining Processes
Volume 10 Forming Processes
Volume 11 Production Engineering
Volume 12 Precision Measurement

Philips Technical Library

HANDBOOK OF PRECISION ENGINEERING

Edited by A. Davidson

Volume 5 Joining Techniques

McGraw-Hill Book Company
New York St. Louis San Francisco Montreal Toronto

English edition published by McGraw-Hill, Inc., 1972.

Library of Congress Catalog Card Number 75-137567

7-015472-4

First published in English by
MACMILLAN AND CO LTD
London and Basingstoke
Associated companies in New York, Toronto, Melbourne, Dublin, Johannesburg and Madras

PHILIPS

Printed in Great Britain by The Whitefriars Press Ltd., London and Tonbridge

Foreword

During the last twenty or thirty years, precision engineering, although not strictly a separate branch of engineering, like shipbuilding and the aircraft industry, has emerged as a specialized technology, involving methods and ideas foreign to conventional mechanical engineering.

Precision-engineered parts (or fine mechanisms, as they are often termed) are very often of limited size, and vary in quantity from one-off to very long runs. But they don't have to be small, of course. For example, the driving mechanism of an observatory telescope, which is fairly large, has to be designed and constructed with extreme precision, whilst an ordinary watch, which is quite small by comparison, is relatively crude as far as tolerances are concerned.

Examples of fine mechanisms are to be found in telecommunications equipment; optical apparatus such as microscopes and comparators, as well as cameras and projectors; office machines like typewriters, accounting and calculating machines; various toys of a technical nature, mentioned because of their often ingenious design and the consummate tool-making that goes into them; electric shavers; and various electronic equipment used in the home.

There is a need, in industry and in training schools, for a Handbook such as this, covering, in a convenient form, the different subjects that are related to the design of fine mechanisms. Much of the material dealt with is not new in itself, of course, but, because it is presented specifically in the context of precision engineering, it should be invaluable to designers, manufacturers and users of precision-engineered parts.

The authors, each of whom is a specialist in his own field, do not claim to have covered the individual subjects fully in the chapters assigned to them. They have presented the main features as clearly as possible, and provided international references from which the reader can obtain more details as required.

The Handbook is divided into twelve parts, of which Volume 1 deals with the general principles to be observed when designing a product, and Volume 2 with the materials available. Theoretical and practical data useful to designers and draughtsmen is given.

Volumes 3–6 discuss production techniques, and production staff will find them to contain comprehensive information on production methods and plant. Volumes 7–12 deal with the practical design of precision-engineered products, and the equipment and components required for this purpose. Designers, manufacturers and users can all profit from them.

The editor would like to take this opportunity of thanking the authors and all who have helped to produce the Handbook.

Eindhoven, 1968 A. Davidson

Contents

Introduction

A. Davidson

There are a number of aspects of manufacturing methods used in precision engineering which are hardly ever dealt with in summaries given in the literature. This provides sufficient reason to devote this section to the techniques of making permanent joints, since these play such an important part in precision engineering. Small and medium sized products, especially in series and mass production, are often mounted or assembled by means of permanent joints. By permanent joints is understood the type of joint which can only be loosened destructively. The methods of mechanical joining, welding, soldering and adhesive bonding as specifically used in precision engineering are described separately.

As a satisfactory survey of wire winding techniques is not to be found in literature, a special chapter is devoted to this subject especially in connection with electric and electronic applications. The winding of wire springs, however, is described in Volume 11 since it belongs rather to the plastic forming techniques.

The methods of applying texts, etc., are discussed in the last chapter. For this there are many techniques which can usually be reduced to those already described in the other volumes. Nevertheless it will be useful to give a survey specially concerned with the application of texts, printings, etc., while various applications and decoration techniques are also dealt with. The chapters of this book are not intended for those intensively engaged on the various techniques. For these people there are usually specialised books and magazine articles in which appropriate reference is made in the list of literature.

The units and symbols used are those of the Practical Unit System (ISO) as explained in volume 1, chapter 1.

Chapter 1

Mechanical Joints (Permanent)

C. A. Nottrot

1.1 Flanging

1.1.1. General

A flanged joint is a permanent joint produced by plastic deformation of the connecting medium. The following considerations play a part in selecting a flanged joint:

It is a simple and cheap operation.

It has hardly any effect on the characteristics of the basic material.

It is possible to join different materials together.

It calls for no special preparation of the materials to be joined, and no after-treatment of the joint.

Machine (tool) costs are low.

However, the strength of the joint is low, and there is a risk of damaging the material to be jointed.

The following components can be used for flanging: tubular rivets (usually countersunk at the end), connecting bushes, rods or pins with countersunk ends.

The materials commonly used are free cutting steel (with or without a lead content), brass and aluminium.

Occasionally, free cutting steel without a lead content shows faults and among a large number of satisfactory joints, there will be one that is inferior. In general, free cutting steel without lead and aluminium are looked upon as the materials most difficult to flange.

The average strength of a flanged joint decreases in the sequence: steel without a lead content, steel with a lead content, brass, aluminium.

With the proper tools, these materials can be used without subsequent annealing. There is a risk of product distortion with annealing so after-treatment may be necessary.

Fixing bushes or tubular rivets are used whenever possible, as they are cheaper than solid rivets.

A satisfactory flanged joint can be obtained by matching the tools as well as the wall thickness (s) and the free height (l) of the edge (see Fig. 1.1).

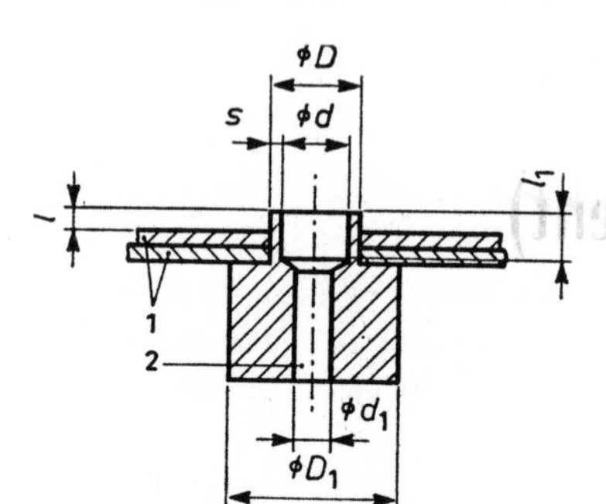

Fig. 1.1. Drawing of a joint before flanging.
1 = flanging plate.
2 = hole or screw thread.

It is clear that a compromise will have to be made. When working with the correct punch profile and a wall thickness of 0·6 mm (0·5 mm for bushes with a diameter $D = 2$–3 mm), and a free edge height (l) of 0·9 mm, a satisfactory flange quality is obtained by either pressure or rotary flanging (spinning). Under these conditions, the quality of the joint depends but little on the material, diameter and flanging time. The nominal limit values of the free edge height are 0·8 and 1·1 mm respectively. If the flange edge is too thin or too low, the bead profile cannot be filled completely. It is then impossible to obtain a sufficiently strong joint.

Too large a flange volume gives rise to burrs. Both faults can also be due to a fault in the construction of the flanging tools.

The most important dimension of the *bush* is the diameter of its hole (d). This dimension is decisive for the position where the tool takes effect, and deformation sets in. Theoretically, a close tolerance would be required, but an E11 fit is found to be sufficient. The internal depth of the flange (l_1) should be at least large enough to keep the lower plane of the central part of the punch clear from the bottom, even when the outer edge of the punch is in contact with the surface surrounding the flange after curling.

As a rule, the diameter of the flange (after curling) will not be greater than the outer diameter $(D)+1{\cdot}5$ mm, the height being about 0·5 mm.

It is best to construct flanged joints so that there are no tensile forces in the flange.

The dimensions of drawn products (soldering lugs, contact eyelets, fixing bushes, etc.) are never stable, and they do not comply with the above-mentioned requirements, since a greater wall thickness involves a higher price. Fixing bushes with a seam are not suitable.

In case of a fit or threaded hole, the diameter should not become less due to the flanging process. Therefore, the edge below the flange is made thicker ($D_1 - d_1$). The pressures occurring during rotary flanging are very small compared with those during pressure flanging. It is necessary, therefore, for this thickening to be heavier for pressure-flanged joints. In certain cases, where this diameter cannot be maintained and pressure flanging is required, the wall thickness will be 0·6 instead of 0·5 or 0·4 mm (at the expense of the outer diameter (D), so that it is possible to use the same tools). The quality and strength of the joints, however, are no longer a maximum (the effect of the smaller wall thickness is greater in rotary flanging than in pressure flanging).

1.1.2 Pressure flanging

In contrast to what is generally believed, the flange is not folded over in pressure flanging: the material is merely forced aside (see Fig. 1.2).

The greater the diameter, the less the deformation. The advantage of pressure flanging is the short flanging time, and that the operation can be carried out in multiple. The principle of the flanging tool is shown in Fig. 1.3.

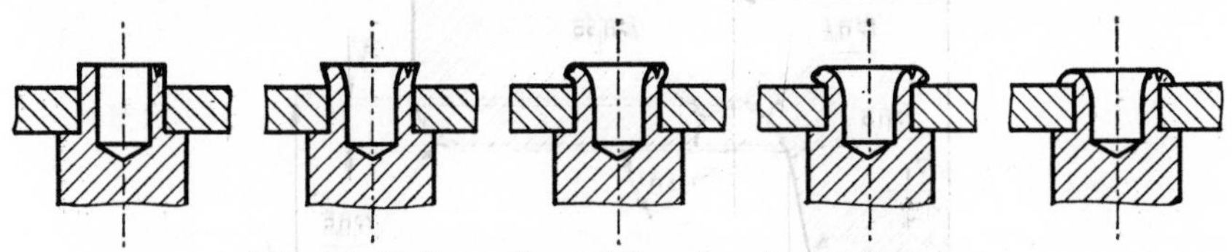

Fig. 1.2. Deformation of the edge during flanging.

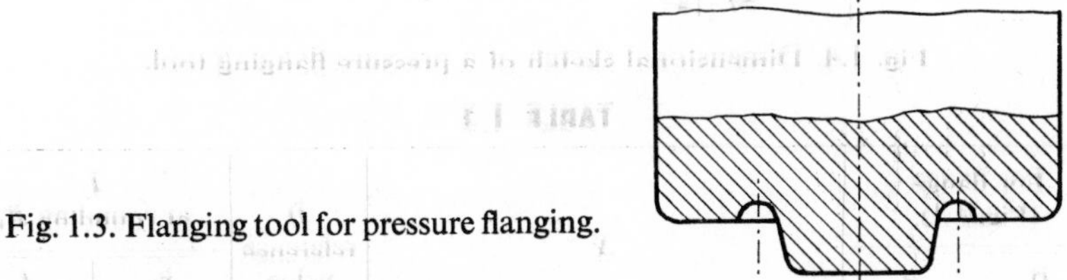

Fig. 1.3. Flanging tool for pressure flanging.

The flanging force required for pressure flanging steel fixing bushes increases rapidly with the diameter (see Table 1.1).

Pressure flanging is usually not employed for diameters exceeding 6 mm. Pressure flanging a 6 mm bush requires lubrication with machine oil containing a dope.

Full annealing means less flanging force (about 20%), but full annealing costs time and money because of the after-treatment (finishing).

When using rivets or bushes in rather brittle materials (resin-bonded paper, bakelite, etc.) it is necessary to use a pressure limitation as regards the

TABLE 1.1

Diameter of fixing bush (mm)	Riveting force (N)		
	free-cutting steel	brass	aluminium
2	5 700	5 000	4 000
4	14 000	13 500	10 000
6	19 000	15000	12 000

stacking tolerances (varying thickness tolerances of the component parts of the product to be flanged). Instead of pressure flanging under a press, flanging can also be done manually (hammer and punch), usually in the following cases:

(a) When the rivets are far apart.
(b) When the components are too high to be stacked under a press.
(c) When there is no room to use a press.

Table 1.2 surveys the average strength of a well-made pressure-flanged joint.

The data of Fig. 1.4 and Table 1.3 serve as a guide for the dimensions of flanging tools for pressure flanging.

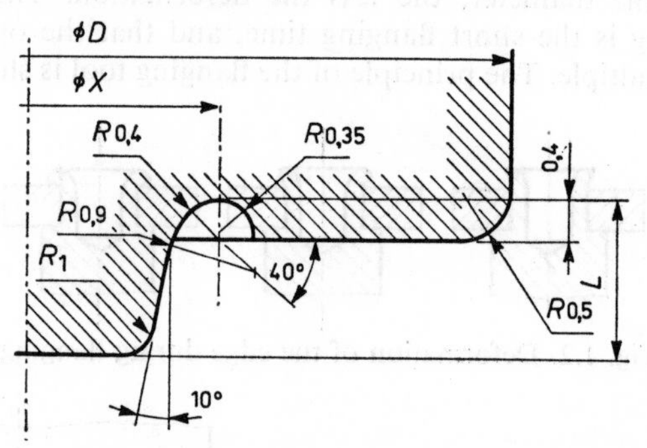

Fig. 1.4. Dimensional sketch of a pressure flanging tool.

TABLE 1.3

For flange (Fig. 1.1) D nominal	d nominal	X	D reference value	L at rounding R_1: R_1	L
2·0	1·0	d nominal + 1·10 mm	8	0·4	1·2
2·5	1·5	d nominal + 1·10 mm	8	0·4	1·2
3·0	2·0	d nominal + 1·10 mm	8	0·4	1·2
3·5	2·3	d nominal + 1·10 mm	8	0·4	1·2
4·0	2·8	d nominal + 1·25 mm	10	0·3	1·55
4·5	3·3	d nominal + 1·25 mm	10	0·3	1·55
5·0	3·8	d nominal + 1·25 mm	10	0·3	1·55
5·5	4·3	d nominal + 1·25 mm	10	0·3	1·55
6·0	4·8	d nominal + 1·25 mm	10	0·3	1·55

TABLE 1.2

	Bush diameter = 2 mm			Bush diameter = 4 mm			Bush diameter = 6 mm		
	free-cutting steel	brass	aluminium	free-cutting steel	brass	aluminium	free-cutting steel	brass	aluminium
Average torque strength (Nm)	0·7	0·5	0·5	2·2	0·8	0·8	4·0	2·0	2·0
Average axial strength (N)	1 500	1 000	800	2 500	1 300	1 200	3 700	1 700	1 700

A very simple apparatus for pressure flanging is the toggle press. The force is transferred by means of a toggle whose outer position is determined by the straight position of the toggle. The pressure on the ram increases as the toggle approaches its straight position, increasing to 1 000–1 500 kg (see Fig. 1.5).

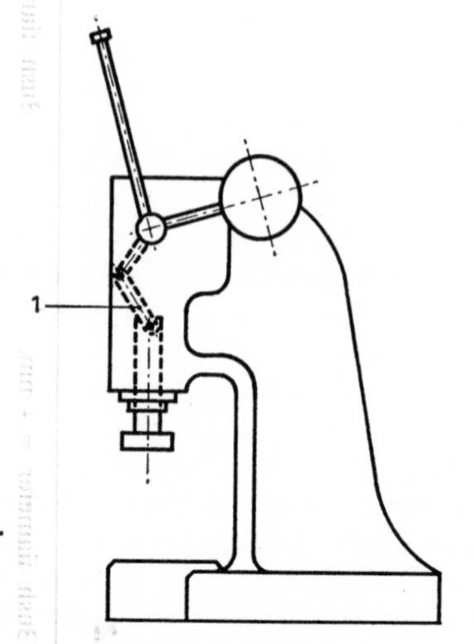

Fig. 1.5. Manually-operated PHILIPS press.
1 = toggle

Fig. 1.6 shows a pneumatic press. By means of air under pressure, a continuously increasing force is exerted on the ram. The correct ram pressure can be adjusted by a reduction valve, to protect the press against damage or

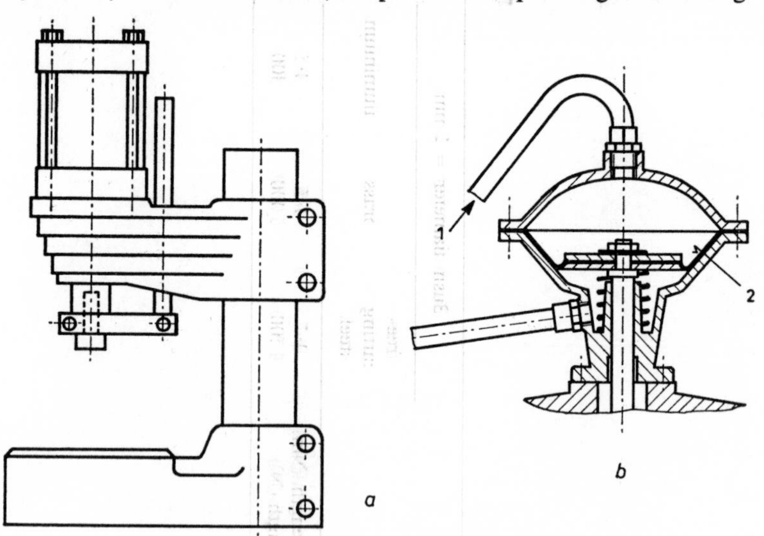

Fig. 1.6. Pneumatic press.
a. Design with cylinder. *b*. Design with membrane.
1 = air pressure. 2 = membrane.

fracture. A type with a cylinder (Bläsi) or with a membrane (Affbe) can be used.

1.1.3 *Rotary flanging*

Rotary flanging is used for:

(a) stacking several plates to be flanged in connection with tolerances.
(b) if the curled flange must meet high appearance requirements.
(c) if the material of the plate to be flanged is fragile (resin-bonded paper, etc.),
(d) for dimensions where the forces used for pressure flanging would be too high (the pressures required are smaller by a factor of 5–10 than for pressure riveting).

A disadvantage is that multiple operation is not possible.

To avoid scoring, the flanging tool must have a well-polished surface and a good lubricating oil used (machine oil containing a dope, for instance), especially when flanging aluminium. The form of the flanging tool is shown in Fig. 1.7. The run-in must be well rounded and show a smooth finish to avoid cutting. There must be no sharp angles at the run-in end.

Table 1.4 gives an overall survey of the axial tensile strength of a well-made flanged joint.

The loosening torque of rotary flanged joints is low. For a 5 mm bush, an average of 10×10^{-1} Nm; for a 10 mm bush, an average of 23×10^{-1} Nm.

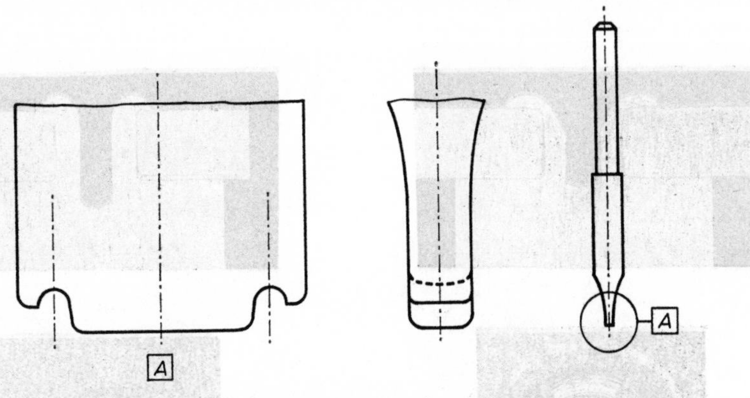

Fig. 1.7. Tool for rotary flanging (spinning).

TABLE 1.4

Fixing bush diameter (mm)	Tensile strength (N)		
	free-cutting steel	brass	aluminium
2	850	750	700
4	2 100	1 700	1 450
8	4 400	4 000	3 000
16	6 200	6 100	5 100

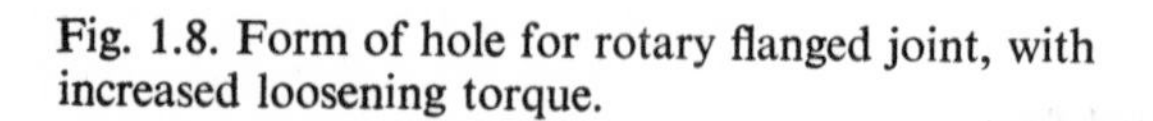

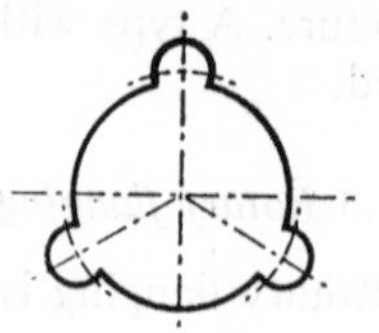

Fig. 1.8. Form of hole for rotary flanged joint, with increased loosening torque.

To increase this loosening torque (2 or 3 times), the shape of hole shown in Fig. 1.8 can be employed.

The required flanging pressures vary from 200 to 500 Nm. It is best to match the number of revolutions to the diameter to be flanged. For instance:

diameter	*revolutions*
ϕ 2– 4 mm :	1200 rpm
ϕ 4– 8 mm :	600 rpm
ϕ 8–20 mm :	400 rpm

For rotary flanging, stops must be used whenever possible, to make sure that the tool travels over the correct distance, and protect product and tool against damage. Figure 1.9 shows the effect of the profile of the flanging punch and the form of the fixing bush.

The data in Fig. 1.10 and Table 1.5 will serve as a guide for the dimensions of tools used for rotary flanging.

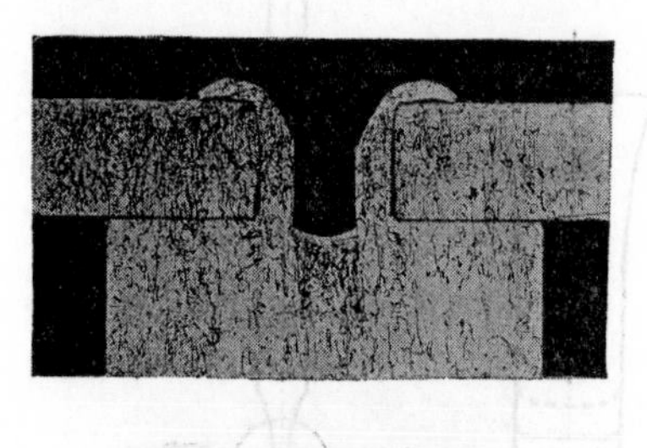

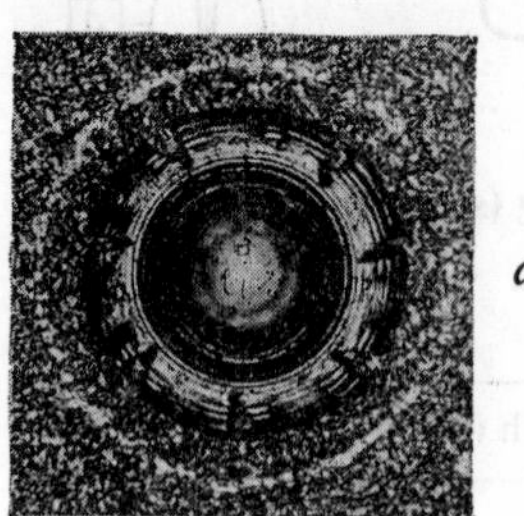

a

b

Fig. 1.9. Effect of profile of flange tool and shape of fixing bush on quality of joint.
a. Bush diameters 2 × 1·2mm (material brass).
b. Bush diameters 2 × 0·8mm (material brass).

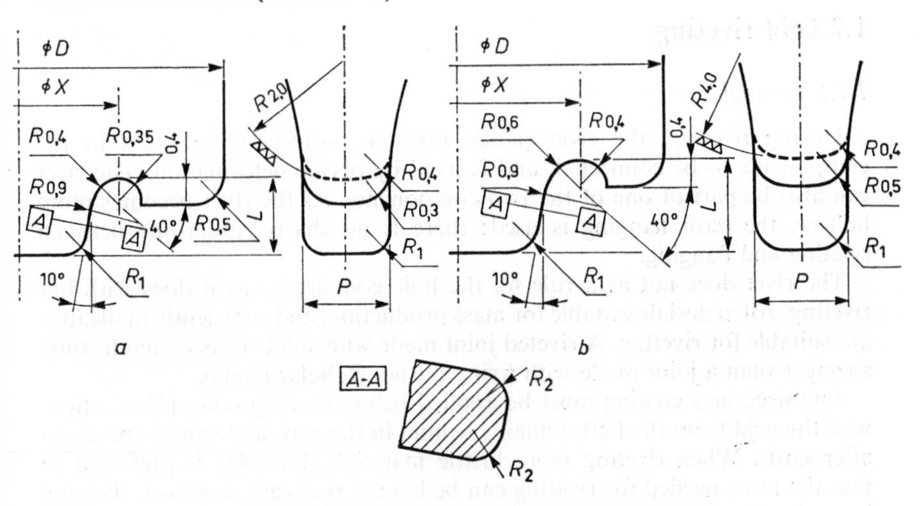

Fig. 1.10. Dimensional diagram of a rotary flanging tool. For shapes of tools *a* and *b* see Table 1.5.

TABLE 1.5

For flange (Fig. 1.1) *D* nominal	*d* nominal	shape of punch	*X*	*P*	*D* reference value	*L* at rounding R_1: R_1	*L* at rounding R_1: *L*	R_2 +0·1
2	1	a	*d* nominal + 1·10 mm	0·9	7	0·4	1·2	0·1
2·5	1·5	a	*d* nominal + 1·10 mm	1·3	7	0·4	1·2	0·2
3	2	a	*d* nominal + 1·10 mm	1·5	7	0·4	1·2	0·2
3·5	2·3	a	*d* nominal + 1·10 mm	1·5	7	0·4	1·2	0·2
4	2·8	b	*d* nominal + 1·45 mm	1·6	7	0·5	1·5	0·2
4·5	3·3	b	*d* nominal + 1·45 mm	1·7	8	0·5	1·5	0·2
5	3·8	b	*d* nominal + 1·45 mm	1·8	8	0·5	1·5	0·3
5·5	4·3	b	*d* nominal + 1·45 mm	1·9	9	0·5	1·5	0·3
6	4·8	b	*d* nominal + 1·45 mm	2·0	9	0·5	1·5	0·3
6·5	5·3	b	*d* nominal + 1·45 mm	2·1	10	0·5	1·5	0·3
7	5·8	b	*d* nominal + 1·45 mm	2·2	10	0·5	1·5	0·3
8	6·8	b	*d* nominal + 1·45 mm	2·4	11	0·5	1·5	0·3
9	7·8	b	*d* nominal + 1·45 mm	2·6	12	0·5	1·5	0·4
10	8·8	b	*d* nominal + 1·45 mm	2·6	13	0·5	1·5	0·4
11	9·8	b	*d* nominal + 1·45 mm	2·8	15	0·5	1·5	0·4
12	10·8	b	*d* nominal + 1·45 mm	2·8	16	0·5	1·5	0·4
13	11·8	b	*d* nominal + 1·45 mm	2·8	17	0·5	1·5	0·4
14	12·8	b	*d* nominal + 1·45 mm	3·0	18	0·5	1·5	0·5
15	13·8	b	*d* nominal + 1·45 mm	3·0	19	0·5	1·5	0·5
16	14·8	b	*d* nominal + 1·45 mm	3·0	20	0·5	1·5	0·5

1.2 Cold riveting

1.2.1 General

During riveting, the rivet passes through corresponding holes in the components to be connected, and is fixed by plastic deformation. The rivet can also be part of one of the two components. As the rivet becomes more hollow, the term flanging is used: there is no sharp border-line between riveting and flanging.

The rivet does not as a rule fill the hole completely, as it does with hot riveting. All materials suitable for mass production, and sufficiently malleable, are suitable for riveting. A riveted joint made with solid rivets is significantly stronger than a joint made with fixing bushes (tubular rivets).

Any necessary coating must be applied before riveting takes place, otherwise the acid from the bath remains behind in the gaps and causes corrosion afterwards. When riveting more brittle materials, less play is preferred so that the force needed for riveting can be less; in that case, however, flanging is preferable. For thin plates, too, the clearance between rivet and hole should be a minimum.

The form of the riveted head should be as simple as possible. The following considerations play a part: good appearance; ease of production; countersunk holes only to be made by extruding thin plates at minimum costs; and the permissible forces on the components to be connected.

Rivets with rounded heads are usually employed. Soft material requires a large head diameter, or a washer can be used instead. Riveting material that is too short, or too long, should be avoided, if satisfactory products are to result.

1.2.2 Impact riveting

The impact press operates on the principle of applying energy stored in a spring, usually mechanically, although there are electromagnetic and pneumatic systems. This is explained in Figs 1.11(*a, b* and *c*).

Figure 1.11(*a*) shows the initial situation where the spring can, if required be given a pre-tension. The spring is tensioned (Fig. 1.11(*b*)) until the shaft

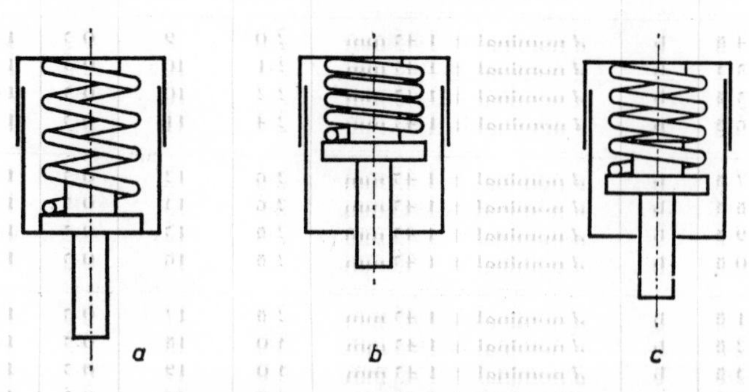

Fig. 1.11. Principle of a mechanical impact press (spring-operated).

comes free, the energy stored in the spring driving the shaft down until it hits a pin, the profiled end of which rests on the product to be riveted (Fig. 1.11(*c*)).

With free-cutting steel, the best results are obtained with a pin diameter of 2–3 mm, at a free riveting height of 0·7–0·9 mm and a minimum impact energy of 25×10^{-1} Nm, for a diameter of 2 mm, and 40×10^{-1} Nm for a diameter of 3 mm. The resulting deformation is about proportional to the impact force applied, no matter whether this takes place in one stroke or in a sequence of strokes. Figure 1.12 shows the *Automator* as an example.

The ratio between the diameters of the riveted components d_2/d_1 in Fig. 1.13 also has a great effect on the resulting deformation. A smaller diameter of the contact area (d_2) gives more distortion, which is greatest directly below the plate in which riveting takes place (d_4). At a larger d_2, there will be less distortion, and it will take place mainly at the end of the riveted product (d_3). For a shank diameter of 2 mm (for free-cutting steel) $d_2 = 4$ mm is quite satisfactory; for 3 mm, $d_2 = 5$ mm can be used, the average loosening torques being 6 and 11×10^{-1} Nm respectively.

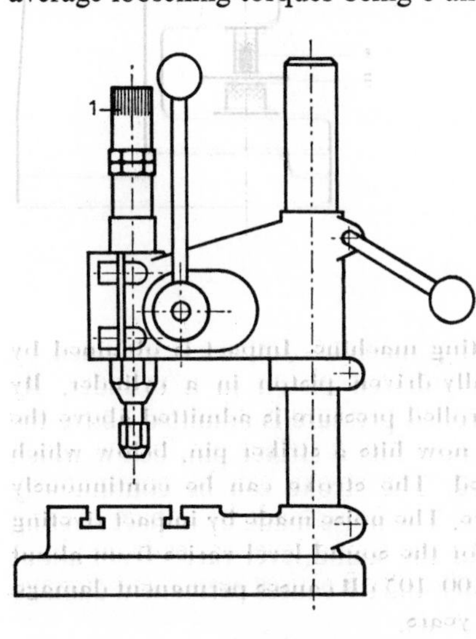

Fig. 1.12. Mechanical spring-operated impact press AUTOMATOR, manufactured by Mermet-Virthner, France. 1 = adjusting screw for spring force.

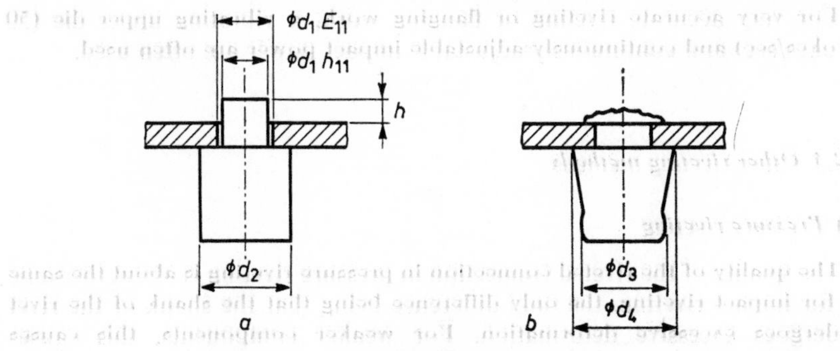

Fig. 1.13. *a.* Product to be riveted. *b.* Riveted joint (deformation).

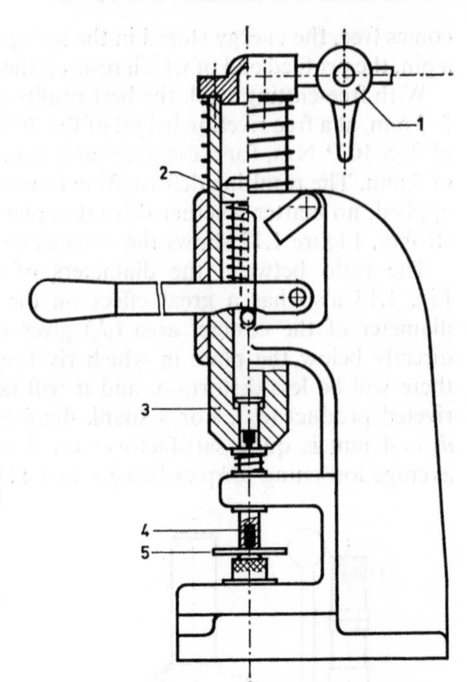

Fig. 1.14. PRESSTAKER pneumatic riveter, made by Giannini Controls Corp., U.S.A.
1 = valve. 4 = striker pin.
2 = driven piston. 5 = product.
3 = cylinder.

Figure 1.14 shows a pneumatic riveting machine. Impact is obtained by releasing the power of a pneumatically-driven piston in a cylinder. By operating a valve, air at a certain controlled pressure is admitted above the piston and drives it down. The piston now hits a striker pin, below which is positioned the product to be riveted. The stroke can be continuously controlled by means of a reduction valve. The noise made by impact riveting machines is annoying in a workshop, for the sound level varies from about 105–120 dB. A constant noise of about 100–105 dB causes permanent damage to the ear after a period of four or five years.

For very accurate riveting or flanging work, a vibrating upper die (50 strokes/sec) and continuously-adjustable impact power are often used.

1.2.3 Other riveting methods

(a) *Pressure riveting*

The quality of the riveted connection in pressure riveting is about the same as for impact riveting, the only difference being that the shank of the rivet undergoes excessive deformation. For weaker components, this causes deformation of the workpiece itself. The fact that pressure riveting produces no noise is an advantage of course.

(b) *Roll riveting*

This method of riveting produces significantly less noise than impact riveting, but the tools are rather expensive and also take up much more space. This can give rise to difficulties, especially when riveting in corners at vertical sides. The shank of the rivet is hardly upset. A good clamping force between rivet and workpiece is vital.

A set of profiled rollers, carried by a main shaft rotating at a high speed (about 1 400 rpm), form the rivet profile. By forcing down these rollers on to the rivets, a smooth, regular head is formed (Fig. 1.15 Steinel).

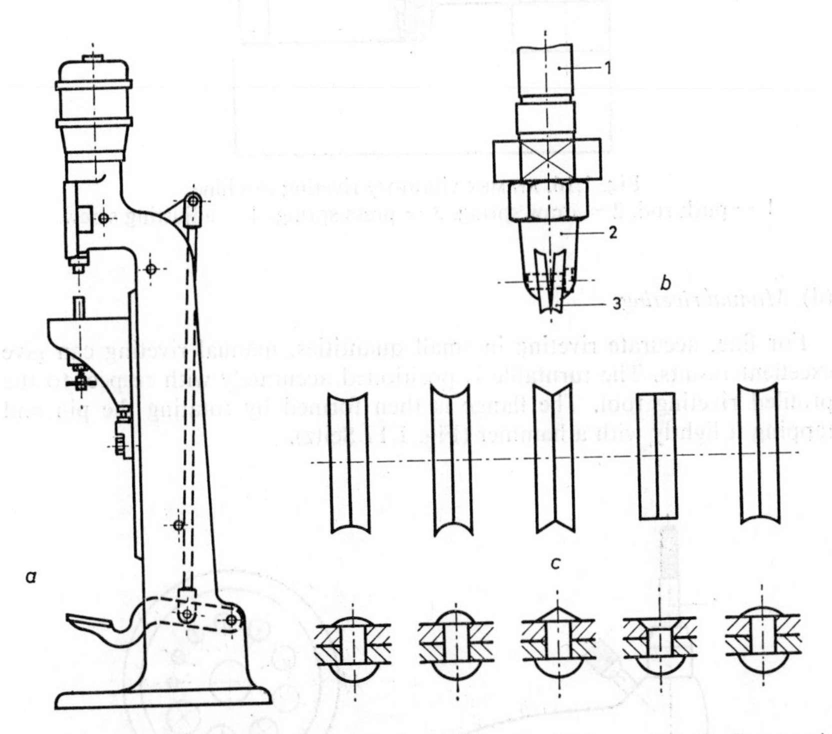

Fig. 1.15. *a.* STEINEL riveting machine for roll riveting (Hahn und Kolb, Germany).
b. Riveting head with riveting rollers.
c. Different models of rivet together with riveting rollers to match.
1 = main shaft. 2 = roller holder. 3 = riveting rollers.

(c) *Vibratory riveting*

A repeating hammer operated by a push rod strikes the die, whose profiled end is forced against the rivet under pressure of a spring. The length of the stroke can be adjusted by an adjustment screw. (Fig. 1.16 Animex).

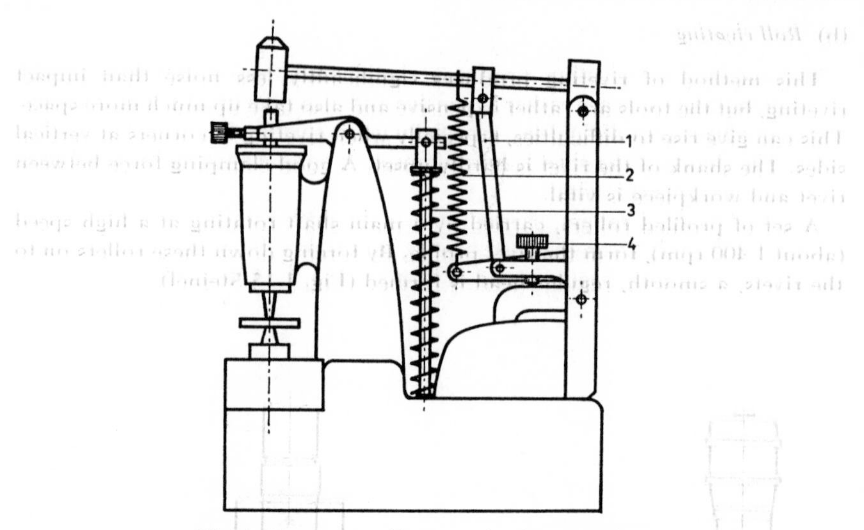

Fig. 1.16. ANIMEX vibratory riveting machine.
1 = push rod. 2 = draw spring. 3 = press spring. 4 = adjusting screw.

(d) *Manual riveting*

For fine, accurate riveting in small quantities, manual riveting can give excellent results. The turntable is positioned accurately with respect to the profiled riveting tool. The flange is then formed by rotating the pin and tapping it lightly with a hammer (Fig. 1.17 Seitz).

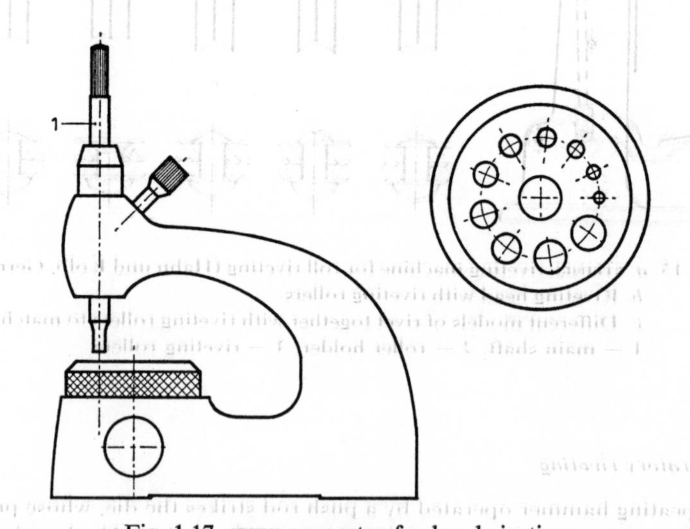

Fig. 1.17. SEITZ apparatus for hand riveting.
1 = riveting pin.

(e) *Progressive riveting*

This rather new method is practically noiseless. The riveting tool is placed at a certain angle on the rivet to the rivet shaft, after which it is forced down on the rivet shank under pressure (see Fig. 1.18).

The pressure is so high that, on the contact line between the riveting tool and the rivet shank, the compressive strength of the material is slightly exceeded.

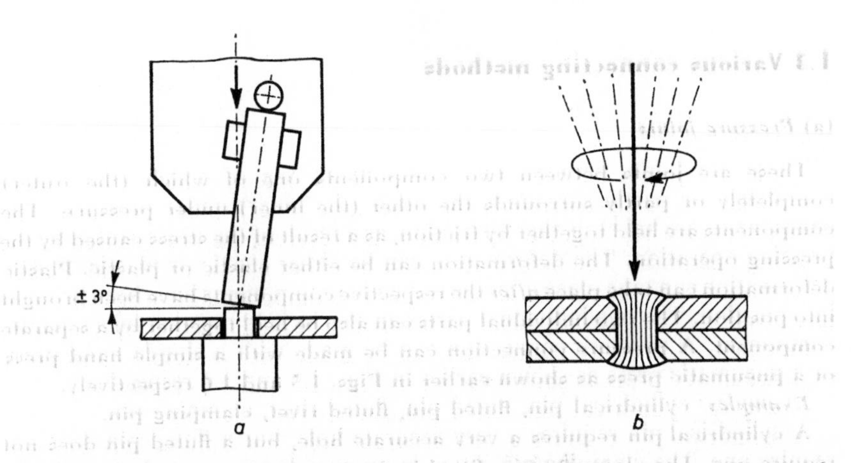

Fig. 1.18. Machine for progressive riveting, made by Bodmer & Co., Switzerland. *a.* riveting principle. *b.* rivet produced.

The material is deformed along a line running from the rivet axis to the rivet edge, and since the deformation after one revolution of the riveting tool is but small, the material has sufficient time to flow completely.

Considerable deformation of material takes place, because the material is rolled further down as the riveting tool progresses at each revolution, and these rolling operations are repeated in rapid succession.

The workpiece itself does not rotate and no lubrication is needed. In this way a mechanically strong joint is obtained, while the riveting pressure and time can be accurately controlled. For instance, low pressure and longer times can be adopted for components liable to damage: or high pressure and shorter times (time saving). The minimum height of the flange can be small (about 1/3 to 1/4 of the usual height).

Multiple riveting, however, is impossible without high costs, and the riveted joint must be within easy reach, and rather accurately centred below the riveting tool. The maximum and minimum diameters of the rivet head are 9 mm and 1 mm respectively.

REFERENCES

H. EDER	*Feinwerktechnische Verbindungen durch plastisches Verformen,* Mitteilung aus der Zentralkonstruktion der Siemens und Halske A.G.
	Fasteners Handbook, Machine Design, Reference Issue, Fastening and Joining (5th edition) September 11, 1969.
E. KENNEL	*Das Nieten im Stahl- und Leichtmetallen.*
RICHTER and v. VOSS	*Bauelemente der Feinmechanik.*
J. SOLED	*Fasteners Handbook,* Reinhold Publishing Corporation, 1957.

1.3 Various connecting methods

(a) *Pressure joints*

These are joints between two components one of which (the outer) completely or partly surrounds the other (the inner) under pressure. The components are held together by friction, as a result of the stress caused by the pressing operation. The deformation can be either elastic or plastic. Plastic deformation can take place *after* the respective components have been brought into position. The two individual parts can also be held together by a separate component. A pressure connection can be made with a simple hand press or a pneumatic press as shown earlier in Figs. 1.5 and 1.6 respectively.

Examples: cylindrical pin, fluted pin, fluted rivet, clamping pin.

A cylindrical pin requires a very accurate hole, but a fluted pin does not require one. The clamping pin, fitted in an even-less-accurate hole, will not loosen due to vibrations. In many cases a knurl can be used. The work finish need not be as accurate as for smooth shafts. When forcing the pin down, it is guided by pilot edges at the hardest end. A cylindrical part in front of the knurl ensures better guidance. The knurl must be applied to the hardest part. Examples of distortion of the outer component after pressing are: attaching curved dove-tail shaped contacts; spread effect of a tap forced into a blind hole (see Fig. 1.19).

Nails are widely used for wood joints: screw nails require 4 times the extraction force of normal nails.

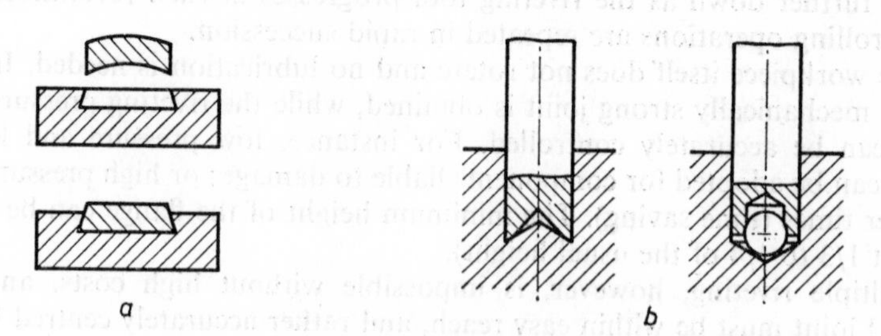

Fig. 1.19. *a.* Fixing silver contacts.
b. Fixing a shaft by spread effect.

(b) *Folded joints*

Flat sheets can be joined by folding over the edges, hooking them together and pressing down the joint so obtained. The strength and tightness can be improved by folding the edge over once more. Sometimes joints are made by means of a separate part (see Fig. 1.20). For folding tools see also Volume 11, Chapter 4.

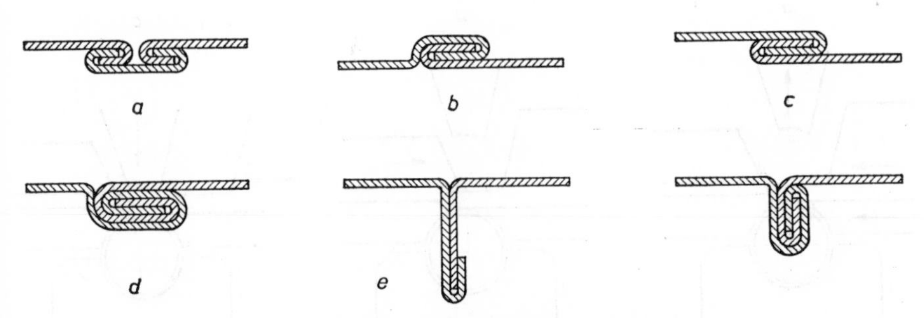

Fig. 1.20. *a* Folded joint with separate component.
b-c. Folded joint without separate component.
d. Flat folded joint.
e. Transverse folded joint before folding.
f. Transverse folded joint after folding.

If tubular products are to be made, they are usually provided with one seam on a single machine. By means of a 4-slide machine and corresponding folding mechanisms, products with different kinds of folded joints can be made according to the principle of Fig. 1.21.

For the machines see also Volume 10, Chapter 4.

(c) *Joints with folded lugs*

Joints can be made, too, by folding over lugs that form part of one of the components to be connected; or the lug can be a separate part. The material to be folded over must be very soft, otherwise resilience would result in loose joints. Fig. 1.22 shows joints made by folded lugs.

This kind of joint is often used in cheap mass production, and was introduced on a large scale in the toy industry. The line over which the lug is folded must fall slightly outside the components, while the bending line itself must be at right-angles to the direction in which the material is rolled. Lugs bent to face each other form a stronger joint than lugs bent in the same direction.

Twisted lugs (Fig. 1.23) require a greater height. By using the proper form a satisfactory joint can be obtained by employing a "key" system. The lugs are usually twisted with a wrench. The lugs can also be folded over with simple pliers or toggle presses.

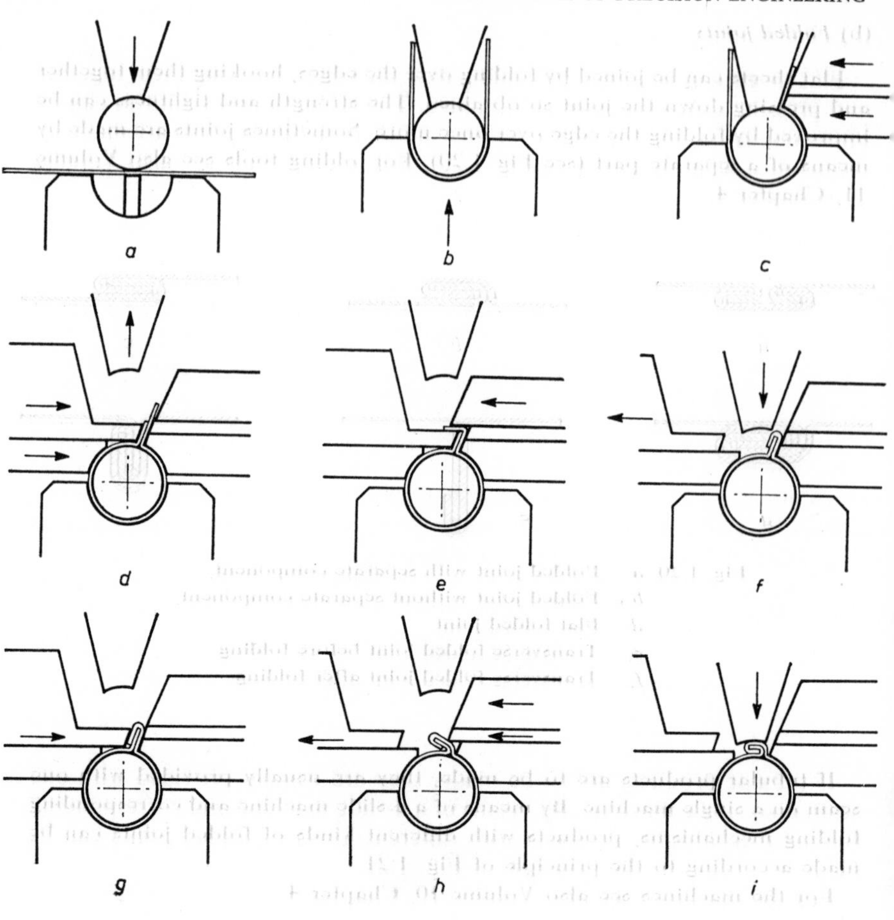

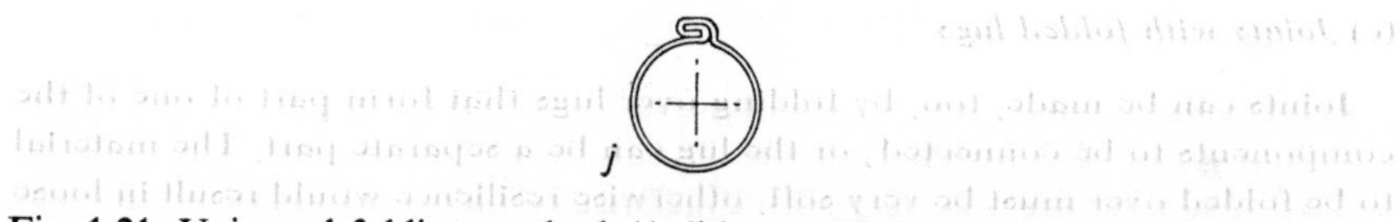

Fig. 1.21. Universal folding method (4-slide machine) for making folded joints. Production order *a* to *i* inclusive; *j* = end product.

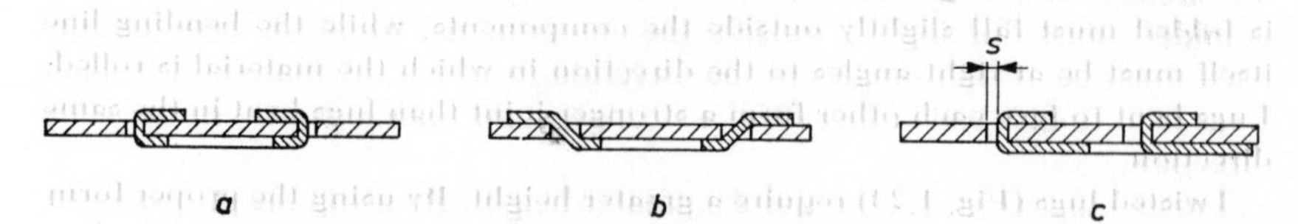

Fig. 1.22. *a.* Rigid joint with folded lugs.
b. Semi-rigid joint with folded lugs.
c. Non-rigid joint with folded lugs (*s* = play).

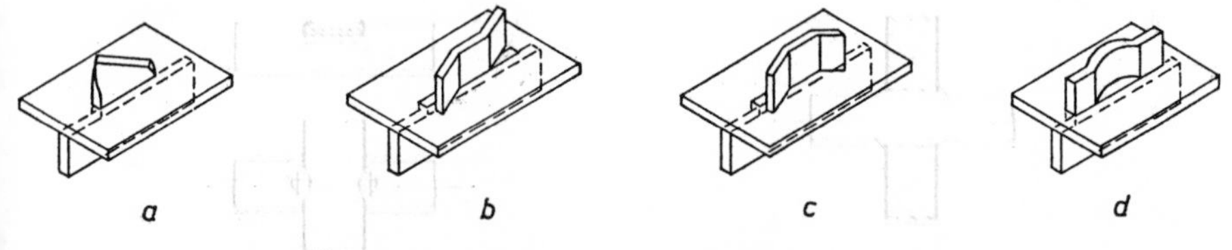

Fig. 1.23. *a.* Joint between two perpendicular components made by means of a wrenched tag.
b. Joint made by two tags wrenched in opposite directions.
c. Joint made by two tags wrenched in the same direction.
d. Joint between two components made by a lug cut loose at the centre.

(d) *Fluting* (notching)

By this is understood the joining of two components by means of impressions made in one component (or both). The impressions can, for instance, be made simultaneously in the components to be joined together (Fig. 1.24). It is also possible to make an impression in a prepared recess of the other component (Fig. 1.25), or by enclosing the other component (Fig. 1.26).

Fig. 1.24. *a.* Fluted joint between a pipe and a rod.
b. Fluted joint between two thin pipes.

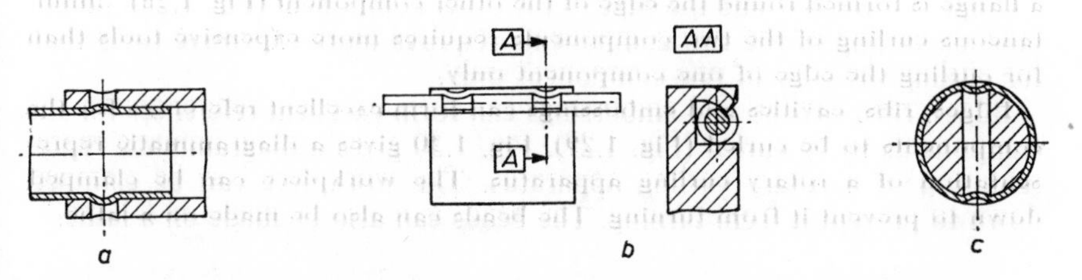

Fig. 1.25. *a.* Joint between a thick pipe and a thinner pipe, made by expansion
b. Joint between a shaft and a plate, by fluting.
c. Joint between a shaft and a thin pipe.

(e) *Curling*

Curling is a sort of flanging of the edges of the workpiece. The two components can either be curling together (Fig. 1.27) or separately, so that

Fig. 1.26. *a.* Joint between a disc and a shaft, by fluting.
b. Joint between a strip and a shaft, by fluting.

Fig. 1.27. *a.* Joint between a pipe and a flat plate.
b. Joint between a pipe and an embossed plate.

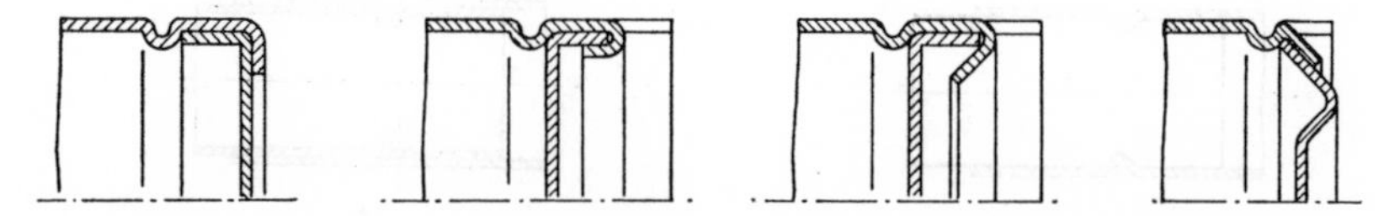

Fig. 1.28. Various ways of joining bottom and cylinder by curling.

a flange is formed round the edge of the other component (Fig. 1.28). Simultaneous curling of the two components requires more expensive tools than for curling the edge of one component only.

Edges, ribs, cavities and embossings can form excellent references for the components to be curled (Fig. 1.29). Fig. 1.30 gives a diagrammatic representation of a rotary curling apparatus. The workpiece can be clamped down to prevent it from turning. The beads can also be made on a lathe.

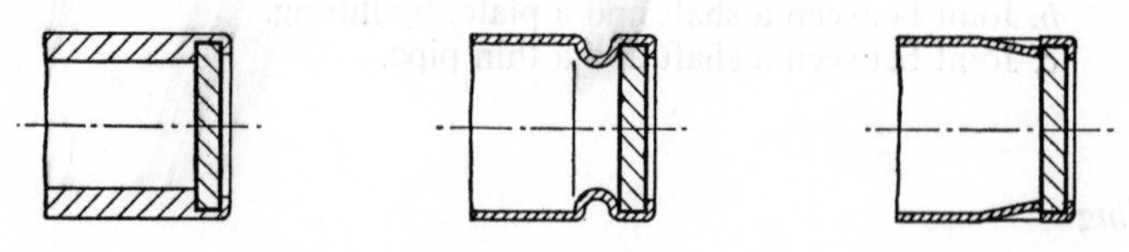

Fig. 1.29. Various joints made by curling with, as a reference, a turned chamber, groove or notch.

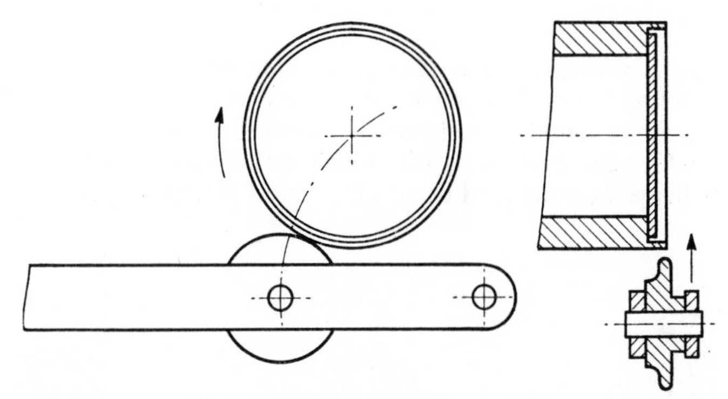

Fig. 1.30. Spinning tool for curling.

(f) *Fluting*

Tubular components can also be joined by fluting. A flute can be made in the two components together (Fig. 1.31), or first in one component, so that the next flute can be forced into the recess of the first component (Fig. 1.32).

Simultaneous fluting is not used for thick material. One or more flutes then have to be made, to keep the other components in their respective positions (Fig. 1.33).

Here, too, a lathe can be used to make the flute.

Fig. 1.31. Joint between two pipes made *a.* by an inward flute, and *b.* by an outward flute.

Fig. 1.32. Joint between two components where the flute is forced into a recess in the other component.

a. outward pressed. *b.* inward pressed.

Fig. 1.33. Enclosing components by fluting.

(g) *Blind riveting*

If riveting is necessary in a place accessible only from one side, the method of "blind riveting" is employed. A blind rivet generally consists of a hollow rivet fitted round a steel pin. The rivet is inserted in a pre-drilled hole, after which the steel pin is withdrawn with a special tool. As a result the rivet is flanged on the far side so that a head is formed. As soon as the

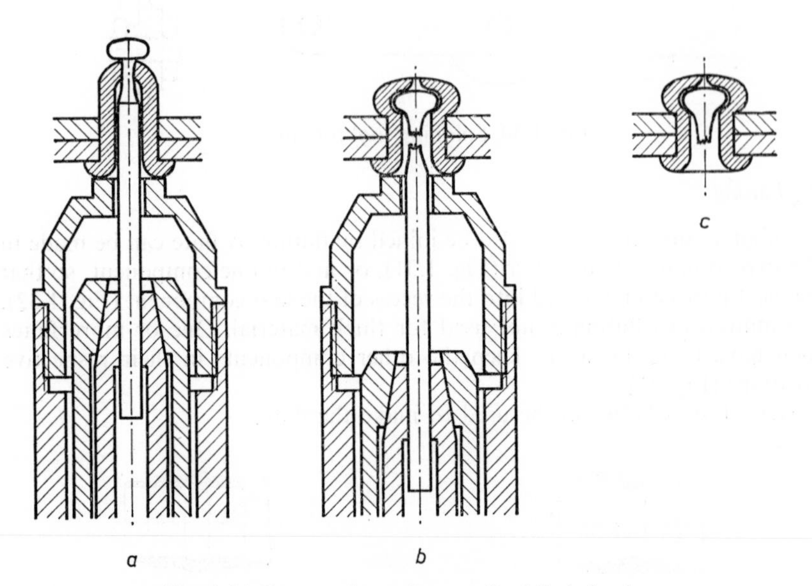

Fig. 1.34. Pneumatic equipment for blind riveting.
a. inserting the rivet. *c.* end product.
b. withdrawing the pen and breaking of the rivet.

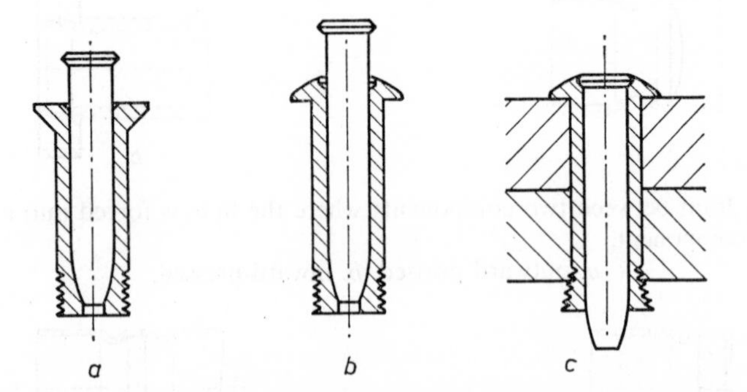

Fig. 1.35. DEUTSCH drive pin blind rivet driven in with the hammer.
a. rivet with sunk head. *c.* end product.
b. rivet with spherical head.

components have been drawn firmly together, the pin breaks off at the appropriate place. The following types are the most common:

Pop rivet (*United Shoe*)

With a pneumatic riveting machine, about 1 000 rivets can be formed per hour (Fig. 1.34).

Deutsch driver pin rivet

These are inserted by hand and driven down with a hammer. They are highly shear-resistant, and provide satisfactory sealing (Fig. 1.35).

Du Pont explosive rivets

These rivets contain chemicals that give a light explosion when heated sufficiently, and thus widen the shank of the rivet to give a spherical shape. There are open and closed types. One man can easily apply 20 or 25 of the rivets per minute (Fig. 1.36).

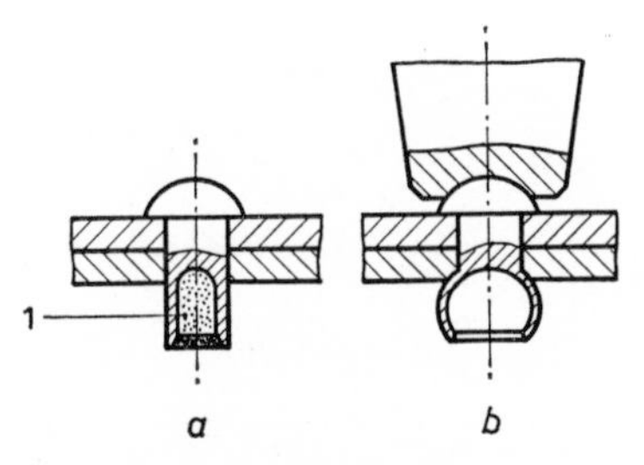

Fig. 1.36. DU PONT explosive rivet (open type).
1 = explosive. a. before explosion.
b. after explosion.

REFERENCES

W. BACHMAN — *Gebohrte Niete in der Feinwerktechnik*, Feinwerktechnik 64, (1960). Heft 8, Seite 293.

H. EDER — *Feinwerktechnische Verbindungen durch plastisches Verformen,* Mitteilungen aus der Zentralkonstruktion der Siemens und Halske A.G.

K. HAIN — *Die Feinwerktechnik.*

E. KENNEL — *Das Nieten im Stahl und Leichtmetallen.*

H. PÖSCHL — *Verbindungselemente der Feinwerktechnik.*

RICHTER and V. VOSS — *Bauelemente der Feinmechanik.*

K. H. SIEKER — *Fertigungs- und stoffgerechtes Gestalten in der Feinwerktechnik.*

J. SOLED — *Fasteners Handbook.* Reinhold Publishing Corporation, 1957.

F. WOLF — *Die nicht-lösbare Verbindungen in der Feinwerktechnik.*

Machine Design, Reference Issue, Fastening and Joining (5th edition). September 11, 1969.

Chapter 2

Welding

J. O. Zwolsman

2.1 Introduction

Metal welding is the process of joining metallic materials by applying heat and/or pressure, with or without the addition of a corresponding material of the same (or approximately the same) melting temperature or traject.

The joining method thus differs from soldering, where use is always made of an additional material, the solder, having a melting point (or melting traject) significantly below that of the metals to be joined.

Welded connections are permanent connections whose binding strength depends on the cohesive power of the molecules. This cohesion does not occur only at the fusion weld, where the welded area crystallizes into a new structure, but prevails to the same degree in the transitional zone of two metals fusing into a solid state, the diffusion weld. The whole mechanism of welding is so complex that an exact approach is impossible. Since Slavianoff's invention (1891) for welding with a metal electrode, electric arc welding has found a very wide range of applications. And it may still be considered surprising that, of the molten welding material, only a few per cent are burned in the arc of 4000 °C, while the remaining metal forms an homogeneous welding bead. And after many years of research and study, the quality of the bead first applied still depends largely on the insight and skill of the welder. By describing typical characteristics of the various welding methods, it is intended here to help the reader select the most suitable method for his welding problem.

In the first place, attention is given to the methods of pressure welding, which are so important in precision engineering. The weldability of metals is explained in Section 2.3 by reference to the resistance welding process. Finally, some methods of quality control are described in Section 2.4.

2.2 Welding methods

2.2.1 Introduction

As each welded joint has specific requirements regarding the kind of material, and the dimensions and shape of workpiece, a rather accurate matching of the welding variables is required for each welding method, which has resulted in a large number of welding systems. Other new systems are still in the developing stage.

The essence of most welding methods is the source, supply, control and dosaging of the necessary welding heat. In many systems, heat is accompanied by a welding pressure, and in some cases this pressure is predominant. If there is no welding pressure, the joint is merely the result of local fusion of the welded components. There are various types of **fusion welding** depending on the welding heat developed.

Oxidation process: oxy-acetylene welding.
Electric arc discharge: open arc welding,
protected arc welding,
submerged arc welding.
Joule conversion of electric energy into heat: electro-slag welding,
Weibel's welding process.
Reduction process: thermit welding.

The category **pressure welding** covers the methods in which pressure only is applied. Depending on the type of pressure applied, these methods can be divided into: cold pressure welding and
friction welding.

The degree of deformation of the material is decisive for the quality of the weld; weldability depends on the hardness and the crystal structure of the material.

On the other hand, there are processes in which heat and pressure are used together, referred to as **hot pressure welding** and divided into:

thermo-compression welding,
resistance welding,
percussion welding, and
stud welding.

Pressure and heat are both used to ensure good metallic contact, and to accelerate diffusion of the atoms or to produce a fusion area.

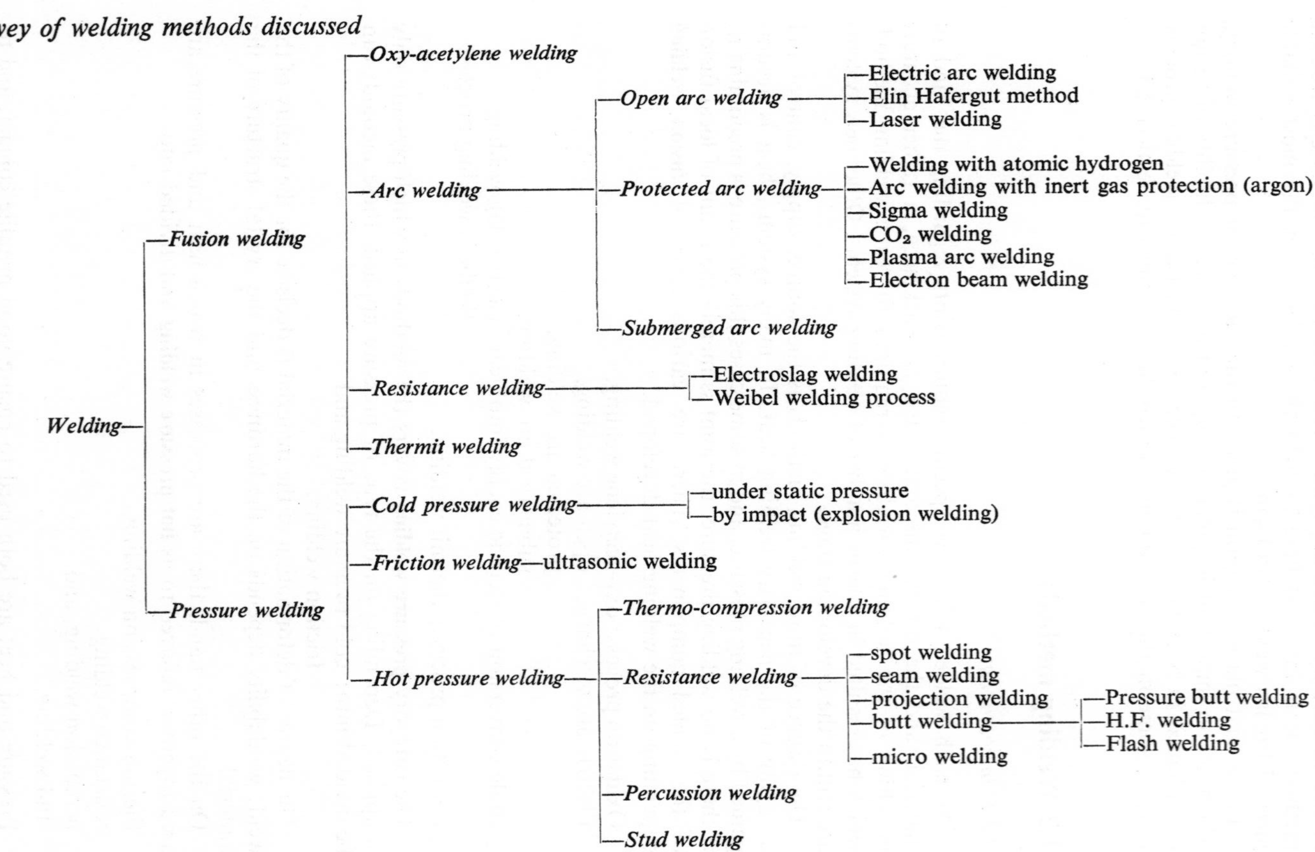

Survey of welding methods discussed
Welding
Fusion welding
Pressure welding
Oxy-acetylene welding
Arc welding
Open arc welding
Electric arc welding
Elin Hafergut method
Laser welding
Protected arc welding
Welding with atomic hydrogen
Arc welding with inert gas protection (argon)
Sigma welding
CO_2 welding
Plasma arc welding
Electron beam welding
Submerged arc welding
Resistance welding
Electroslag welding
Weibel welding process
Thermit welding
Cold pressure welding
under static pressure
by impact (explosion welding)
Friction welding—ultrasonic welding
Hot pressure welding
Thermo-compression welding
Resistance welding
spot welding
seam welding
projection welding
butt welding
Pressure butt welding
H.F. welding
Flash welding
micro welding
Percussion welding
Stud welding

The thermo-compression welding method develops completely in the solid state, and the welding time is not critical. The other three welding methods, however, resistance, percussion and stud welding, require accurate attention to time, pressure and heat. The welding time is very short compared with the systems mentioned earlier. Finally, there are a few other welding systems, still partly under development, namely:

ultrasonic welding,
plasma arc welding,
electron beam welding and
laser welding.

These methods are dealt with later under the heading "Special welding methods". The last three clearly demonstrate the aim of welding engineers to supply the welding heat to the workpiece in ever closer concentrations.

The coverage in this chapter is intended to give the reader an insight into the essential factors of the welding process, so that he can, on the one hand, place the process in the whole scale of welding methods, and, on the other hand, form a clear picture of the effects welding variables have on a certain weld. In order to give a clear survey of existing welding methods, the methods used in heavier engineering are also (briefly) included.

2.2.2. *Oxy-acetylene welding* [2, 3]

Welding process

A mixture of oxygen and acetylene ignited in a gas burner gives an adjustable and directed flame of high temperature, which is used for fusing the material of the workpiece with that of the added welding wire (Fig. 2.1). As the heat supply does not depend on the supply of material, the welding specifications can easily be complied with.

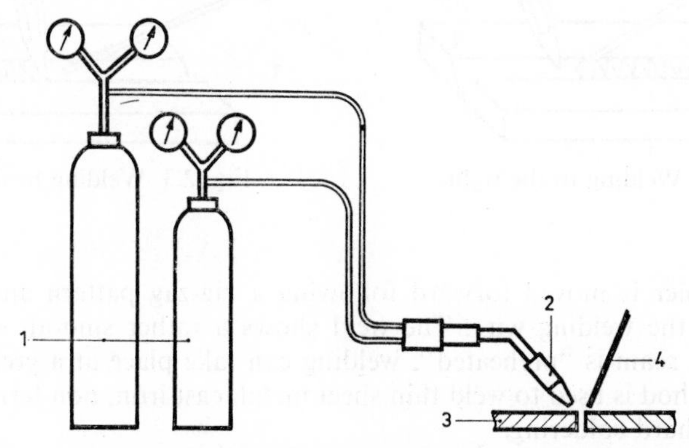

Fig. 2.1. Oxy-acetylene welding.
1 = gas cylinder. 2 = burner. 3 = workpiece. 4 = welding wire

Welding flame

The flame size is determined by the diameter of the jet used in the burner. The supply of oxygen and gas can be controlled separately on the burner. To ignite the flame, first the oxygen valve is opened and then the acetylene valve, the acetylene supply being reduced until a bright cone becomes visible at the end of the burner. When so adjusted, the flame has reached its maximum temperature (3100 °C), and any chemical attack on the weld is reduced to a minimum.

Welding wire

For steel and non-ferrous metals, bare drawn and annealed wire is used.

For a sheet thickness up to 5 mm, the wire diameter = sheet thickness.
For a sheet thickness >5 mm, the wire diameter = $\frac{1}{2}$ sheet thickness.

Cast iron is welded with a cast iron rod.

Welding methods [3]

Taking the burner in the right hand and the welding wire in the left, a choice can be made between:

Welding from left to right (Fig. 2.2).
The welding wire follows the burner, tracing a zig-zag pattern as the burner moves backwards. The weld cools slowly, shrink stress is low and the penetration depth is great. Suitable for heavy sheet metal work.

Welding from right to left (Fig. 2.3).

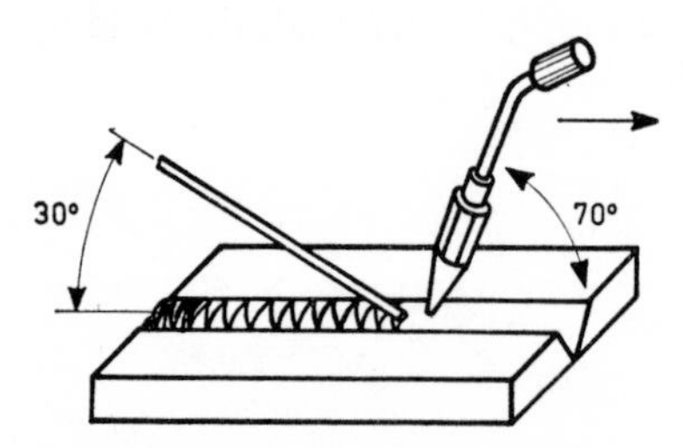

Fig. 2.2. Welding to the right.

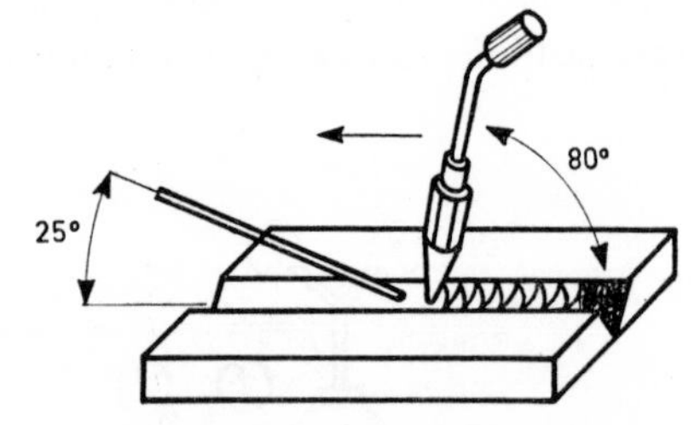

Fig. 2.3. Welding to the left.

The burner is moved forward following a zig-zag pattern and is aimed towards the welding wire. The weld shows a rather smooth surface. As the open seam is "preheated", welding can take place at a greater speed. This method is used to weld thin sheet metal, cast iron, non-ferrous metals and for hard soldering.

The choice of method depends mainly on the dimensions, shape, position and kind of material used for the workpiece, and the welder's experience.

Micro welding technique

It is remarkable how one of the oldest welding methods is now being applied in modern micro-welding. In this technique, use is made of an oxygen-acetylene or oxygen-hydrogen flame which can be very finely adjusted and reaches a temperature of 3500 °C.

With burner jets having an opening of 75–750 μm, it is possible to weld (or to solder) wire upwards from 25 microns. Kinds of wire: gold, platinum, iron, nickel, copper.

Furthermore, the micro-flame can be used for welding, soldering, melting, cutting or tempering, and for all other jobs requiring a pin-point flame.

Special mention should be made of the "Water-welder" manufactured by a British firm. This electro-chemical instrument breaks down distilled water by electrolysis into the elements oxygen and hydrogen. These gases are then automatically mixed in a stoichiometric ratio in the gas generator and fed to the micro-jet, where they are burnt to produce a temperature of 3300 °C (Fig. 2.4).

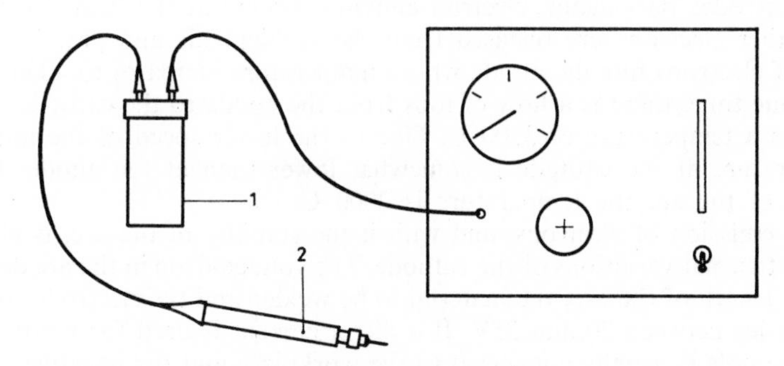

Fig. 2.4. Water welder.
1 = barrel containing methyl alcohol for drying the gas mixture.
2 = burner holder.

2.2.3. *Arc welding*

2.2.3.1. *Open arc welding*

(a) *Electric arc welding* [1, 2, 3]

Welding process

An electric arc is struck between the electrode and the workpiece and produces so much concentrated heat that the material of the workpiece and of the electrode fuse together (Fig. 2.5).

The coating of the electrode can be selected to have a favourable effect on welding.

The supply of material and heat supply are interdependent, so a choice must be made from electrodes of different dimensions and structures.

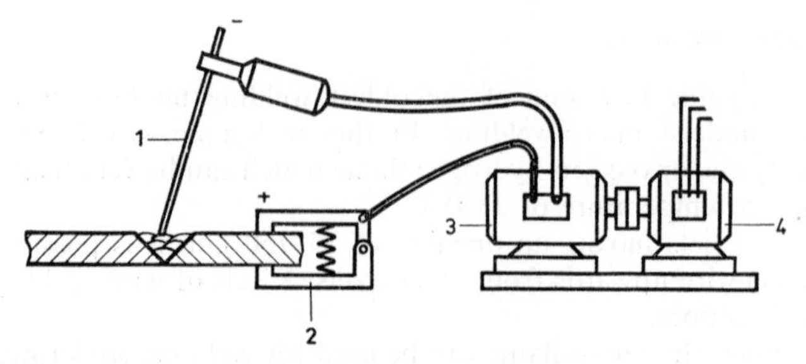

Fig. 2.5. Electric arc welding.
1 = electrode. 2 = earth clamp. 3 = generator. 4 = motor.

The metal arc

A voltage is applied between two electrodes, and if these two electrodes are brought into contact with one another, there will be a flow of current which, owing to the contact resistance, will heat the place of contact. When the electrodes part again, electron emission occurs at the heated cathode, and other electrons are released from the ambient air and gas. The total flow of electrons hits the anode whose temperature increases to 4000 °C. At the same time, there is a flow of ions from the anode to the cathode, which reaches a temperature of 3500 °C. Due to the lower speed of the ions, the temperature at the cathode is somewhat lower than at the anode. In the centre of the arc the temperature is 2000 °C.

The emission of electrons, and with it the stability of the arc, is affected by the thermal variations of the cathode. The voltage-drop in the arc depends on the length of the arc, the material to be welded and the electrode coating, and varies between 20 and 35V. If a direct current is used for welding, the positive pole is usually connected to the workpiece and the negative pole to the electrode.

Material transfer [3]

Independent of the direction of the current and the welding position material is transferred drop-by-drop from the electrode to the workpiece.

The electrode can be regarded as a bundle of parallel conductors attracting each other electrodynamically. As soon as the electrode metal softens, constriction takes place and the material is separated from the electrode in the form of a drop (Pinch effect) (Fig. 2.6).

As soon as the drop of metal comes in contact with the weld pool, it is absorbed due to the surface tension of the pool. This means that overhead welds can be made.

Magnetic blow effect [3]

Two parallel conductors carrying currents flowing in opposite directions repel each other. If the electrode takes up a position perpendicular to the workpiece, the magnetic force will give the metal arc a direction opposing

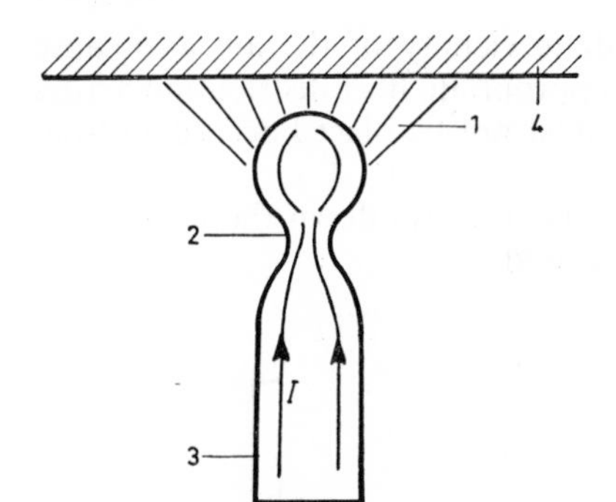

Fig. 2.6. Pinch effect in overhead welding.
1 = arc. 3 = electrode.
2 = constriction. 4 = workpiece.

that of the current (Fig. 2.7). On the other hand, however, if welding takes place at the edge of a magnetic workpiece, a magnetic force bends the arc inwards (Fig. 2.8).

This phenomenon is known as *boundary effect*. Especially when thin sheet steel is welded, does this edge effect give rise to difficulties. The two effects causing the arc to deviate can neutralize each other. By locating the electrode obliquely, the position of the arc, which began to deviate when the current was switched on, is corrected.

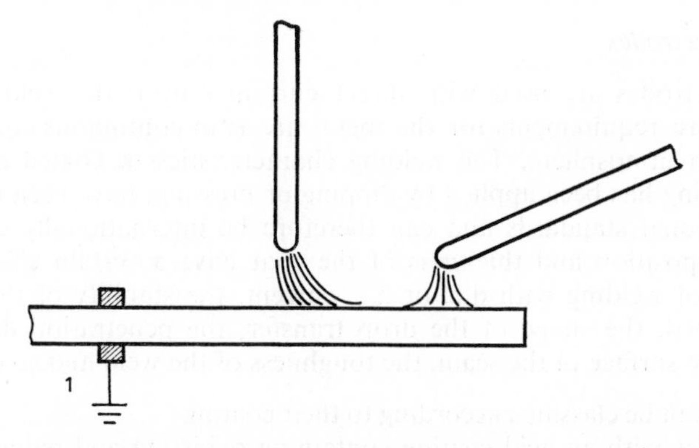

Fig. 2.7. Magnetic blow effect caused by the current.
1 = earth clamp.

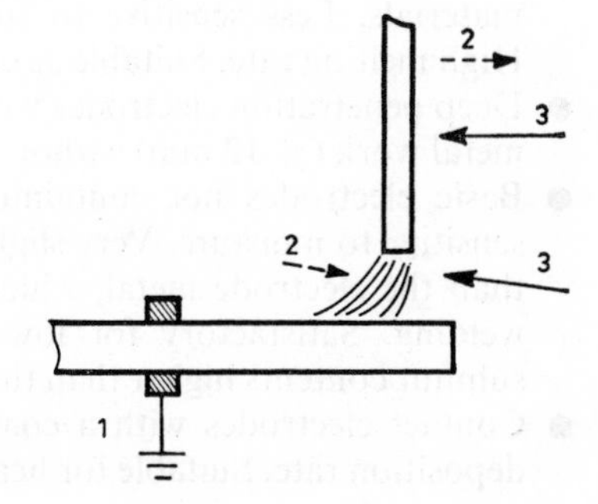

Fig. 2.8. Boundary effect caused by the magnetic field in a ferromagnetic workpiece.
1 = earth clamp.
2 = deviation due to the current.
3 = deviation due to the boundary effect.

Large quantities of iron in the immediate vicinity of the arc also cause it to deviate. For instance, to ensure an appropriate heat distribution across the weld of Fig. 2.9, the earth terminal must not be placed according to a, but according to b.

The magnetic hysteresis in the welded metal slows down the blow effect considerably, when alternating current is used.

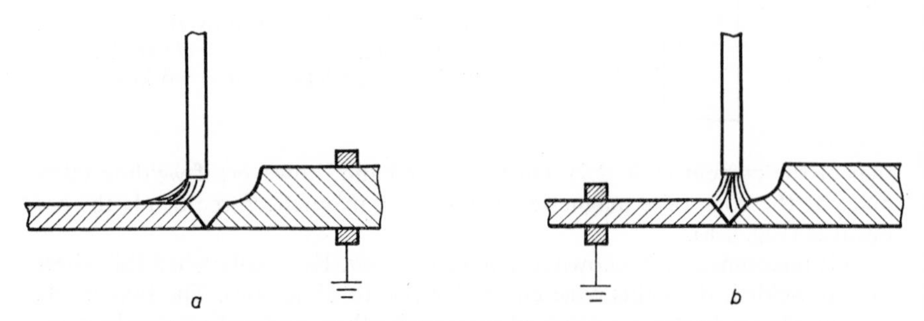

Fig. 2.9. Positioning the earth clamp to obtain a well directed arc. *a.* bad positioning of earth clamp. *b.* good positioning of earth clamp.

Welding electrodes

Bare electrodes are used with direct current only if the weld need not satisfy severe requirements for the metal arc is in continuous contact with the ambient atmosphere. The welding characteristics of coated electrodes, whose coating has been applied by dipping or pressing, have been laid down in international standards and can therefore be internationally compared.

The composition and thickness of the coat have a certain effect on the possibility of welding with d.c. or a.c. current, the stability of the arc, the melting speed, the shape of the drop transfer, the penetration depth, slag forming, the surface of the seam, the toughness of the weld and so on.

Electrodes can be classified according to their coating:

- Electrodes with an acid coating containing oxidizing and reducing metal powders. Suitable for horizontal welding with wide arc.
- Electrodes with a rutile coating containing titanium oxides and organic materials. Less sensitive to sulphur than electrodes with acid coatings. High melting rate. Suitable as contact electrodes for thin sheet metal.
- Deep penetration electrodes with a heavy coating. Suitable for heavy sheet metal work ($\leq$ 12 mm) without preparation.
- Basic electrodes not containing iron oxides in the coating. Short arc, sensitive to moisture. Very slight tendency to freeze. The slag melts slower than the electrode metal, which makes the electrode suitable for vertical welding. Satisfactory for low alloy steels and steels with carbon and sulphur contents higher than those of steels that can easily be welded.
- Contact electrodes with a coating of (conductive) metal powders. High deposition rate. Suitable for heavy welding.

- Core wire electrodes. They consist of a core wire surrounded by a metal braiding, the whole being covered by the coating. Any length of this electrode can be wound on a wheel, thus opening the way to a continuous welding process.

Summarizing:

Bare electrodes and electrodes with a calcareous-basic or thin titanium oxide coating melt in large drops and show a tendency to shortcircuit. The arc voltage is low, the power consumption and melting power low. The welded joint is rough.

Electrodes with an acid coating or a thick coating of titanium dioxide melt in fine droplets. The melting energy is high, they give a smooth joint and require a high welding voltage and high welding power.

Current source

Arc welding can be done with either alternating current or direct current.

Welding with *alternating current* (a.c.) requires no expensive equipment, shows no magnetic blow effect, produces a less stable arc than d.c. current and is less suitable for the application of bare electrodes.

The current can be drawn directly from the mains, and the required welding voltage adjusted by choke coils.

The load on the mains is, however, very unequal and efficiency is low, for the surplus voltage involves losses in the choke coil. Better welding is obtained with a welding transformer whose no-load voltage is kept at about 75V for safety reasons. The transformer has been given a steep characteristic in order to minimize voltage and current variations when the length of the arc changes.

Eddy current losses are accordingly high. Control is effected by changing the winding ratio at the primary or secondary side, using choke coils or varying the magnetic field by shifting the windings with respect to each other.

Welding with *direct current* (d.c.) supplied by a converter results in a symmetrical mains load by using a 3-phase converter, a stable arc and easy ignition, the use of all types of electrodes, including bare ones, a higher permissible welding voltage and a chance of magnetic blowing.

The welding converter consists of a 3-phase or d.c. motor coupled to a welding generator. The generator may also be driven by a petrol engine. The voltage-current characteristic of the generator follows a rather steep course. The welding current rectifier consists of a 3-phase transformer with rectifier, thus uniting the characteristics of a welding converter and those of the voltage transformer. The rectifier must satisfy severe requirements.

(b) *Elin Hafergut method*

Under a copper rail, an acid electrode with a heavy coating is placed on the workpiece, the length of the electrode being 1–2 m. For d.c. welding, the electrode is connected to the negative terminal, and the workpiece to the positive terminal (Fig. 2.10).

The arc is ignited by touching the end of the electrode with a carbon or copper rod, after which the electrode burns up automatically.

The copper rail serves to keep the electrode in position. The Elin Hafergut method is used to weld thin sheets of mild steel with a low C content, in places not easily accessible.

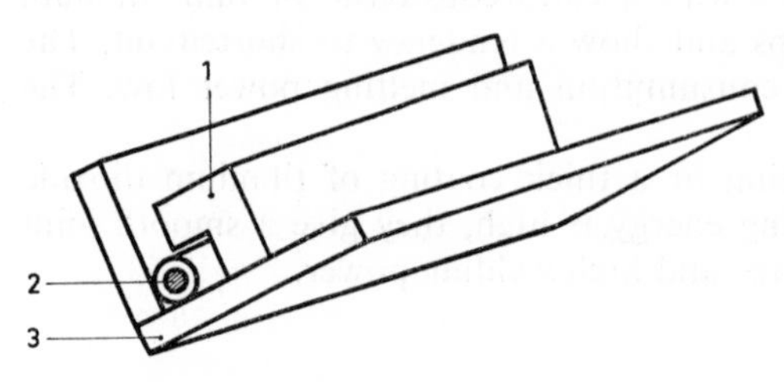

Fig. 2.10. Elin Hafergut welding.
1 = copper rail.
2 = electrode (negative terminal).
3 = workpiece (positive terminal).

2.2.3.2 Protected arc welding

(a) *Welding with atomic hydrogen* [2]

Welding process

Using a variable transformer, an arc is drawn between two tungsten electrodes, surrounded by a supply of hydrogen. The hydrogen is ionized in the arc and takes heat from the electrode points which, due to the gas protection, do not melt (Fig. 2.11). The heat is released at the edge of the disc-shaped arc during the transition from $2H \rightarrow H_2$, which produces a temperature of 4000 °C.

The hydrogen also protects the weld pool against effects from the air.

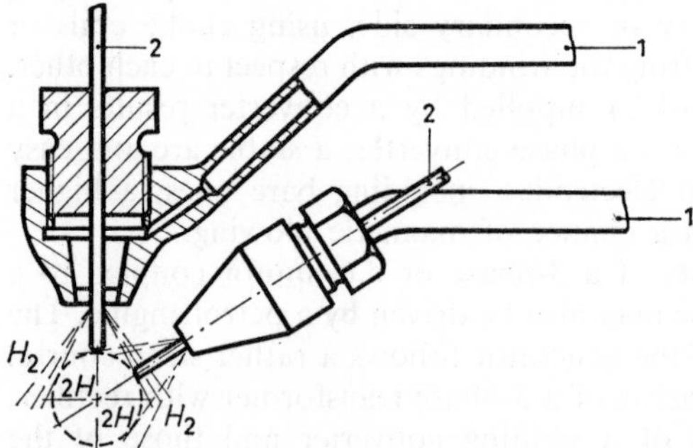

Fig. 2.11. Arc of atomic hydrogen.
1 = hydrogen and current supply.
2 = tungsten electrode.

Equipment

The welding apparatus consists of a variable transformer, a hydrogen cylinder with a gas valve and a welding burner or torch with supply cables and hoses.

The transformer supplies 15–120 A, and the welding voltage is 80V.

The open voltage, which is also the ignition voltage of the arc, is 300V. A safety switch reduces the open voltage of the transformer to 40V.

Application

The method is still used for repairing heavy machine tools, welding components of excavating machines, rails and other workpieces subjected to heavy wear, but atomic hydrogen welding is being replaced more and more by argon arc welding and sigma welding*, both discussed later, to avoid embrittlement of the weld by hydrogen and because of the unwieldy welding torch.

(b) *Argon arc welding* [2]

As early as 1930, a patent was granted in America on arc welding with a protective curtain of helium or argon, but it was only during World War II that this method was further developed, and found wide applications for the welding of aluminium in the aircraft industry and shipbuilding.

Welding process

The arc, surrounded by a curtain of neutral argon or helium gas, is drawn between a non-melting tungsten electrode and the workpiece (Fig. 2.12). The gas curtain protects the weld pool against deteriorating effects from the oxygen and hydrogen in the air.

Since the electrode does not melt, an extra welding rod is usually required for the supply of material.

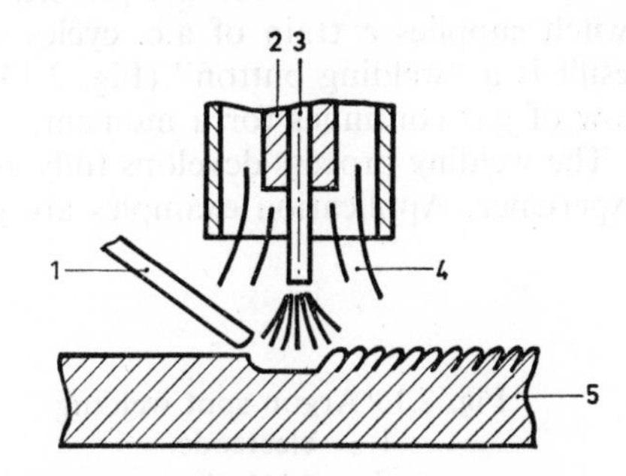

Fig. 2.12. Argon arc welding.
1 = welding rod.
2 = electrode holder.
3 = electrode.
4 = protective gas.
5 = workpiece.

Welding current

D.c. or a.c. current is used, depending on the materials to be welded.

Direct current is eminently suitable for welding metals with a thin-flowing layer of oxide, like stainless steel, copper, brass and nickel alloys.

The negative pole is connected to the electrodes and the positive pole to the workpiece, where the temperature is then a maximum (see 2.2.3.1 (a)). The joint is narrow and deep. A positive electrode produces an arc with an oxide-purifying character, by means of which all non-ferrous metals (including aluminium) can be welded. The joint is wide and shallow. Still

* Sigma welding stands for shielded inert gas metal arc welding, and is currently replaced by mig (metal inert gas) welding.

electrodes plus poling is seldom applied, because of the heavy thermal load of the electrode (see also sigma welding 2.2.3.2 (c)).

In the positive half-cycles, a.c. current produces the oxide purifying effect: in the negative half-cycles, the electrode cools off. Superimposed on a weak high-frequency (h.f.) current, the a.c. current causes the arc to ignite even at a distance of 4 mm. Moreover the h.f. current has a stabilizing effect on the arc. The welding temperature lies between those of the two welding methods mentioned here in Sections 2.2.3.1 (a) and 2.2.3.2 (c). A.c. current is widely used for the welding of light metals. When welding brass, a smoother joint is obtained, for less zinc evaporates than with direct current. Thin plate (≈ 1 mm) is welded with alternating current.

Protective gas

The helium gas formerly used has been replaced by the cheaper argon. Moreover, argon is heavier and less volatile and is more suitable for the welding of thin sheets. Consumption varies from 5–18 litres/min.

Electrode material

Tungsten containing 1·2% thorium oxide needs a significantly lower temperature for electron emission than pure tungsten. Th-W then has a longer life. At under-load conditions, the arc is not quite stable.

Spot welding with argon arc [9, 10]

The sheets to be welded are pressed together with a welding gun. A time-switch supplies a train of a.c. cycles superimposed on a h.f. current. The result is a "welding button" (Fig. 2.13). After the arc has disappeared, the flow of gas continues for a moment.

The welding process develops fully automatically and requires no welding experience. Application examples are given in Table 2.1.

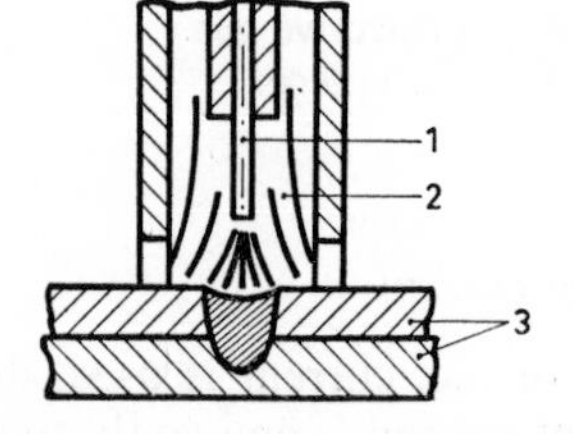

Fig. 2.13 Argon spot welding
1 = electrode.
2 = protective gas.
3 = workpiece.

TABLE 2.1

Material	Sheet thickness (mm)	I x t (A·s)	Welding time (s)
Al sheet Al 99·5	1–2	450–110	3–6
MS sheet Ms. 63	1–2	480–1 600	4–10
Thomassenst. St. VII–23	0·8–2	240–620	2–5

Note: I = current (A): t = time(s).

Spot welding can also take place with a melting electrode and is classified as such under sigma or mig welding.

(c) *Sigma welding* [2]

Sigma welding can be regarded as a continuation of argon arc welding. A metal flame arc surrounded by a neutral gas stands between the melting welding wire and the workpiece (Fig. 2.14). The wire, at the same time serving as the electrode, is wound off a reel at constant speed.

The fusion rate and welding speed are greater than with argon arc welding.

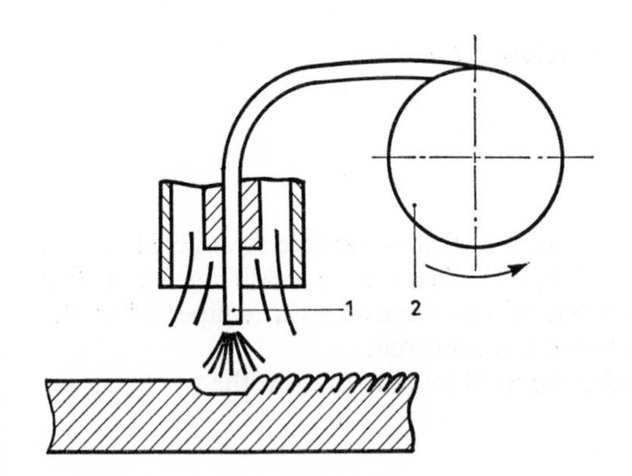

Fig. 2.14. Sigma welding.
1 = welding wire, likewise electrode.
2 = reel with wire.

Polarity

Only direct current is used for welding, the welding wire forming a positive pole in most cases. This gives a stable arc, uniform melting of the wire, small risk of short-circuit, and a purifying effect at the weld by disturbance of the oxide layer.

For metals with a high fusion oxide layer, sigma welding is the best method. Due to the high welding power, thicker sheets can be welded than with argon welding.

Applications: non-ferrous metals and light metals, high alloy steels and carbon steels.

(d) *CO_2 welding* [11]

The metal arc between the melting welding wire and the workpiece is surrounded by a gas-curtain of CO_2, which is much cheaper than the expensive argon.

The method has the advantages of sigma welding, namely, a high melting capacity and a high welding rate.

To obtain a non-porous joint, and to protect the weld from oxidation, the welding wire contains reducing constituents. The droplets of metal in the arc and the losses due to sputtering are greater than if argon were used. The arc is less stable and the joint has a rougher appearance.

The weld arc must be short and constant, and this requires more skill from the welder.

To keep the welding voltage constant, the transformer is given a flat E-I characteristic.

Applications: construction and fine-grain steels.

(e) *Plasma arc welding*

For a description of this welding method see Section 2.2.9 under the heading "Special welding methods".

(f) *Electron beam welding.* See Section 2.2.9.

2.2.3.3 *Submerged arc welding* [2]

Welding process

The arc between a bare welding wire and a workpiece is covered with a layer of powder which has a metallurgical effect on the weld (Fig. 2.15). The welding powder is supplied in front of the electrode via a pipe; after the welding process, the remaining powder is removed.

A copper rail is positioned under the weld seam to prevent the weld powder from running away.

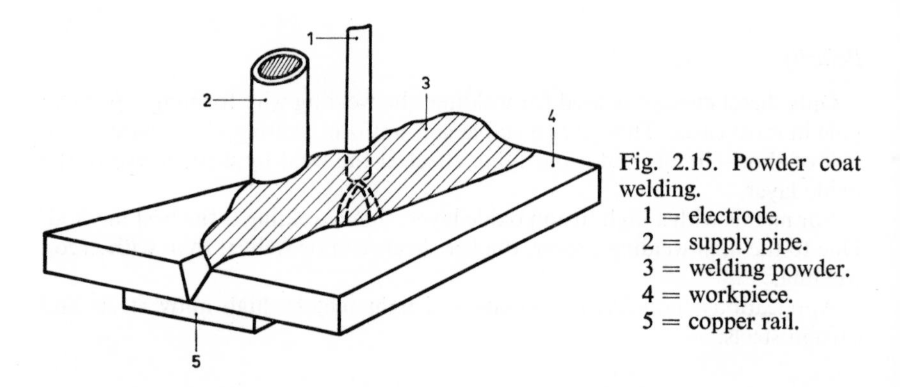

Fig. 2.15. Powder coat welding.
1 = electrode.
2 = supply pipe.
3 = welding powder.
4 = workpiece.
5 = copper rail.

Heat development

The heat development is not due to the electrical resistance of the powder material, but to the arc igniting at a low voltage between 25 and 60V. A high current density of 20 to 200 A/mm^2 can be attained, and consequently a high welding power. The composition of the welding powder and of the welding wire influence the ionization of the arc. Both d.c., a.c. (including 3-phase) current can be used for welding. If 3-phase current is used, the arc is formed between two welding wires, so that two welds of a different level can be applied simultaneously.

Metallurgical aspect

Because of the gradual development of the welding process, and the slow cooling of the weld, all impurities of a specific weight lower than that of the weld pool are given the opportunity of rising to the surface, where they can easily be removed.

Application

Although originally used only for heavy construction in boiler and vessel building, submerged arc welding is now employed in lighter engineering too.

Apart from carbon steels, high-alloy steels can be welded in the same way.

2.2.4 *Resistance welding*

2.2.4.1 *Electroslag welding* [10, 11]

Round about 1950, a welding method was developed in the Paton Institute at Kiev (USSR), by means of which heavy sheets with a thickness up to 60 mm could be vertically welded, without any need for expensive equipment.

Welding process

A continuously supplied bare welding wire is led into a pool of molten powder (Fig. 2.16).

The energy required for the deposition of the wire is generated by the electric current passing through the fluid slag: it causes the weld wire and the edges of the sheets to fuse. An arc occurs only at the beginning of the welding process to melt the powder; after that, the arc is extinguished.

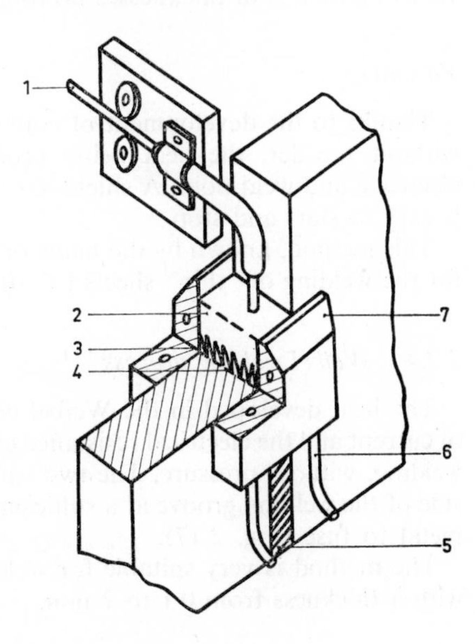

Fig. 2.16. Electroslag welding.
1 = wire supply. 5 = weld.
2 = molten slag. 6 = cooling water.
3 = molten metal. 7 = copper block.
4 = welded metal.

This is a form of resistance welding without pressure. For heavier welding jobs, 2 or 3 welding wires can be supplied simultaneously.

Direction of welding

Copper blocks prevent the molten material from flowing away. Thus, it is only possible to use the method in a vertical direction. The welding process should not be interrupted.

Welding powder

The welding powder is of similar composition to that used for submerged arc welding and contains mainly oxides of Si, Al, Mn, Ca and Mg. The powder consumption per kg of molten metal is 50 g, which is very little. The pool has no effect on the structure of the welded material, but merely serves to protect and heat the weld pool.

Quality

The long heating and cooling times of the weld are conducive to the growth of a coarse-grained dendritic structure of lower strength and toughness than can be obtained by arc welding. The grain structure is refined by heat treatment.

Application

Electro-slag welding is used in heavy industry, for boiler construction, shipbuilding and engineering. Due to the great melting capacity per electrode, 18–24 kg/mm, wall thicknesses up to 60 mm can be welded.

Variant

Thanks to the development of core electrodes, it is now possible to weld without powder, the heat being produced by a continuous arc between electrode and weld pool. A shield gas is led over the weld pool. The process is easy to start and stop.

This method, known by the name of electro gas welding, is used especially for the welding of "thin" sheets ($\leqq$ 10 mm).

2.2.4.2 Weibel welding process [3]

The heat developed in the Weibel welding process is produced by a flow of current and the electrical resistance of the material. It is a kind of resistance welding, without pressure. The two carbon electrodes are drawn along each side of the welding groove at a sufficiently low speed for the two edges of the metal to fuse (Fig. 2.17).

The method is very suitable for welding thin sheets of non-ferrous metal with a thickness from 0·1 to 2 mm.

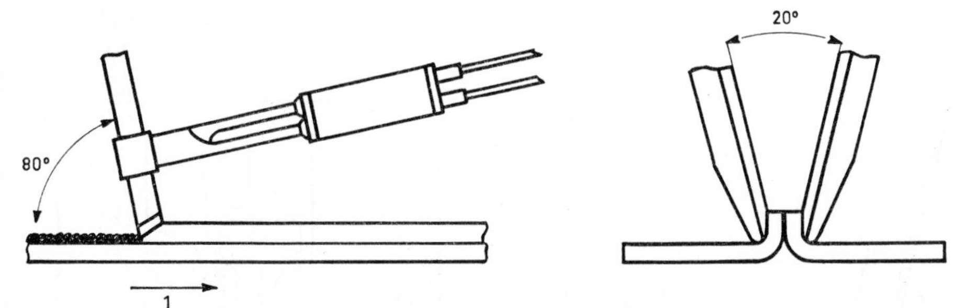

Fig. 2.17. Weibel's welding process.
1 = welding direction.

2.2.5 Thermit welding [2]

At the beginning of the twentieth century the thermit welding process began to gain firm footing in Europe, especially for the welding of railways and heavy machine parts. Although **thermit pressure welding** is the oldest branch of the thermit welding method, **thermit fuse welding** is now finding the widest field of application.

Welding process

The location of the weld is surrounded by a collar of wax which is in turn surrounded by a sand mould with runners and heating openings. The wax is removed by pre-heating and the sprue is left free. The thermit mixture is ignited in the crucible, after which the overheated metal is poured into the mould. After the cooling process, the sand mould and surplus metal are removed (Fig. 2.18).

Preparation and after-treatment are very time consuming. The **thermit quick welding method** developed in 1955 enables two men to complete a prepared rail weld in 15 min. using standard moulds.

Heat development

The heat development is based on the chemical exothermic reaction between a mixture of metal oxide and a reducing agent, when this mixture is ignited.

The thermit mixture consists of 75% iron oxide and 25% aluminium powder, with additions of ferromanganese and ferrosilicon. The aluminium reduces the iron oxide to molten metal and serves to form the liquid slag. To improve the welding characteristics, various other elements can be added.

The oxidising heat of the aluminium exceeds the reduction heat removal of the iron oxide to such an extent that each kg of thermit mixture yields 524 g of molten iron with a temperature of 3000 °C, according to the following reaction:

$$Fe_2O_3 + 2Al \rightarrow Al_2O_3 + 2Fe + \text{heat (759·6 kJ)}$$

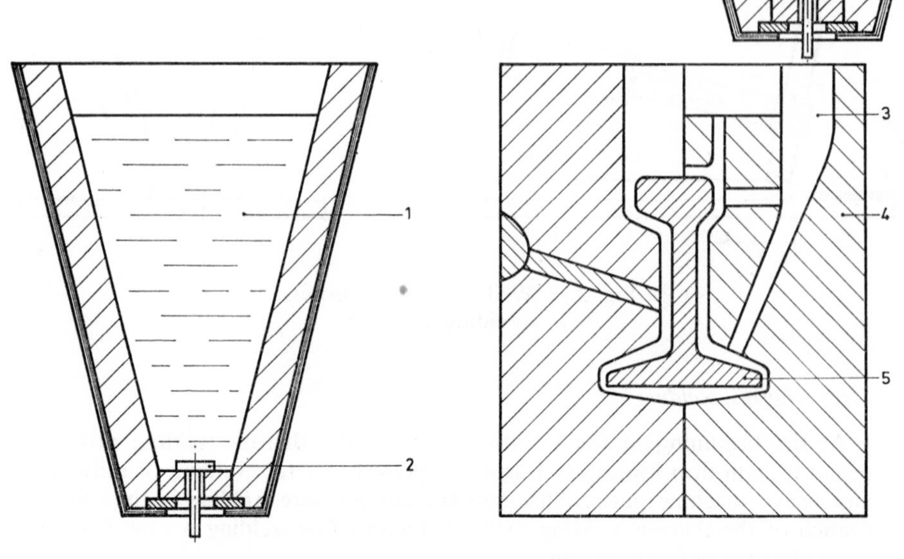

Fig. 2.18. Thermit welding.
1 = thermit powder. 4 = sand mould.
2 = dosing valve. 5 = workpiece (rail).
3 = runner.

In reality, the temperature is approximately 2100 °C due to the addition of cold metal, heat conduction and radiation.

2.2.6 Cold pressure welding

2.2.6.1 Under static pressure [15, 16]

Welding process

The two components to be welded are forced together under great pressure, so that at the points of contact the oxide layer breaks and the pure metal penetrates into the mutual surfaces. By continued deformation of the metal and expansion of the contact area, the two metals form a very close combination, in which the two structures are adapted to each other and an exchange of free electrons can take place. A cross-sectional view does not usually show a dividing line. The entire plastic deformation develops in the solid phase.

The resulting reinforcement of the metal gives the weld a strength far greater than the non-deformed metal.

Welding methods

For the lap weld and the butt weld, plastic deformation takes place according to Figs. 2.19 and 2.20.

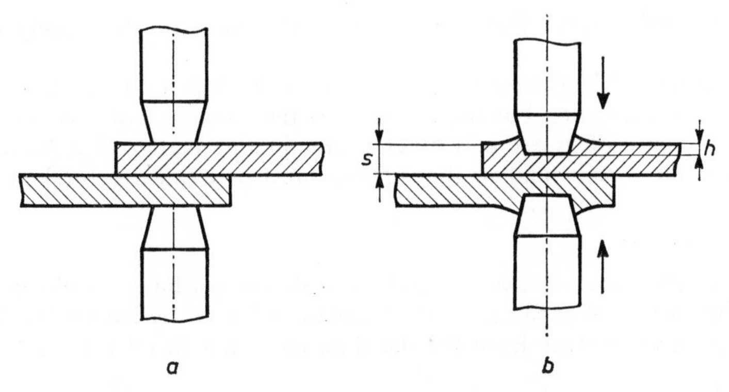

Fig. 2.19. Overlapping pressure weld.
a. before welding. *b.* after welding.

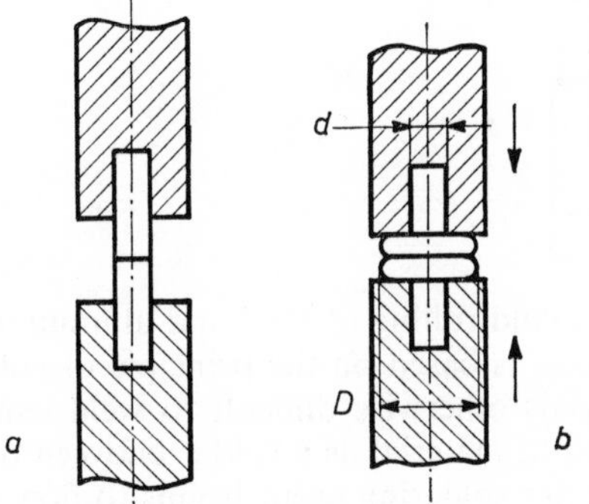

Fig. 2.20. Pressure butt weld.
a. before welding.
b. after welding.

To produce a weld, a certain minimum degree of distortion ε is required.

Lap weld: $\varepsilon_o = \frac{h}{s} . 100\%$

Butt weld: $\varepsilon_s = \frac{D^2 - d^2}{d^2} . 100\%$

The greater the ratio between the hardness of the oxide layer and the hardness of the metal, the lower ε will have to be. See Table 2.2 [16].

TABLE 2.2

Metal	Degree of deformation		Hardness ratios: oxide layer / basic metal	Hardness of basic metal $\times 10^5$ N/m^2
	ε_0	ε_s		
Aluminium	60	160	4·5	16
Copper	84	180	1·3	48

A value of ε higher than required has little effect on the quality of the weld.

In contrast with friction welding, the heat developed during the welding process plays no part; but the hardness of the metal is a decisive factor.

The presence of grease and other contaminating matter impedes a good fusion of the metals, and so the surfaces must be cleaned beforehand.

Different metals

The contact areas between metals of different hardness should undergo the same degree of expansion, which can be achieved by having the harder metal protrude further from the die than the softer metal (Fig. 2.21).

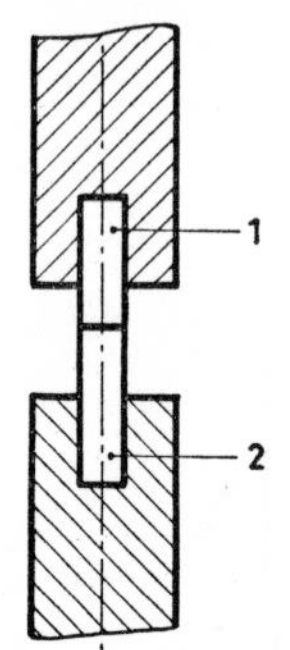

Fig. 2.21. Pressure butt welding of two metals of different hardness. 1 = soft metal. 2 = hard metal.

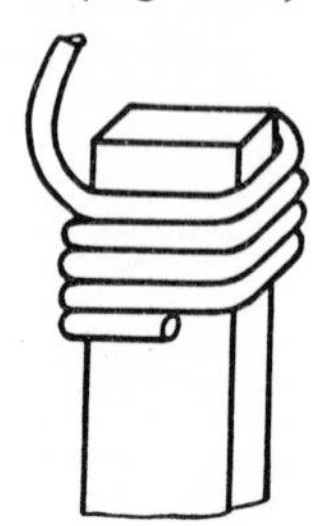

Fig. 2.22. Wire wrap welding method.

The cold rolling of steel and aluminium, or aluminium and tin solder, for instance, is based on the principle of cold pressure welding. The welding of materials otherwise difficult to weld can be facilitated by applying a layer of a third material as a solder between them. As the pressure increases, first the softer soldering agent begins to flow and consequently hardens, then the harder metal follows.

Connections such as Fe–Al–Fe, Fe–Cu–Fe, Al–Pb–Al, Fe–Pb–Fe can be cold pressure soldered.

Applications

- Solder contacts on copper lamellae.
- The copper cap on a transistor can be welded by inserting indium foil between.
- Aluminium cover on aluminium housing.
- The seam of welded Al rail represents a constant electrical resistance during heavy current loads.
- The welding of plastics.
- Wire connections in electronic components and systems according to the wire-wrap method. The wire is tightly wound round the rectangular connecting pin (Fig. 2.22). A cold pressure weld occurs on the sharp edges of the pin [17].
  Wire material: copper, brass, nickel iron and nickel.
  Sheet material: copper, beryllium copper, steel, brass, phosphor bronze and nickel-silver alloys.

2.2.6.2 *Cold pressure welding by impact (explosion welding)* [18, 19]

Welding methods

With nitrocellulose as the explosive powder, a welding pressure is built up in the very short time of 30–50 ms (Fig. 2.23). Because of the heat developed by friction, a lower pressure is required than for the cold welding process described above. Furthermore, there is no real distinction between cold pressure welding at a stamping speed of 10 to 30 mm/min. and explosion welding with stamping speeds varying form 400 to 900 mm/s. In this process, too, a plastic deformation of the components to be joined is the first requirement for obtaining a good weld. The set-up of Fig. 2.24 is used for welding strips or sheets.

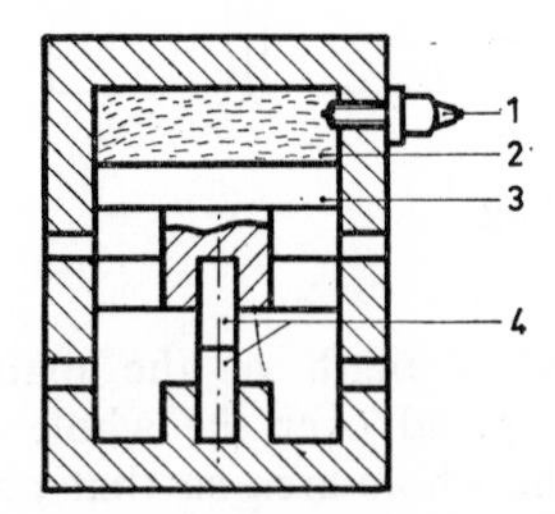

Fig. 2.23. Explosive butt welding.
1 = sparking plug.
2 = explosive.
3 = piston with die.
4 = workpiece

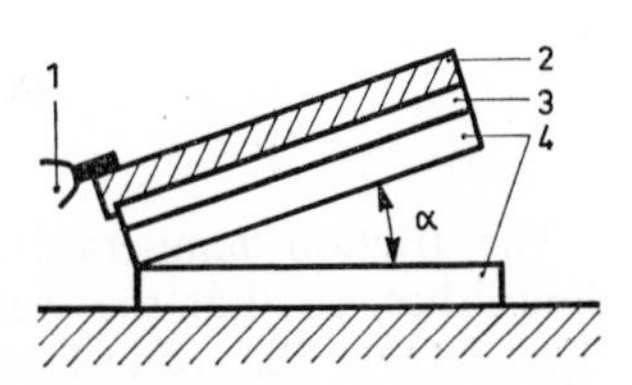

Fig. 2.24. Explosion sheet welding.
1 = sparking plug. 3 = buffer.
2 = explosive. 4 = workpiece.

α = opening angle.

An explosive load is uniformly applied on the upper plate. A buffer plate protects the plate against damage.

Strips of Cu, Al, Ta, Mo, Fe, Ms and Re, with dimensions of 125×25 mm^2 and thicknesses varying from 1 to 2 mm, are welded with an opening angle of $\alpha = 5°$. The contact area of the welded metal shows an undulation whose length and amplitude depends on the angle α. Each weld involves a reinforcement depending on the type of metal welded and the angle α.

Application

Explosion welding was accidentally discovered in America during the formation of metal workpieces by means of an explosive method, when the workpiece fused with the metal mould, thus producing a weld. Today, explosion welding is used by Lockheed Aircraft, National Northern Corporation and North American Aviation Rocketdyne Division. Metal layers serving as anti-corrosion layers can be applied in places difficult of access, inside thick-walled tubes for instance.

2.2.7 *Friction welding* [6,27,28,29]

Welding process

The cylindrical components to be welded together are positioned axially with respect to each other, one part being made to rotate (Fig. 2.25a). The stationary clamped component is then forced against the rotating component.

Due to the high concentration of pressure, welds are produced at the contact points, but they are instantly torn apart again (Fig. 2.25b).

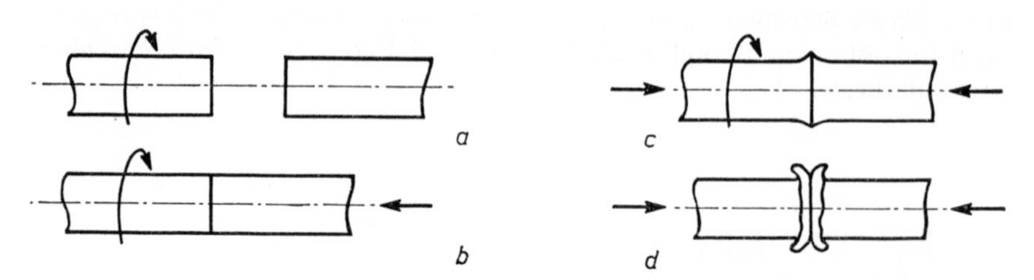

Fig. 2.25. Friction welding.

The friction temperature increases and the strength of the material diminishes, so that the weld points begin to spread over the whole area (Fig. 2.25c). The oxide layer is broken over the whole area so that purely metallic contact can now take place. The rotating movement is stopped and the axial pressure increases (Fig. 2.25d).

Pressure and heat play an important part in the welding process. Due to the heat development the pressure required is lower than for cold pressure welding.

Plastic deformation and expansion of the weld area, and the appearance of a weld collar, are characteristic. Due to the local heat development, the weld collar is smaller than the one formed in cold welding. The process is self-regulating; as soon as the temperature increases and the material softens, friction decreases. The main problem, therefore, is to stop rotation at the correct moment.

Welding characteristics

Rotational speed is not critical. For ferrous metals 2 000 to 4 000 rpm, for copper and aluminium 4 000 to 6 000 rpm. The friction pressure varies from 300 to $2\,000 \times 10^5$ N/m^2. A low pressure applies to copper, aluminium and mild steel, but a high pressure is used for alloyed steel, bronze and nickel steel.

Welding pressure is from 600 to $4\,000 \times 10^5$ N/m^2, with the same distribution as for the frictional pressure.

Welding time: from 3 to 20 sec.

A rapid braking of the rotating shaft, in 1·5 to 3 sec., is required to prevent the weld from fracturing locally. Welding power for normal welding times is 1175×10^4 W/m^2; for shorter welding times it is $2\,940 \times 10^4$ W/m^2.

Application

The friction welding method is still in its initial stages. The first publications appeared in the USSR in 1957 and in the USA in 1960, so the field of application is as yet rather limited.

In electronics, the method is suitable for welding aluminium and copper: in cryogenics, for welding tubes: in the car industry, for welding valve stems, pins and shafts.

In the field of precision engineering, ultrasonic welding is now strongly developing. As far as characteristics are concerned, it can be classified with friction welding, but because of its particular character, ultrasonic welding will be discussed in Section 2.2.9, "Special welding methods".

2.2.8 Hot pressure welding

2.2.8.1 Thermo compression [20,21,22,26]

If two metals are brought into intimate contact with each other, atom diffusion takes place at a temperature-dependent rate. The diffusion results in a perfect bonding, with electrical and metallurgical characteristics equal to those of the basic metal.

The original division plane cannot be seen even after considerable magnification.

The atomic structures of the two metals should be equal to one another, or they should at least be in accord. If the differences are too great, a layer of some other material on one of the two surfaces can promote diffusion.

The heat is externally applied, and the whole area round the weld brought to a constant temperature, which must be accurately adjusted and be about half the melting temperature of the metals to be joined. Pressure and time are less critical. The intimate contact between the metal surfaces calls for preparation immediately before the welding process. Thermo-compression, known by many names in English literature, such as diffusion bonding, heat pressure bonding, thermo-compression bonding, etc., is used particularly in semiconductor manufacture.

In the ever-increasing miniaturization of electronic components, the quality of the connections has a great effect on the electronic characteristics. For instance, a recrystallized zone of a weld or the solder of a soldered connection can have an interfering effect.

Welding methods

For the welding of connecting wires to transistor crystals or vacuum-deposited layers on a substrate, use is made of a die or tool that presses the wire on to the plate with a force of $\approx 700 \times 10^5$ N/m^2. For a 20μ wire, this varies from 80 to 100 mN. The wire is supplied via a capillary tube and compressed air (Fig. 2.26).

The ambient temperature is maintained by a gas. Heat can also be supplied via the tool, but this is a more critical method. A gas mixture (95 N, 5 H_2)

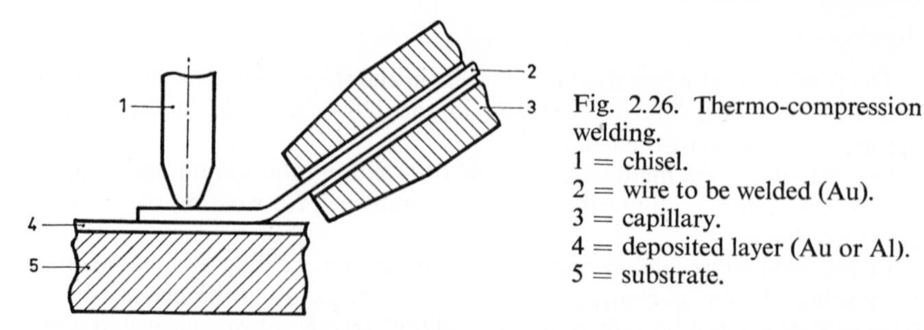

Fig. 2.26. Thermo-compression welding.
1 = chisel.
2 = wire to be welded (Au).
3 = capillary.
4 = deposited layer (Au or Al).
5 = substrate.

protects the weld against corrosion. A vertically-positioned capillary can also serve as the tool. Before bonding, the wire is cut-off and bent over the capillary nozzle (Fig. 2.27).

Instead of cutting the wire, it can be burnt off with a gas flame and the resulting "nail head" used to make the connection (nail head bonding) (Fig. 2.28).

In the methods mentioned above, the whole area round the weld is heated. Wells Electronic Inc., Indiana, have put a nail head bonding apparatus on the market, the tool of which consists of two electrodes insulated from each other. During the welding process a current of 1–300As flows from one electrode to the other via the wire head, so that, due to the Joule effect, the weld is heated locally (Fig. 2.29). The heating time of 100–500 ms suggests

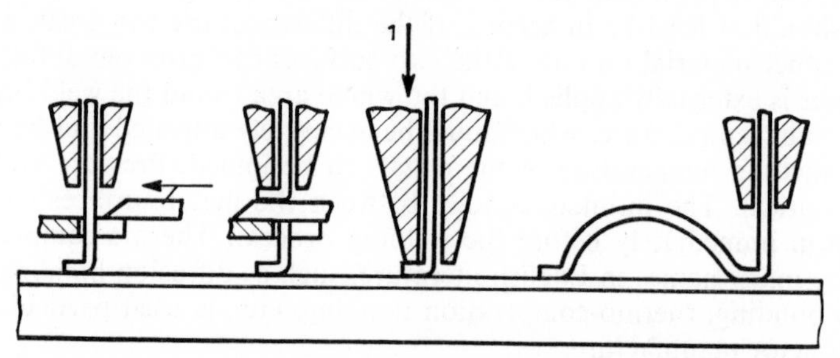

Fig. 2.27. Thermo-compression welding: cutting off the wire.
1 = welding pressure.

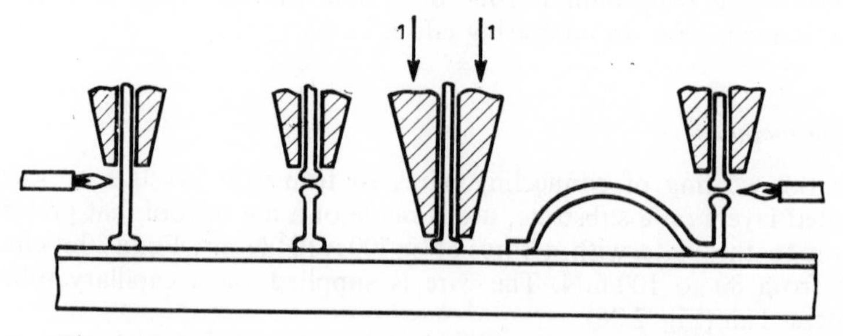

Fig. 2.28. Thermo-compression welding: burning off the wire.
1 = welding pressure.

that, not only is diffusion taking place, but there is also an element of recrystallization. Moreover, the bonding temperature is no longer under full control. Otherwise, the process has developed according to Fig. 2.28.

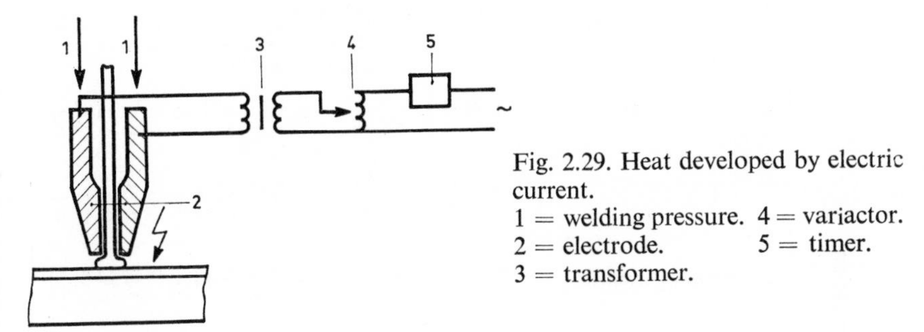

Fig. 2.29. Heat developed by electric current.
1 = welding pressure. 4 = variactor.
2 = electrode. 5 = timer.
3 = transformer.

Application

In the manufacture of transistors, diodes, vaporized components of crystal circuits, etc., the bonding process is used on a wide scale for establishing connections.

Materials like aluminium, gold, lead, platinum, silver and tin are suitable for the process. Their main characteristics are friction coefficient (+), hardness (−), elasticity module (−), atom size (+), proportionality (+) or disproportionaliιy (−) to the tendency to bonding.

Table 2.3 surveys the characteristics of different materials as regards their suitability (γ) for bonding [26].

2.2.8.2 Resistance welding

Introduction

A resistance weld is made under the influence of pressure and heat. With a current flow (I), the heat developed (W) is concentrated in the weld, depending on the electrical resistance (R), and time (t) according to Joule's Law: $W = I^2Rt$ (Fig. 2.30). Usually, R occurs in the contact area between

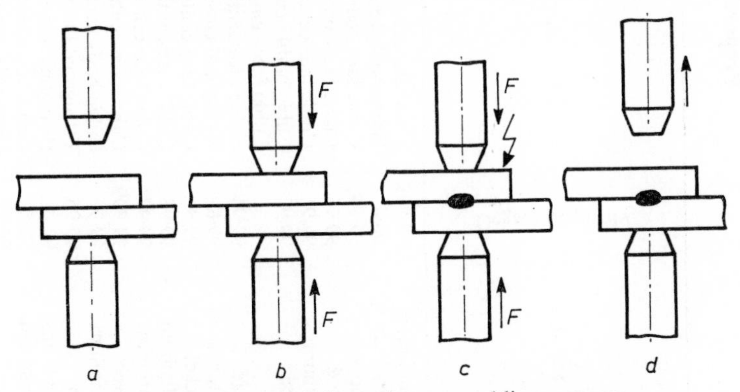

Fig. 2.30. Stages of the resistance welding process.
F = Welding pressure.

TABLE 2.3

Properties of metals and their suitability for thermocompression

Metal	Degree of purity	Crystal structure	Crystal dimensions $\times 10^{-10}$ m	Modulus E (normal) $\times 10^9$ N/m^2	Modulus E (measured) $\times 10^9$ N/m^2	Tensile strength (measured) $\times 10^6$ N/m^2	Cold hardening coefficient	Hardness (Knoop)	Bonding coefficient		Friction coefficient	
									γ	γ max.	Static	Dynamic
Aluminium	99·999	Cubic face-centred	2·82	69·3	*	80·5	0·56	15	1·8	7·0	0·35	0·28
Beryllium	98·8	Hexagonal close-packed	2·25	300	385	273	0·23	270	0	5·0	0·57	0·47
Copper	99·99	Cubic face-centred	2·55	126	*	*	*	25	0·70	4·4	1·1	0·91
Gold	99·9	Cubic face-centred	2·88	80	*	*	*	75	1·84	9·0	2·6	1·8
Iridium	99·9	Cubic face-centred	2·71	526	*	*	*	470	0	0	0·25	0·21
Iron	99·9	Cubic bodily centred Cubic face-centred	2·52	200	*	*	*	61	0·28	4·0	0·53	0·43
Lead	99·94	Cubic face-centred	3·49	16·1	29·3	14·7	0·71	7	3·5	8·3	1·24	1·0
Magnesium	99·8	Hexagonal close-packed	3·20	45·5	27·3	210	0·49	48	0·02	0·04	0·63	0·38
Nickel	99·9	Cubic face-centred	2·49	210	*	*	*	184	0·48	4·0	1·5	1·18
Palladium	99·98	Cubic face-centred	2·75	119	131	*	1·0	170	0·75	17	2·05	1·05
Platinum	99·99	Cubic face-centred	2·77	154	*	207	0·49	65	0·86	5·4	1·3	1·0
Silver	99·999	Cubic face-centred	2·88	77	77	175	0·36	28	0·82	2·5	0·85	0·63
Tantalum	99·9	Cubic body centred	2·94	77	*	*	*	230	0·4	6·0	0·24	0·17
Tin	99·96	Tetragonal body centred	3·16	47·5	45·5	16·8	0·32	7	1·0	1·4	0·73	0·65
Titanium	99·4	Hexagonal close-packed	2·93	112	98	637	0·31	240	0·21	0·8	0·72	0·5
Vanadium	99·5	Cubic body centred	2·71	140	196	86·1	0·26	330	0·2	1·2	0·75	0·43
Zinc	99·89	Hexagonal close packed	2·75	98	57·4	57·4	0·38	56	0·2	0·45	0·7	0·5
Zircon	99·5	Hexagonal close packed	3·19	94·8	48·3	721	0·65	230	0·1	0·4	0·56	0·45

* = not available

the components to be welded together. If a welding zone or welding lens occurs, it is fuse welding. If the welding process develops in the solid state of the material, we are dealing with a diffusion weld, whose form shows similarity with a hot pressure weld.

Usually, both these aspects can be discovered in a resistance weld. In English literature we often come across the expression "braze weld", by which is understood the weld made by means of a third non-ferrous metal with a melting point lower than that of the basic metals, but higher than 425 °C. Although often a chemically or galvanically applied layer is used to improve the weldability of a metal, it is usually incorrect to speak of a soldered weld as described above; the term will therefore not be used from now on.

Owing to the strong heat concentration, and the corresponding high temperatures, the welding time (t) is many times shorter than for hot pressure welding. In precision welding engineering especially, the welding time is measured in msec.

The welding pressure (F) ensures that the oxide layer and contaminations on the welding surface are broken and removed. The same pressure ensures a complete welding process during melting or softening of the material, so that no arc is drawn. Consequently, the corrosive effect of the surrounding atmosphere is reduced to a minimum.

Heat development and temperature distribution

The heat development (W) and the temperature distribution (θ) taking place in the weld during current passage depend on the local material resistances (R_1, R_3, R_5, R_7) and contact resistances (R_2, R_4, R_6) (Fig. 2.31). If a weld is symmetrical in shape, dimensions and material with respect to the contact plane of the two components, the (θ) distribution for a certain welding time (t) resembles Fig. 2.31.

To achieve this optimum distribution, R_4 must be at maximum.

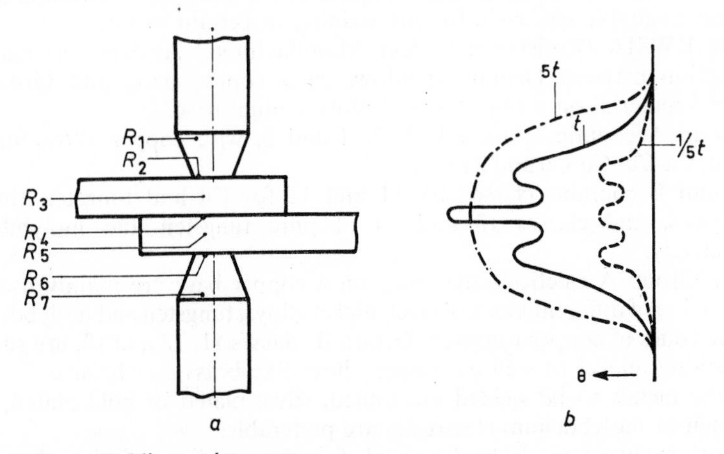

Fig. 2.31. *a*. Welding resistances.
b. Temperature distribution (θ) resulting from the passage of current

If R_4 is low, however, as in the case of brass, this resistance can be increased by applying a thin layer of metal (tin, for instance), or by providing one of the workpieces with projections (see "Projection welding" on p. 62). Also the welding pressure (F) has an effect on R_4, and also on R_2 and R_6 ($R^2 \propto 1/F$). In order to keep W low at R_2 and R_6, a large F will be used, but in that case, R_4 decreases. A compromise will have to be found.

As regards the resistances R_3 and R_5, the general rule is to weld soft metals under low pressure and hard metals under high pressure.

For the remainder R_2 and R_6 can also be influenced by the choice of electrode shape and material (see next paragraph on Electrodes). If welding takes place at low values of R_2 and R_6, a great part of the welding head will be lost owing to conduction. In this case a short welding time will be selected.

For metals with a high value of R_2 and R_6, it is often better to have W take place more gradually, which involves a longer t.

If, after a certain time t, a temperature distribution is obtained as in Fig. 2.31b, it is found that at a shorter time, say $1/5\ t$, the shares of R_2, R_4 and R_6 in the temperature increase are equal. On the other hand, however, after about $5t$ the temperature throughout the distance between the electrodes will be equal.

Electrodes

The electrodes have three functions:

- To transfer the welding pressure.
- To conduct and concentrate the current in the weld.
- To remove the welding heat from the metal surface.

They must therefore have the following characteristics: a suitable hardness and wear resistance; an homogeneous crystal structure; a high electrical conductivity; and a low tendency to alloy with the metals to be welded.

A metal cannot satisfy all these requirements. The only solution is to find the most suitable electrode for the welding of certain metals.

The RWMA (Resistance Welder Manufacturers' Association) makes a distinction between Group A: alloys on a copper base, and Group B: sintered material on a tungsten and molybdenum base.

Group A contains classes 1, 2, 3, 4 and 5, with copper alloys such as Cr-Cu, Ca-Cu, Co-Cu and Zr-Cu.

Group B contains classes 10, 11 and 12 for Cu and tungsten sintered electrodes, and classes 13 and 14 of pure tungsten and molybdenum respectively.

The Group A electrode materials on a copper base are mainly used for the welding of different kinds of steel, nickel alloys, tungsten and molybdenum.

Electrodes of copper-tungsten, Group B, classes 11, 12 and 13, are suitable for welding nickel as well as copper alloys like brass and bronze.

If the metals to be welded are tinned, silver-plated or gold-plated, pure tungsten or molybdenum electrodes are preferable.

Thorium-tungsten electrodes used for the welding of tinned metals, aluminium and brass give remarkably good results. The tip of the electrode

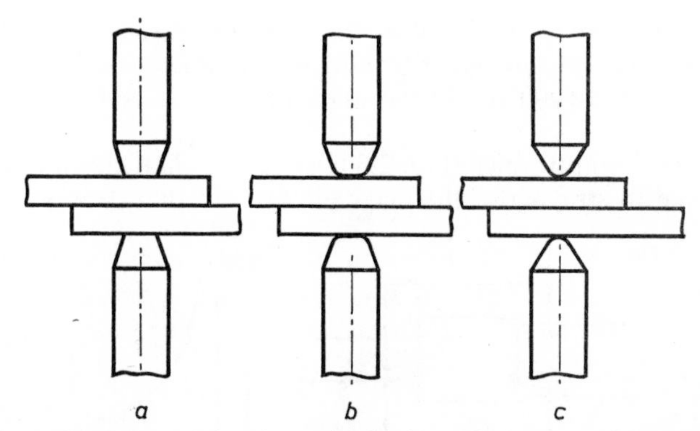

Fig. 2.32. Shape of the electrode tip depends on the weld material.

Fig. 2.32(a) A flat tip is used for bare steel and non-ferrous metals. During the welding process the pressure, and consequently the current density, remains constant.

Fig. 2.32(b) the tip is rounded for welding metals with an oxidized surface. The current density is maximum at the beginning of the welding process, so the oxide layer is easily penetrated.

Fig. 2.32(c) A sharp rounded tip is used for rusty metals and for light metals, owing to the heat concentration.

is selected according to the surface condition of the metals to be welded (Fig. 2.32).

Energy supply

The energy is not supplied continuously: on each welding pulse, a certain quantity of energy is fed to the weld.

The energy can be dosaged in two ways.

(a) According to the transformer timer method.

(b) According to the transformed capacitor discharge method.

(a) *Transformer timer method*

When S switches over (Fig. 2.33), timer TS passes one or several cycles

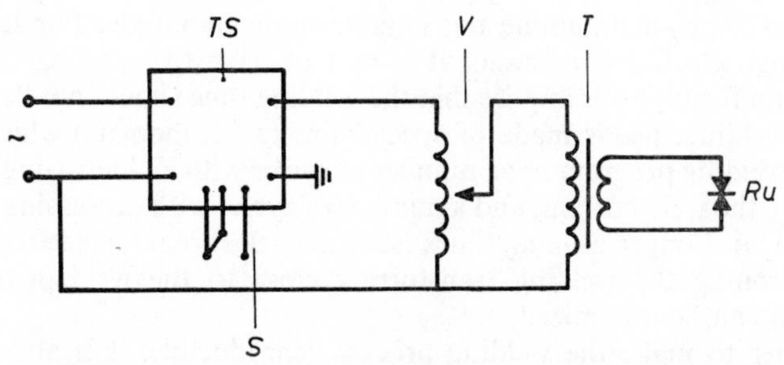

Fig. 2.33. Diagram of transformer welding method.

of the mains voltage via variac V or the auto-transformer to welding transformer T, in which the voltage is transformed to a few volts and the current to a value of 100–10 000 As. In the weld, the current and voltage drop are in phase.

If T is a core transformer (Fig. 2.34a), the *E-I* characteristic will form a steep line due to straying of the magnetic flux. In that case, we speak of a

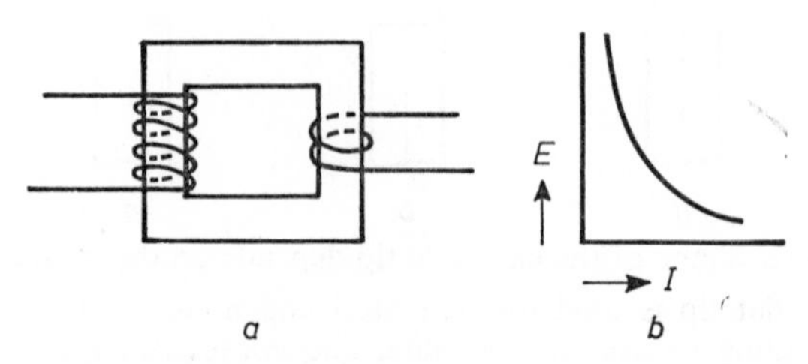

Fig. 2.34. Core transformer with soft *E-I* characteristic.
a. core transformer. *b.* voltage–current characteristic.

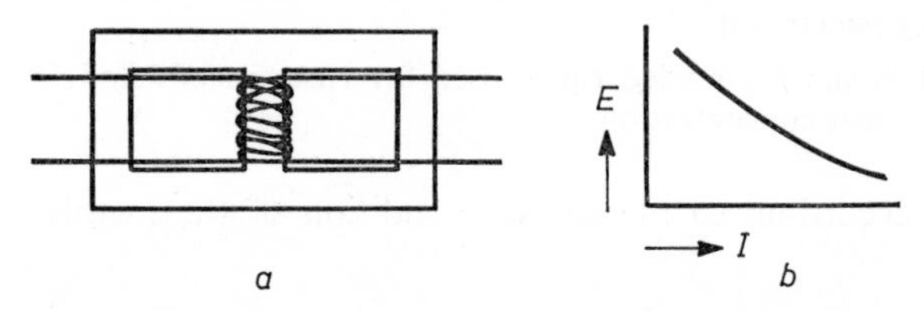

Fig. 2.35. Jacket transformer with hard *E-I* characteristic.
a. jacket transformer. *b.* voltage–current characteristic.

soft transformer. When the welding voltage *E* varies, the product *E-I* will remain constant. For the welding of oxidized or tin-coated and nickel-plated metals it is best to use the soft transformer. The case transformer with a flat *E-I* characteristic (Fig. 2.35) belongs to the hard type. For bare and harder types of metal, for which the voltage remains more constant during the welding process, this transformer, with its low internal resistance, is used.

The welding time can be adjusted to a $\frac{1}{2}$, 1 or several cycles, (1 cycle corresponds to 20 ms) maintaining the same current amplitude. For hard metals with a high electrical resistance it is best to have the welding temperature increase uniformly; this implies that the welding time should not be short. For heavier welding, use is made of special timers (tempomats) which serve to adjust a welding programme; a number of cycles with an increasing amplitude (upslope), the weld current, and a number of cycles with decreasing amplitude (downslope). Longer welding times (several cycles) require electrode cooling. By positioning the welding transformer close to the weld, losses due to induction can be minimized.

In order to make the welding process reproducible, it is also important to keep the resistance of the secondary circuit as low as possible. Variations of the initial weld resistance R_u then have but little effect on the heat development.

What was said above is explained by means of the welding table relating to the simplified model of Fig. 2.36.

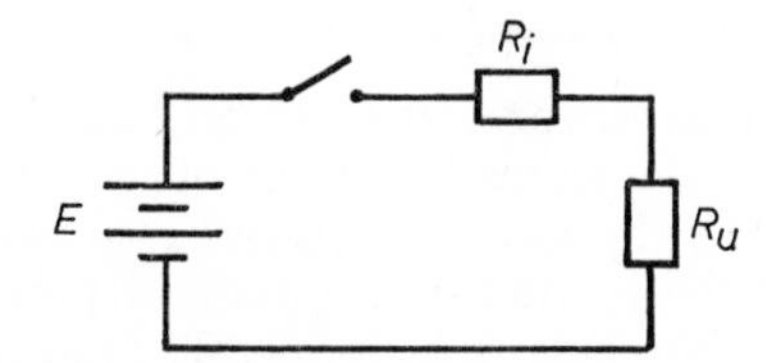

Fig. 2.36. Simplified weld model.

Here R_u represents the welding resistance and R_i the sum of the remaining resistances.

Voltage E corresponds with the open secondary voltage of the welding transformer.

By means of formulae used in electrology, it is easy to calculate that the heat development W in R_u when the current is conducted can be written as follows:

$$W = E^2 \frac{R_u}{(R_i+R_u)^2} . t.$$

Say that $E = t = 1$.

In Fig. 2.37, W is plotted as a function of R_u with R_i as a parameter. A variation of R_{u0} near the top of the W curve has less effect on the heat development than when R_{u_0} is positioned on the upward slope of the W curve.

Moreover, if the welding process takes place between R_{u_0} and R_{u_1}, the heat development at $R_i = R_{i_1}$ will vary less ($\triangle W_1$) than for $R_i = R_{i_2}$ ($\triangle W_2$).

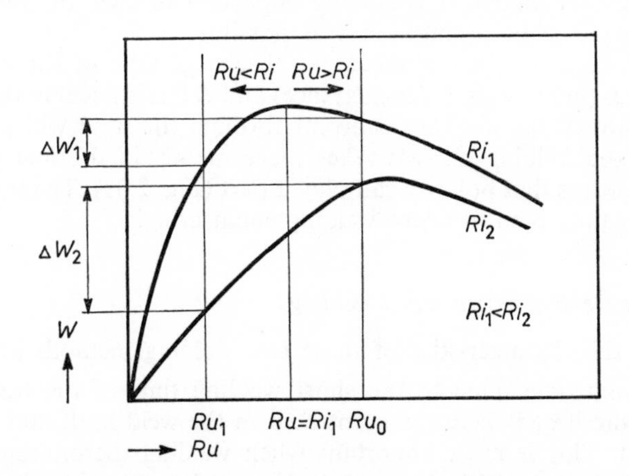

Fig. 2.37. Heat development (W) in the weld as a function of the welding resistance (R_u) at various values of R_i.

R_{u_0} = welding resistance before the welding process.

R_{u_1} = welding resistance after the welding process.

The general rule, therefore, is to make R_l equal to (or less than) R_{u_0} which means that the current conductors should be of heavy construction and the points of contact silver-plated.

(b) *Transformed capacitor discharge method* [43]

A capacitor battery (Fig. 2.38) is charged to a voltage of 100–1 000V from the mains by means of a variable transformer R_t and a rectifier G. When the electrodes exercise pressure on the workpieces to be welded together, switch S is placed in position 2, and C discharges across weld R_u via a pulse transformer T. Transformer T changes the charge voltage V

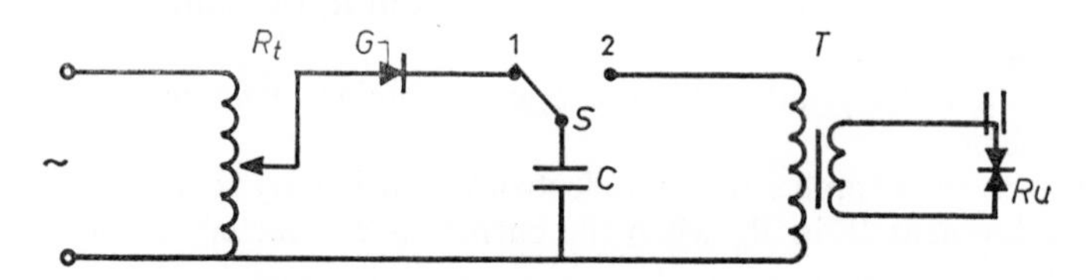

Fig. 2.38. Diagram of the traco welding method.

into the welding voltage E_u ($\leq$20V) and increases the discharge current. Of the total energy $\frac{1}{2}$ CU^2 in the battery, about half is fed to the weld when the battery of capacitors is discharged. The losses depend on the resistances and inductances in the discharge circuit. The iron packet of T can be made of ceramic with iron powder. In contrast with the transformer welding method, just dealt with in (a), R_u has no effect on the power consumption of T. For the remainder, it is equally important to keep the impedance of the secondary current circuit as low as possible.

The welding time is a few msec, the discharge time of the battery, and depends on L (induction), C (capacitance) and R (resistance) in the discharge circuit. Although the discharge current through the weld will pass zero at least once, the welding process takes place mainly in the first part of the cycle. This means that polarity can play a part (Fig. 2.39). The traco welding method is mainly used in micro welding engineering.

Transformer welding versus traco welding

Some of the characteristics of these two welding methods are:

- The welding time. Due to the short welding time of the traco welding method, the heat is better concentrated in the weld itself and less heat is dissipated. This is most important when welding thermo-sensitive components and super-conductive metals. For hard metals with a high electrical resistance, in the presence of an oxide layer or contamination, a more gradually developing welding process could be more suitable. The possibility of adjusting the welding time offers better matching.

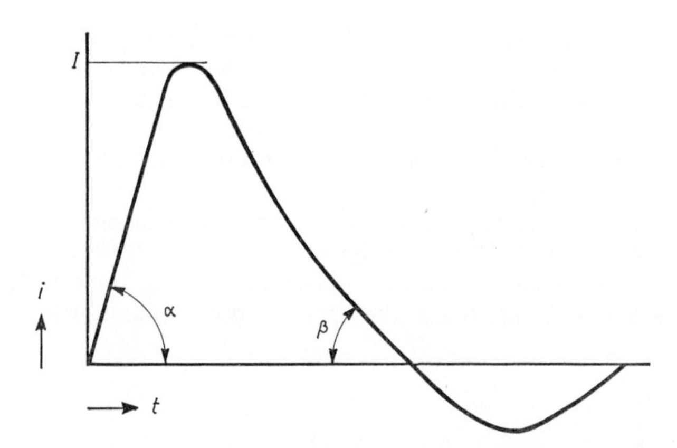

Fig. 2.39. Current (i) variation in the traco welding method.
$a = f(L)$. $\beta = f(R)$. $i_{max} = If(C, U)$.

- The initial voltage of the traco welding method is higher than for the transformer welding method. Consequently, there is a greater chance of sparking and the welding head has to meet heavier demands. The traco welding process is the more explosive of the two, and consequently somewhat more critical.
- The Peltier effect is more pronounced with the traco method than with the transformer method. Especially in micro-welding does this effect become noticeable at times. Depending on the electrothermal voltage of the components to be welded and the current direction, extra heat is developed in the contact area (see Fig. 2.40) at the points A and C, or heat is taken away at B and D [38].

Resistance welding of different materials[37]

To describe the temperature distribution in the weld, a symmetrical welding model was used earlier. If, however, the metals to be welded together have different properties, the welding process changes. For instance, the position of maximum temperature is shifted to the metal with the highest

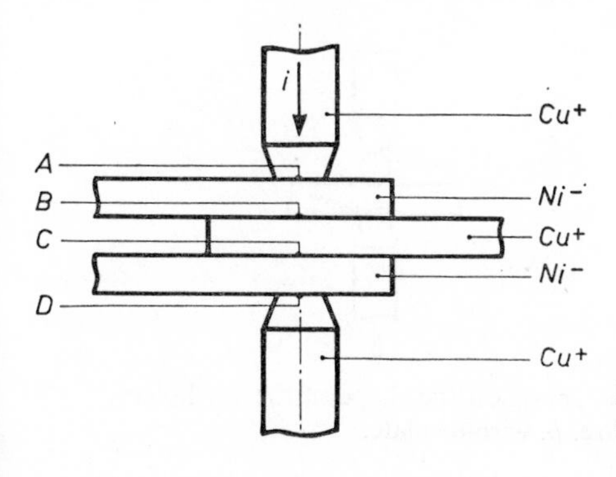

Fig. 2.40. Peltier effect on the welding heat.

specific resistance. In the case of Fig. 2.41, the nickel will recrystallize, whilst, at the position where the weld is to be made, less heat is developed. Conditions can be improved by making the weld according to Fig. 2.42.

This implies that the *dimensions* play an important part, as well as the *electrode materials.* For electrode A, for instance, we choose a tungsten alloy and for B a copper alloy.

The welding process for wire-to-wire (Fig. 2.43a) is more critical than for wire-to-plate (Fig. 2.43b). Also the shape of the workpiece is important.

The welding variables should be so coordinated that, during the welding process, both metals approach the melting point simultaneously. If the

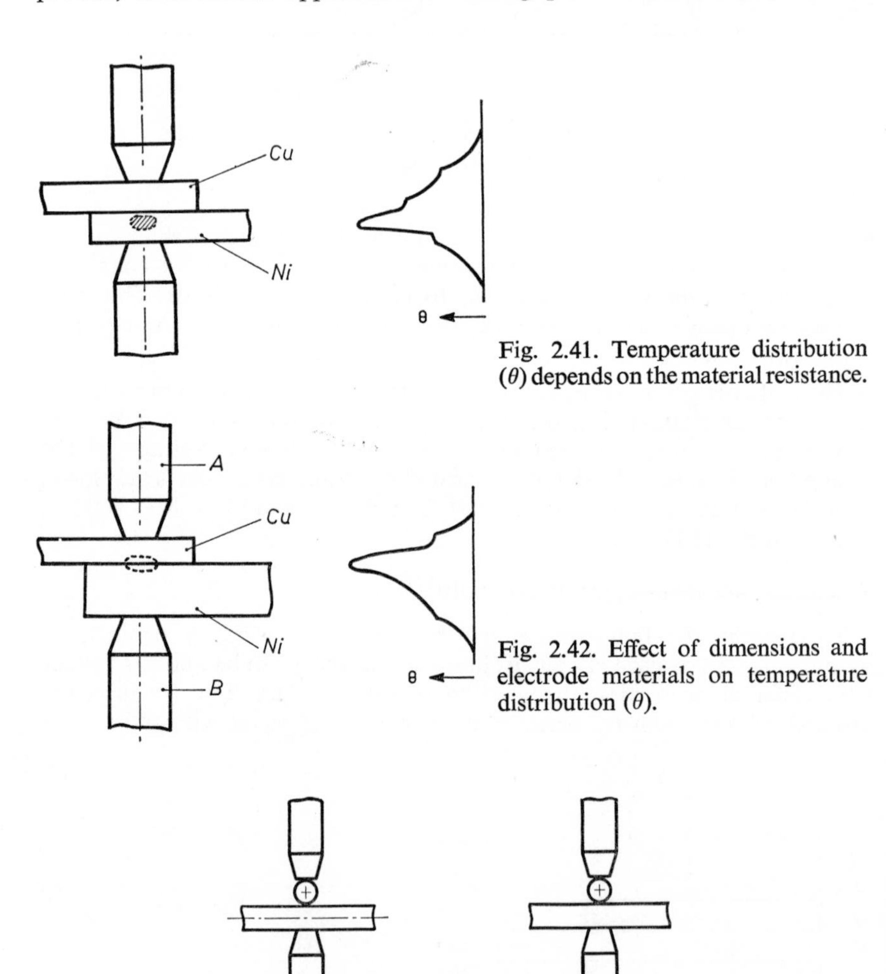

Fig. 2.41. Temperature distribution (θ) depends on the material resistance.

Fig. 2.42. Effect of dimensions and electrode materials on temperature distribution (θ).

Fig. 2.43. Transition resistance depends on the shape of the workpiece. *a.* wire-to-wire. *b.* wire-to-plate.

metals have different *coefficients of expansion*, the weld may be subjected to high internal stresses when cooling off, and this may result in a fractured weld.

For materials with good heat conductivity, such as copper alloys, it will be difficult for a melting zone to occur during the welding process. The weld is made while the metal is in the solid state. To ensure satisfactory bonding, the *lattice structures* of the metal should not differ too greatly. If a melting zone occurs in the weld, it is no longer the original crystal structure but the *solubility* of the two metals that has an effect on the quality of the weld.

The weldability of the metals can be considerably improved by chemically or galvanically applying a layer of some other metal.

Resistance welding methods

(i) *Spot welding*

The electrodes serve to provide the welding pressure, the welding current and to concentrate the welding heat (Fig. 2.44). The combination of pressure and heat concentration sets heavy demands for the tip of the electrode. Electrodes can be cooled to give them a longer working life.

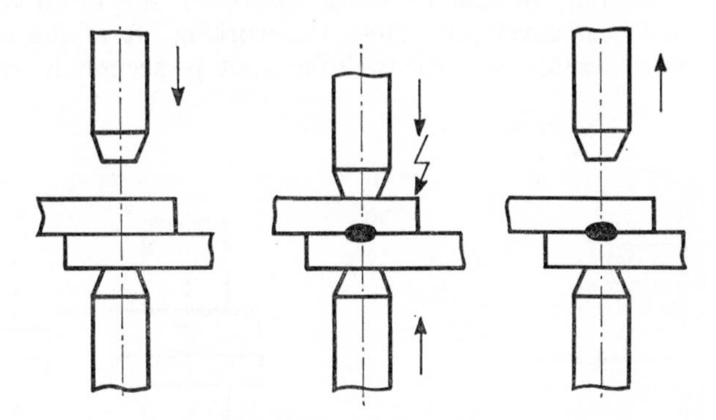

Fig. 2.44. Spot welding.

For the welding of light metals, aluminium for instance, two current pulses are sometimes used, so that, during the preliminary pulse, the electrode sinks slightly into the metal. This increases the contact area between electrode tip and metal, so that the thermal load on the tip is less during the second heavier welding pulse.

During welding, the workpiece material expands laterally, which leaves an impression after the material has cooled off. The depth of this impression depends on the material properties and shape of the electrode.

Spot welding finds a wide field of application in the:

- Automobile industry for body work.
- Aircraft industry: riveting is largely replaced by welding.
- Building industry: for the welding of steel reinforcements and concrete skeletons made of profiled steel.

- Precision engineering industry: for measuring tools and the manufacture of instruments.

(ii) *Seam welding*

Instead of pointed electrodes, rotating electrode discs are used to transfer the pressure and welding power (Fig. 2.45).

The welding current is supplied by a number of pulses per unit on angular rotation.

The field of application includes:

- Canning.
- The car industry: petrol tanks, sheet metal constructions.
- Coach building.
- The aircraft industry: pressurized cabins, fuel tanks and wing construction.
- Radiators.

(iii) *Projection welding* [40]

To increase the welding resistance of the material locally, a projection is made in the material. Instead of using electrodes, the metal itself now concentrates the heat developed. Thus, the working life of the electrodes is increased considerably and the welding spot is accurately positioned (Fig. 2.46).

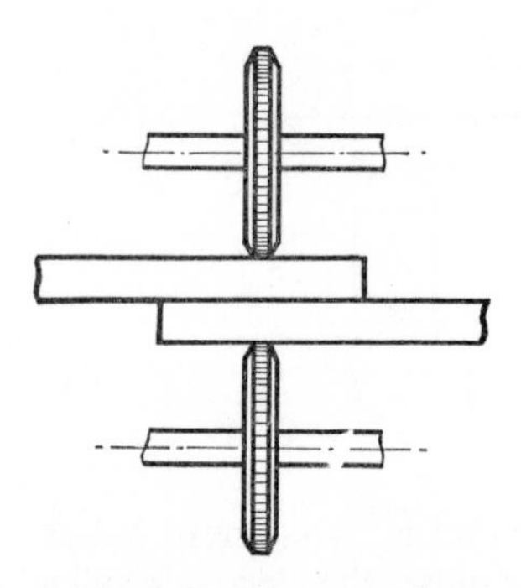

Fig. 2.45. Seam welding

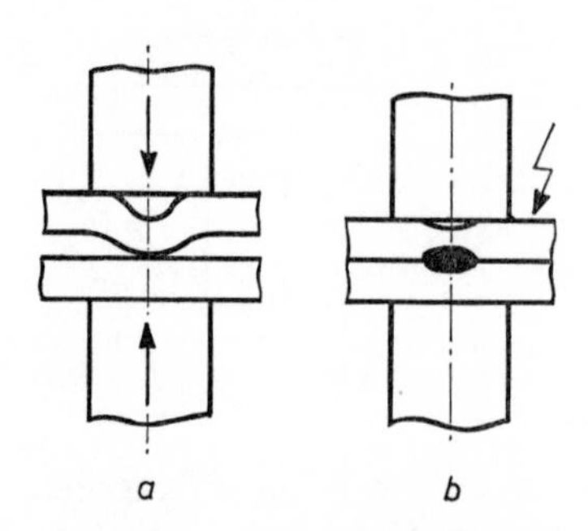

Fig. 2.46. Projection welding.

Projection welding is especially suitable for metals with high heat conductivity or with low contact resistance, such as brass, aluminium, copper and silver-plated, gold-plated and cadmium-plated metals.

If the metals to be welded together are of different hardness and electrical conductivity, the projections are made in the hardest metal with the lowest conductivity.

For two plates differing in thickness, and with the projections made in the thinnest plate, the welding time is $s_2/s_1 \times t$ if t represents the welding time for two plates of equal thickness s_1. (Fig. 2.47).

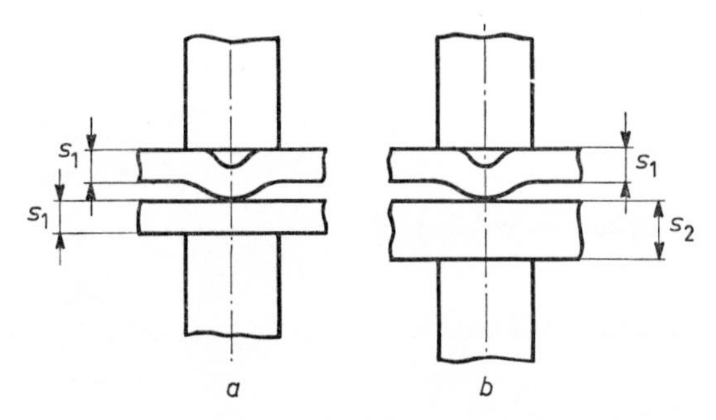

Fig. 2.47. Required welding time depends on the sheet thickness ratio s_2/s_1.
a. welding time *t*. *b*. welding time $s_2/s_1 . t$.

Application includes the mass production of pressed, stamped or drawn components, such as the connections between transistor housings and the base plate.

(iv) *Butt welding*[2]

(1) *Pressure butt welding*

In this rather old method of resistance welding, the electrodes only serve to supply the current and, if necessary, to build up a pressure. (Fig. 2.48).

In mechanical engineering, butt welding is used in the first place for the welding of rod-shaped workpieces. After the pressure has been built up, the current supply follows. The heat development is first concentrated at the place of contact, but then begins to spread over the whole area between the electrodes. During welding, the metal is upset. Due to the heat development, the degree of deformation of the material need not be the same as is necessary for cold pressure welding. Another form of butt welding is used for welding pipes (tubes) axially (Fig. 2.49).

This is a continuous welding process in which a HF welding current up to 450 kHz is used. The method is also known as "High frequency welding" or, with reference to pipe welding, as "Upset tube welding"[39].

Because of the tendency for a high frequency current to flow along the

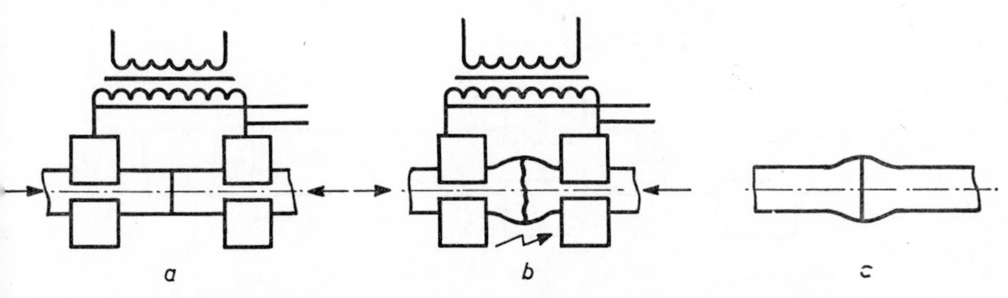

Fig. 2.48. Successive stages in the pressure butt welding process.

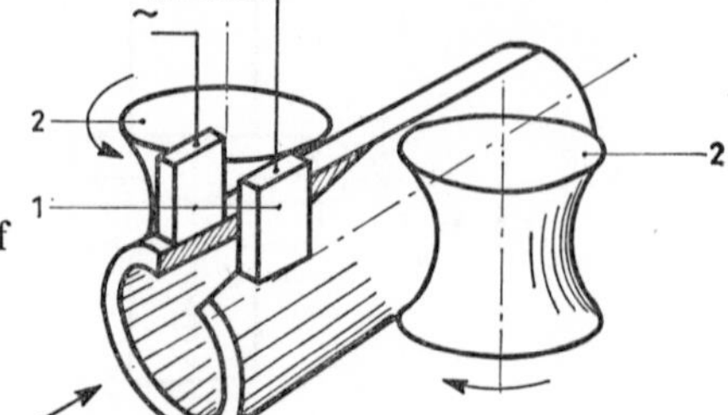

Fig. 2.49. High frequency welding of tubes.
1 = electrodes.
2 = pressure and transport rollers.

surface of the conductor (skin effect) and to follow the route of least inductance, the position between the electrodes and the points where the edges of the pipes come together are mainly heated.

Owing to the expensive current converter, the prices for pipe welding equipment are very high. Apart from pipe welding, the high frequency method is used for the welding of strips. For instance, a thin strip of tape saw to a wider strip of a cheaper steel; the welding of gauze material, profile material; or for welding a helical strip round a cylinder by means of current pulses or a continuous current.

(2) *Flash butt welding*

Is a method of resistance welding in which heat is more developed in the welding plane than is the case with pressure butt welding.

Before the ends of the workpieces are brought into contact, the voltage of the transformer is applied to them. As soon as contact is made, the workpiece is given an axial vibration, to produce a metal arc which melts the metal at the surface and throws it out under a shower of sparks (Fig. 2.50a). Then, the two ends are pressed together and the welding current switched off.

Once the weld has cooled off, it looks like Fig. 2.50b. The different welding stages are completed automatically. In contrast with pressure butt welding, the welded surfaces require no previous treatment. If the areas are too large to draw an arc, the ends are slightly pointed.

The weld shows less thickening than in pressure butt welding. Flash butt

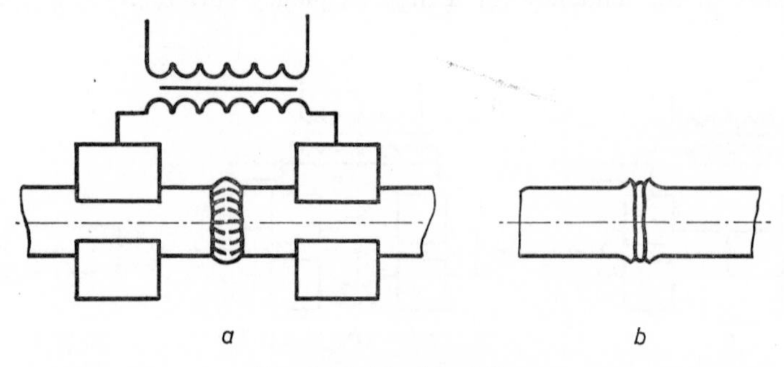

Fig. 2.50. Flash welding.

welding is widely employed for welding crankshafts in engineering, and is also very suitable for welding profiled steel, light metal window frames, drills and milling tools.

Stud welding (Section 2.2.8.4) as used in precision engineering is a modified version of flash welding.

(v) *Micro-welding techniques*[41,44,43]

In the aim towards miniaturization of electronic systems, it is necessary, in welding the connections, that the right amount of energy is always supplied to the welding spot. Especially in America has this outline problem grown into a special branch of resistance welding, known as micro-weld engineering.

It is employed in space research and rocket building, and in many electrical systems like computers, measuring equipment, telephony, radio and television engineering, etc. In all cases, the main points are the welding of:

(1) Small components like transistors, capacitors, wire-wound resistors, diodes, resonators and so on.

(2) Small packages of components, known as modules as regards the internal connections, that is, wire-to-wire connections or wire-to-strip.

(3) Integrated circuits, whose terminals are welded to a layer of metal deposited on a substrate of glass, ceramic or silicon.

For wire-to-wire or wire-to-strip welding the principle of the welding process does not differ from that described in the introduction of 2.2.8.2. The heat development depends both on the contact resistance and the material resistance.

However, for welding a wire or strip to a layer deposited on a substrate, the electrodes are placed side by side (Fig. 2.51). In this situation, the contact resistance is less important than the material resistance.

During the first part of the welding time, the current flows through the strip (3) and as its temperature increases so does the resistance, after which the current tries to find a cooler way and penetrates deeper into the deposited film material (4). The welds are then formed at the location of the electrodes, where the current density is a maximum (series weld).

If the upper material (3) has a high resistance with respect to the lower material (4) the weld becomes critical. A smaller distance between the

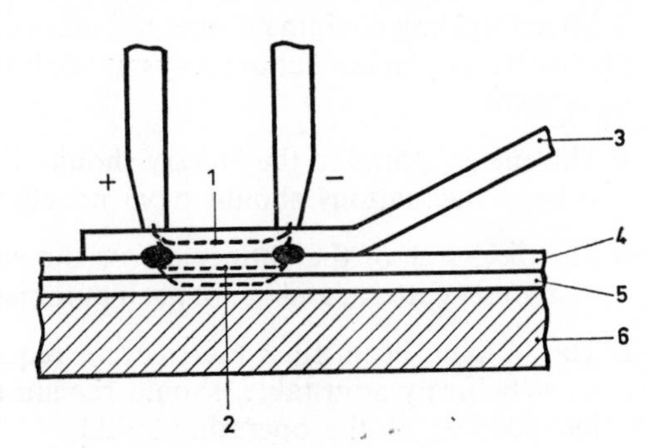

Fig. 2.51. Welding process on a deposited layer.
1 = first current passage.
2 = next current passage.
3 = Au or Ni.
4 = NiCr or Au.
5 = Ni.
6 = substrate, together with chromium layer for the attachment.

electrodes is then desirable (Fig. 2.52). Another solution is to apply mylar foil between the components. If the welding resistance is too low with respect to the resistance of the whole welding circuit, the current will be little affected by R throughout the welding process. This implies that the heat development $W = I^2Rt$ increases with R and results in what is known as "blow out".

Sputtered welds are avoided by using equipment (Texas Instruments Inc. and Weldmatic) that feeds back the welding voltage E to the welding transformer, in order to keep the welding voltage constant. The heat development $W = E^2/R$ will then fall off as R increases.

If the electrodes are positioned close together, the weld will be formed between the electrodes (parallel gap weld) (Fig. 2.52). The electrode distance and the size of the electrode tip depend on the material to be welded and the film thicknesses.

If the electrodes are separated only by an insulating layer forming a rigid body together, the term split electrode (Fig. 2.53) is used. The thickness of insulation δ is about 1·5 times the thickness of the material to be welded. A uniform pressure distribution over the two electrode tips is, however, critical, so this method is less suitable for the welding of hard materials.

Fig. 2.52. Electrodes close together (parallel gap weld)

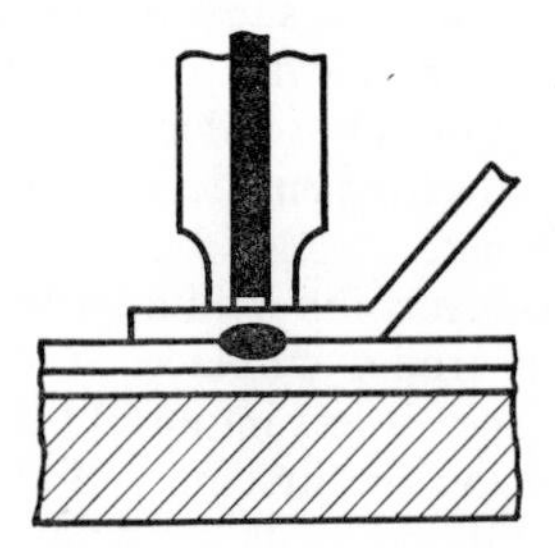

Fig. 2.53. Welding with a split electrode

Micro-welding equipment operates according to the transformed capacitor discharge. The manufacturer pays special attention to three aspects of this equipment:

- The energy stored in the battery should depend only on the setting; mains voltage fluctuations should have no effect.
- The discharge of the battery across the weld should take place according to a preset pattern, and adapt itself automatically to the welding conditions.
- The welding head must meet severe requirements. The welding pressure must be finely adjustable, should remain constant during welding and be independent of the operating speed.

2.2.8.3 Percussion welding[57,58,59]

Welding process

The name percussion, meaning thrust, does not completely cover the welding process; the metal flame arc is far more characteristic during the welding pulse.

A capacitor C (Fig. 2.54), charged via rectifier G and transformer T from the mains, is connected direct across the weld, without a transformer. The workpieces to be welded together are themselves the electrodes. A pointed wire d approaches plate p at a velocity V. When d is at a few microns distance from p, an arc is formed which causes d to fuse at a greater rate than V.

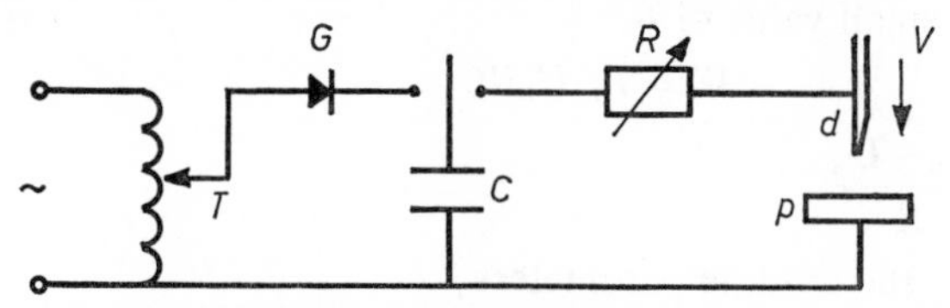

Fig. 2.54. Circuit for percussion welding.

The arc remains intact until C has lost too much of its charge and V exceeds the fusion rate, after which d is forced into a pool of molten metal produced by the wire and the plate. (That is why this welding method is counted among pressure welding). The arc extinguishes and the weld cools off.

Heat development

As soon as the arc ignites, voltage U (Fig. 2.55) falls off from charge

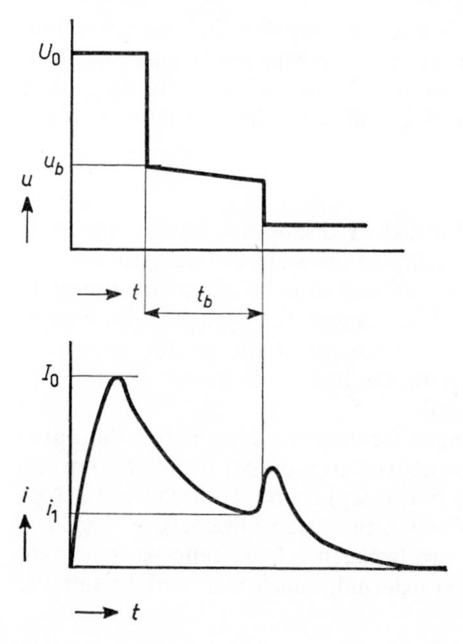

Fig. 2.55. Voltage variation in the percussion welding process.

Fig. 2.56. Current variation in the percussion welding process.

voltage U_0, 70–200V, to the arc voltage $U_b \approx 15$V, the ionising voltage of metal; U_b will drop slightly during the arc time t_b.

The weld current i (Fig. 2.56) drops from the maximum value I_0 to i_1 over a period t_b. As soon as the arc extinguishes, the welding resistance decreases and i increases again, after which C discharges further across the weld. The welding process takes place during arc time t_b. The heat development is then:

$$W \equiv \int_0^{t_b} U_b i \, dt \, (1 - e^{-t_b/RC})$$

$$= I_0 . U_b . R . C$$

For a not-too-small value of t_b,

$$W = I_0 . U_b RC$$

Here $I_0 = \dfrac{U_0 - U_b}{R}$

This gives us the welding parameters.

U_0 = charging voltage of the battery RC, 70–200V.
R = resistance in the discharge circuit, 0·1–10 ohms.
C = capacitance of the battery, 1–10 000 μF.
t_b = arc time, 5–10 ms, which is dependent on the feed rate V.
U_b lies in the vicinity of 15V, that is, the ionising voltage of metal vapours.

The ionising voltage of gases is about 7V. An electric arc is found easier to produce when a shield gas is used.

Application

The percussion welding method is used mainly for fine wire welding; wire-to-wire (head to head connection); wire to connecting pins and lead-throughs; pins to plates and, in the semiconductor industry, for Mo, Ti in combination with wire of a copper or nickel alloy or wire of a ferrous metal.

Special methods

The periods at which the arc is formed or disappears usually varies. If the arc time t_b is not constant, the quality of the weld will vary too.

To obtain a more stable weld, the wire can also be placed on the plate to be welded under a slight pressure. The moment the capacitor discharges, the wire will be withdrawn against the pressure of the spring, under the influence of an electro-magnetic force. As the battery discharges further, the wire is fed back to the workpiece again.

The firm of Cannon are marketing a welding machine under the name 'Percussive Arcwelder" which brings the wire to a pre-set distance from the other workpiece. Then an RF (radio frequency) energy is discharged across the weld, so that the arc is ignited. Finally, the battery discharges[58,61].

This apparatus will also draw an arc between a fixed tungsten electrode and the weld. The heat source is then external, which is a method suitable for the welding of insulated wire.

(a) *Percussion versus resistance welding*

In contrast with resistance welding, the contact resistance plays no part in percussion welding. To obtain a strong electrical field, only one of the parts to be welded needs to be pointed. Combinations otherwise difficult to weld, like copper-brass, copper-molybdenum, copper-steel, can now be realized.

The welding pressure is only important to the extent that, after the arc has extinguished, the weld remains closed. The welding mechanism can therefore be of light construction. Care should be taken that, after the thrust, there is no resilience; in other words, the thrust must be damped. The heat penetration into the sheet metal is low, so thin plates can be welded.

On the other hand, the corroding effect of the ambient air during a period t_b will be greater than with resistance welding.

The welding process develops in a rather violent manner. The electromagnetic effect causes the liquid wire metal to sputter, which results in an uneven weld, while the area round the weld itself is covered with particles of welding material.

2.2.8.4 *Stud welding*[62,63,64]

Introduction

This welding process could be classified as one special method of percussion welding. For the welding of bolts and pins, however, stud welding takes up a place of its own.

The welder needs no special skill to operate the simple automatic equipment.

Welding process

There are three different forms of stud welding, all of which use the metal arc for heat development.

(a) *Electric arc stud welding*

The energy is supplied by a welding converter, rectifier or transformer.

The transformer has recently come into use; welding with direct current dates from an earlier time and lends more stability to the welding process. With a welding gun the somewhat pointed bolt and the ceramic ring are pushed against the plate, after which the process develops as shown in Fig. 2.57. The ceramic ring serves to protect the weld and is removed afterwards.

There are also semi-conductive rings which initially keep the bolt at a certain distance from the plate. The ring melts away and forms a protective slag on the weld.

Welding time = 0·1–1·5 s, bolt diameter = 5–25 mm, plate thickness = 3–6 mm.

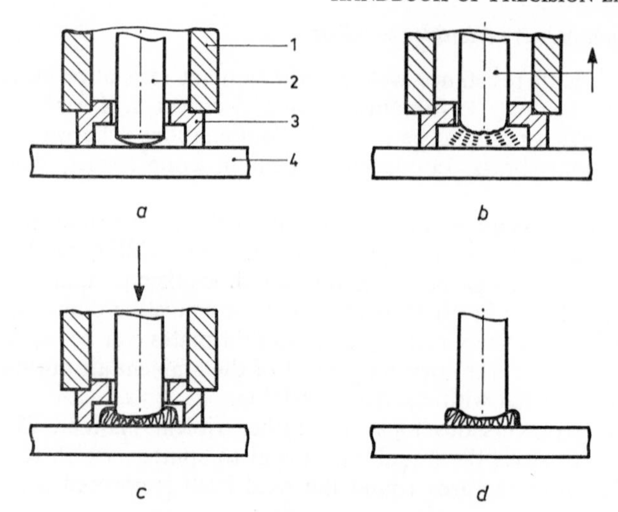

Fig. 2.57. Electric arc stud welding.
1 = gun. 2 = bolt. 3 = ring. 4 = plate.

a. Bolt placed against the sheet, after which the current passes.
b. Welding tool lifts the bolt and draws an arc. Tip of bolt and plate surface fused together.
c. Bolt is forced into a pool of molten metal.
d. Ceramic ring is chopped off and the weld finished.

It is possible to weld thin sheet metal, as the heat development is short-lived and heat removal remains low.

(b) *Capacitor discharge and stud welding*

For the welding of thin pins and bolts provided with a narrow welding tip, a capacitor battery is used as the energy source. As soon as the battery discharges over the welding tip, the latter melts and forms the connecting phase between pin and sheet (Fig. 2.58).

There is no protective ring. Welding time = 1–6 ms.
Bolt diameter = up to 6 mm.

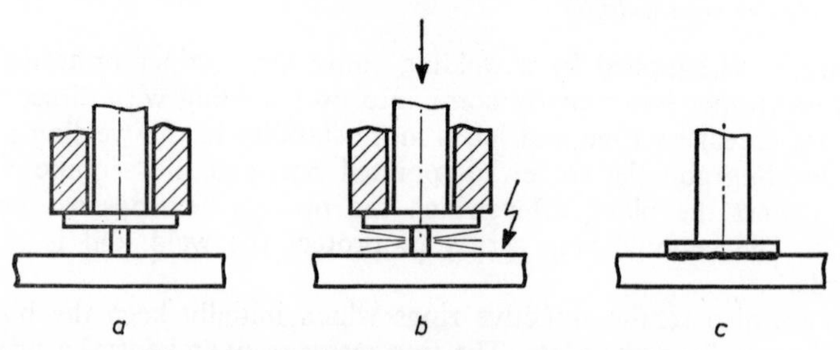

Fig. 2.58. Capacitor discharge stud welding.

a. Pin tip placed on the plate.
b. Battery discharges, the tip melts and the arc is drawn.
c. Arc extinguishes, the weld cools off and the pressure is removed.

(c) *Drawn-arc capacitor discharge stud welding*

If the welding tip must be avoided, the arc is drawn by withdrawing the bolt momentarily during the capacitor discharge (Fig. 2.59).

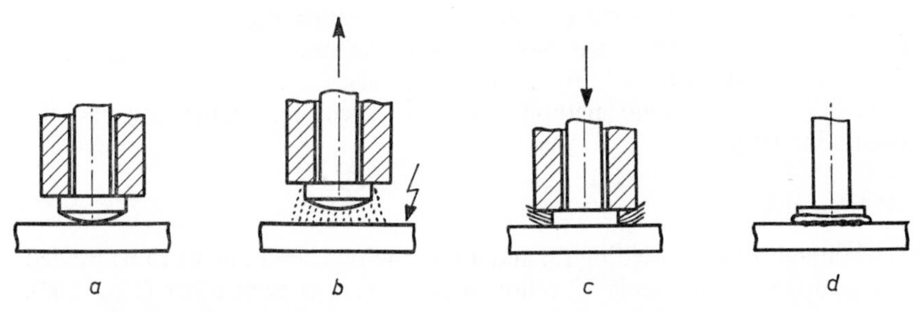

Fig. 2.59. Drawn-arc capacitor discharge stud welding. Welding phases the same as those in Fig. 2.57.

The welding phases are the same as those in Fig. 2.57.

Owing to the short welding time (6–15 msec), no protective ring is needed. During that time sufficient energy can be dissipated only for bolts with diameters up to 6 mm.

Table 2.4 surveys the welding criteria of the three stud welding methods[64].

TABLE 2.4

Stud welding method	a	b	c
Stud diameter (mm) 1·5–3	I	G	G
3–6	B	G	G
6–12	G	I	I
12–25	G	I	I
Cross-section of stud up to 32 mm²	B	G	G
>32 mm²	G	I	I
Stud material:			
Carbon steel	G	G	G
Stainless steel	G	G	G
Alloyed steel	M	B	M
Aluminium	M	G	G
Brass	B	G	G
Sheet metal:			
Carbon steel	G	G	G
Stainless steel	G	G	G
Alloyed steel	M	G	B
Aluminium	M	G	G
Brass	B	G	G
Sheet thickness (mm) <0·4	M	G	G
0·4–1·6	M	G	G
1·6–6	G	G	G
>6	G	G	G

Meaning of the letters: G = good; M = moderate; B = bad; I = impossible.

2.2.9 Special welding methods

2.2.9.1 Ultrasonic welding

In recent years, a great deal of research has been carried out on ultrasonic welding, so it has become possible to join metals that, because of their thermally unfavourable characteristics and the presence of a tough oxide layer, were difficult to weld by thermal methods.

Today, the ultrasonic method is used in precision engineering and the plastic industry.

Welding process

Clamped between welding tip and anvil, the two components to be welded are brought to ultrasonic vibration with respect to each other (Fig. 2.60).

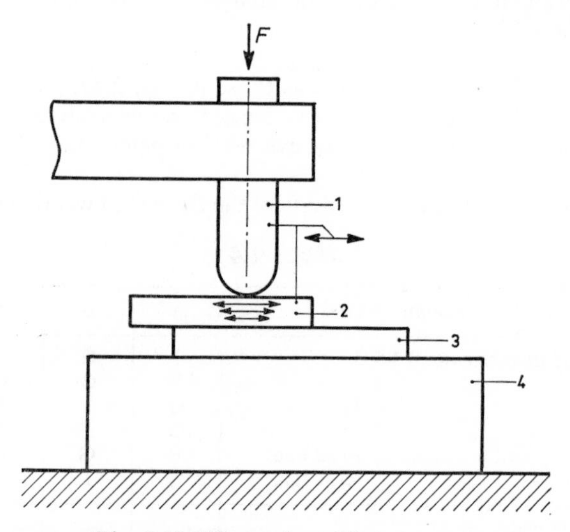

Fig. 2.60. Ultrasonic welding process.
1 = welding tip. 2 = vibrating lip. 3 = lip lay stable on 4. 4 = anvil.

Due to the resulting friction, the layer of oxide will break at various points, so that a purely metallic contact is obtained and atomic binding predominates. During the welding process, the points of contact expand to form lines in the direction of vibration, after which the lines finally form a contact area. If the welding time is too long, the whole area of the two components between welding tip and anvil undergo intensive deformation. See also Volume 4, Chapter 1, Section 1.4.

According to the mechanism of the weld, this process is classified with friction welding, although heat development remains below the melting temperature of the metal to be welded, so there is no recrystallization. On the other hand, the heat development is predominant for the ultrasonic welding of plastics.

Equipment

An a.c. voltage of about 20 kHz, produced by an electric generator, is converted into mechanical vibrations with an amplitude of 1 μm by means of a piezo-magnetic converter or transducer. An amplitude transformer soldered to the converter causes the welding tip (or sonotrode) to vibrate at a greater intensity of 6–10 μm. The workpiece is clamped between the welding tip and the anvil (Fig. 2.61). The welding time is set with a time switch.

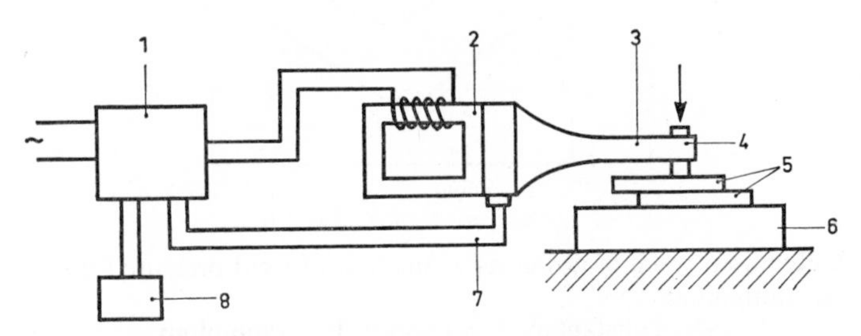

Fig. 2.61. Diagram of ultrasonic welding apparatus.
1 = generator for 20kHz. 4 = welding tip. 7 = feedback.
2 = transducer. 5 = workpiece. 8 = timer.
3 = amplitude transformer. 6 = anvil.

As the characteristic frequency of the system, which depends on the material to be welded, can vary during the welding process, a piezo-electric pick-up of the amplitude transformer gives feedback to the generator, so that the welding energy remains constant throughout the welding process.

The vibrating frequency is just above the hearing threshold, that is, upwards from 20 kHz. Owing to the low efficiency of the converter, too high a vibration factor will not ensure a satisfactory welding effect. Although this method of welding causes no permanent hearing damage there is a chance that some people may be sensitive to ultrasonic vibrations, and will as a result suffer from headaches or sickness. In that case, medical advice should be taken.

The compressive force F indicated by the arrow = 2 to 200 kg, and the generator power P = 20 to 1 000W.

Variables

To obtain a good weld, the variables, represented by the welding energy W, are selected with regard to dimensions, kind of material and surface condition of the components to be welded.

$$W = 4F.\mu.\delta.v.t$$

where F = welding pressure,
μ = friction coefficient,
δ = vibration amplitude,
v = vibration frequency and
t = welding time.

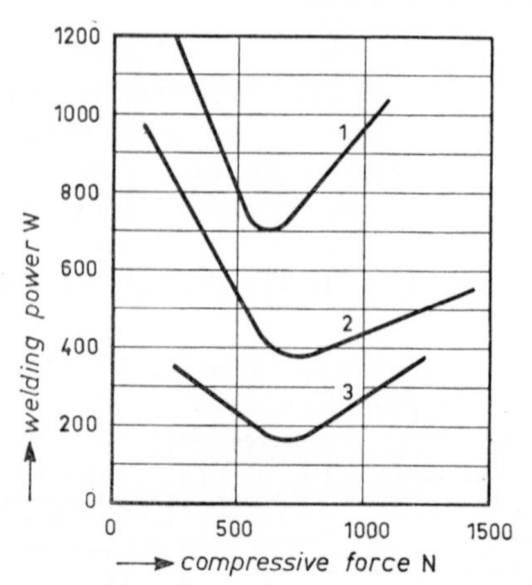

Fig. 2.62. Required welding power as a function of weld pressure for constantan, copper and aluminium.
1 = constantan. 2 = copper. 3 = aluminium

Pressure

For most materials, a minimum required power corresponds with a very specific pressure (Fig. 2.62)[69,70]

A variation in energy implies a variation in the vibration amplitude δ, but it should be noted that, because of load variation, the nodes and anti-nodes of the vibration pattern of the welding tip are displaced. This means that a node is not always present at the location of the tip, and consequently a variation of the energy supplied need not cause a corresponding variation of the welding power. The pressure increases with the hardness and thickness of the components to be welded.

Time

The welding time plays a part, and depends on the hardness of the materials. For aluminium, this time is not critical. If t becomes too great, harder materials will show fatigue phenomena[70,71], so that the breaking strength is reduced. (Figs. 2.63 and 2.64).

Frequency

Each welding apparatus operates with a frequency matched to the efficiently loaded acoustic system. Load variations can cause the characteristic frequency to change: a feedback system matches the generator to the altered frequency. The frequency is then not critical for the weld.

Friction coefficient

If the workpiece has a smooth finish, more energy will be used for friction in the contact area, and the weld will be much stronger.

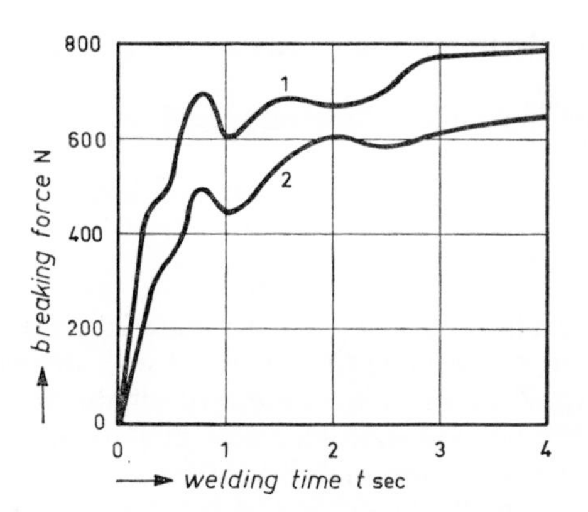

Fig. 2.63. Breaking force as a function of time, for aluminium with a plate thickness of 0·8 mm.

	1.	2.
electric power	280 W	340 W
pressure	800 N	270 N

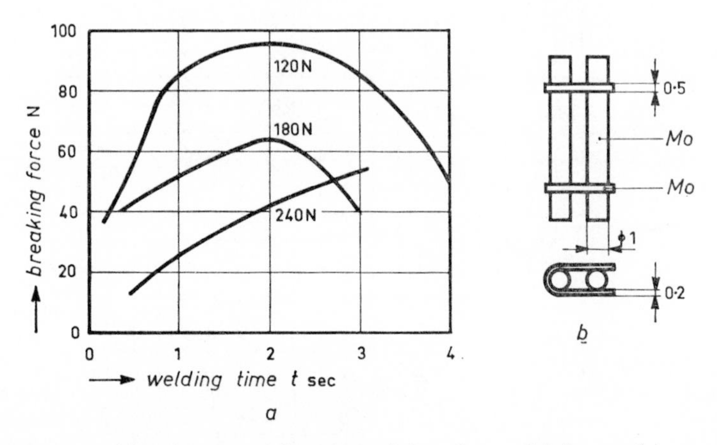

Fig. 2.64. *a*. Breaking force as a function of time for molybdenum (*a*), according to the dimensions in *b*., at various pressures.
Parameters: welding pressures N
(examples 120 N, 180 N and 240 N).

A few general rules

Materials with great differences in hardness are difficult to weld.
Soft materials are easier to weld than hard ones.
For good energy transfer, the welding tip should not be too smooth. If necessary, the shape of the welding tip must be adapted to the workpiece.

The thickness of the upper plate is decisive in the first place for the welding energy applied.

The influence of lower plate thickness is far less.

Application

(a) *Metals*

For a survey of ultrasonically weldable materials, see Fig. 2.65[70]. The movement made by the welding tip is parallel with the welding plane.

- Al, Cu, Au, Ag, Zn, Sn and brass are very suitable for welding. Thermal methods entail difficulties due to electrical and thermal conductivity, the oxide layer and the formation of brittle alloys.
- Insulated wire to a plate: the welding tip penetrates the wire insulation.

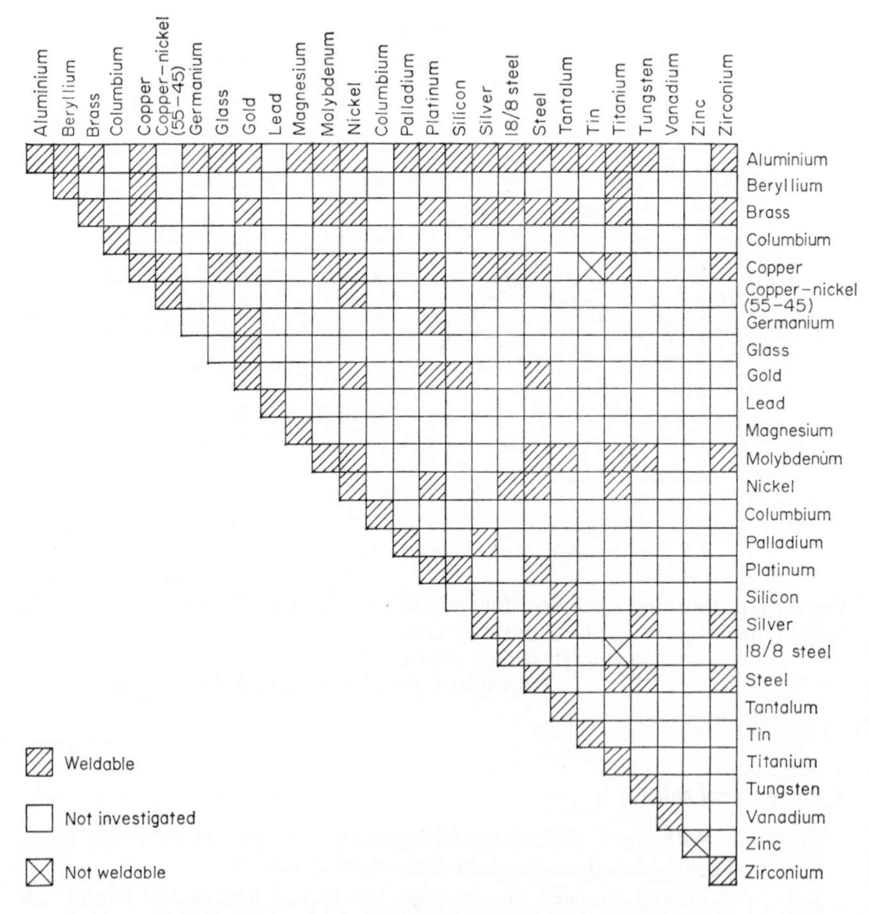

Fig. 2.65. Survey of ultrasonically weldable materials.

- Al foil to Al plate, or Al foil to glass.
- Gold wire to a nickel film deposited on a glass substrate.
- In electron tube manufacture: no brittleness in tungsten and molybdenum connections: no non-permissible transfer of electrode material to the components to be welded.

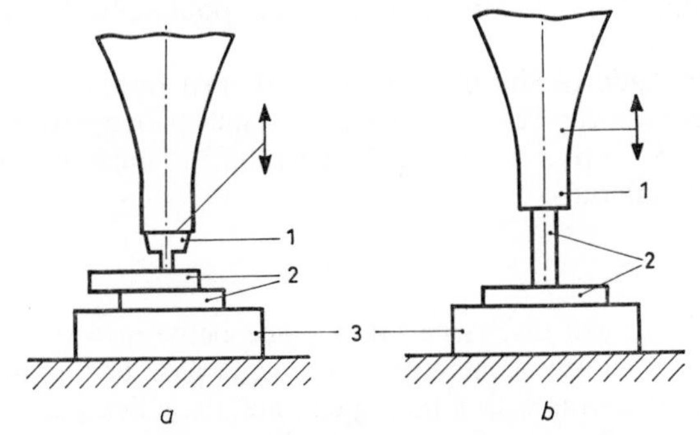

Fig. 2.66. Ultrasonic welding of plastics.
a. contact welding
1 = welding tip. 2 = workpiece. 3 = anvil.
b. remote welding.
1 = welding tip; 2 = workpiece.

(b) *Plastics*

As plastics have no oxide film impeding welding, the hammer method is adopted for these materials (Fig. 2.66a). Friction on the welding face produces heat, causing the materials to fuse with each other.

It is also possible to weld from a distance. A plastic rod clamped at one end is then welded to a plate of synthetic material (Fig. 2.66b).

Not all plastics (it depends on the elasticity module) are ultrasonically weldable (see Table 2.5)[73].

TABLE 2.5

Material	Contact welding	Distance welding
Polystyrene	good	good
Nylon	good	moderate
Acrylate	good	moderate
Polypropylene	good	bad
Hard PVC	moderate	moderate
Flexible PVC	moderate	moderate

2.2.9.2 *Plasma arc welding*[75,76,1]

Over the last few years, more and more materials, satisfying stringent thermal requirements, are used in aircraft and rocketry. A directed heat source at high temperature is required, to work these materials, so the development of the plasma arc was a matter of great interest. Fundamental investigations had been going on in Germany ever since the twenties, and after the Second World War America, in particular, recognized the possibilities of the plasma arc.

With the concentrated plasma arc, it has been possible to achieve a temperature of 50 000°C and an exit velocity of 9 000 m/s. Internal energy is released when, at the edge of the flame, the gas ions return to the molecular state via the atomic state. This reaction can take place spontaneously, or be created by applying powder.

It is characteristic of the plasma arc that heat development takes place without the occurrence of chemical reactions such as occur with the acetylene-oxygen flame, so the plasma arc welding method is most suitable for chemically-sensitive materials.

Plasma arc

A direct current arc discharge takes place between a tungsten electrode and an anode, in an atmosphere of compressed gas. The no-load voltage of the direct current source with a falling characteristic lies above the operating voltage of the burner.

The thermal Pinch effect, caused by the inflow of the cold gas in the direction of the centre of the flame arc, compresses the hot gas.

The magnetic Pinch effect narrows the ions and consequently the plasma beam. These two effects result in a larger current density and a high plasma temperature.

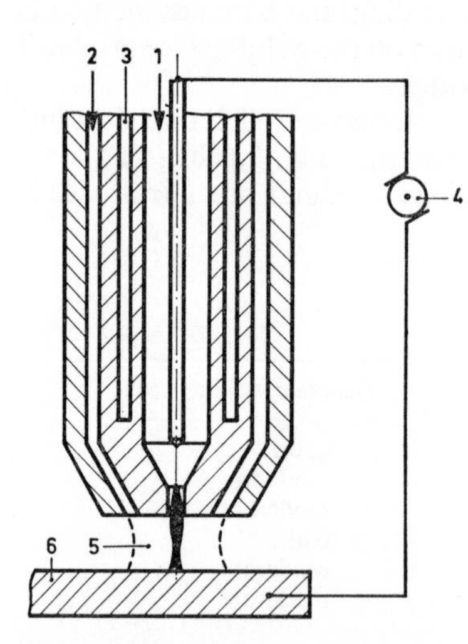

Fig. 2.67. Transferred plasma arc.
1 = arc gas.
2 = shielding gas.
3 = cooling.
4 = current source (DC).
5 = arc.
6 = workpiece.

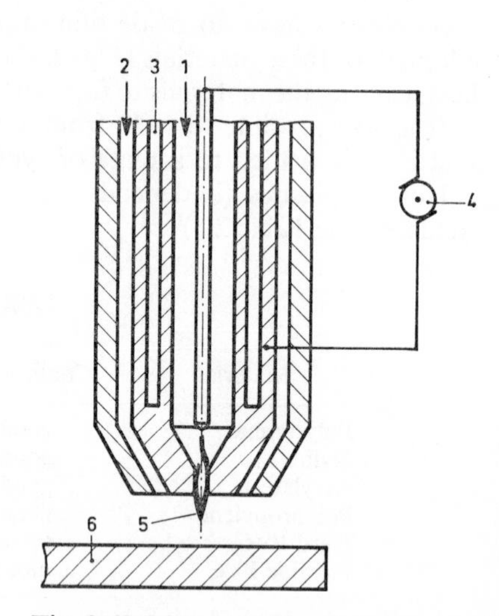

Fig. 2.68. Non-transferred plasma arc.
1 = arc gas.
2 = shielding gas.
3 = cooling.
4 = current source (DC).
5 = arc.
6 = workpiece.

If the conductive workpiece serves as the anode, it is referred to as a transferred plasma arc (Fig. 2.67): if the workpiece itself is not conductive, it is a non-transferred plasma arc (Fig. 2.68). By having the compressed gas transfer metal powder, the indirect plasma arc can be employed to apply metallic films.

Arc gas

Gases causing no corrosion to the electrode, and having no corrosive effect on the burner jet, can be used as pressure gas or arc gas. Other important factors are the energy content or enthalpy. The enthalpy of monatomic gases increases practically linearly with the temperature. At the same temperature, diatomic gases absorb more energy, for during ionization the conversion into atoms first takes place (Fig. 2.69)[75].

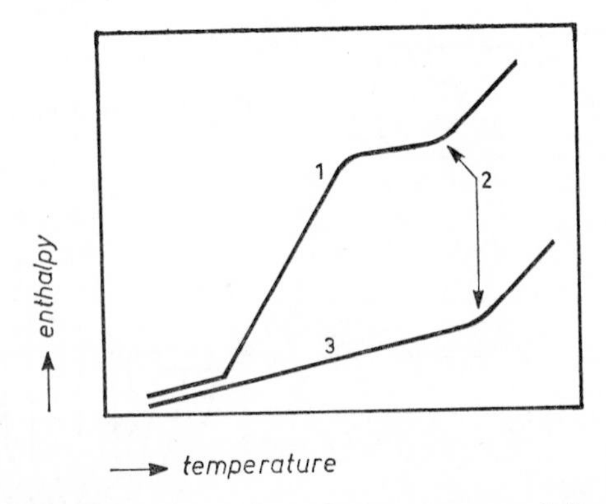

Fig. 2.69. Enthalpy as a function of temperature and diatomic gases. 1 = diatomic gas; 2 = beginning of ionisation. 3 = monatomic gas.

Table 2.6 gives a few values of a plasma burner working with different kinds of compressed gas.

Due to its low reaction speed, argon has a low enthalpy, but therefore protects the weld against oxidation. A mixture of 86 He and 14 Ar gives a rather good efficiency.

TABLE 2.6 [75]

Operating values of plasma burner (Thermal Dynamics Corp.)

Gas	Power absorption of the burner kW	Operating voltage V	Plasma temperature K	Plasma enthalpy kcal/kg	Thermal efficiency %
N_2	60	65	7 600	9 900	60
H_2	62	120	5 400	76 600	80
He	50	47	20 300	51 000	48
Ar	48	40	14 700	4 600	40

Hydrogen and nitrogen, however, are cheaper and have much higher enthalpy values, but hydrogen is very corrosive and thus has a corrosive effect on the weld. A mixture of nitrogen with about 10% hydrogen gives satisfactory results.

Shielding gas

Because of its low temperature, the outer layer of the gas will not ionize; consequently, the diameter of the plasma arc increases, and the molten material could be blown away by the high speed and greater mass of the gas. A typical plasma weld showed the form of a wine glass (Fig. 2.70). By applying a low-pressure gas, it is possible to give the beam an extra convergence and to reduce the "blow" effect. In the case represented by Fig. 2.71, the actual shielding gas is applied round the converging gas jet[76].

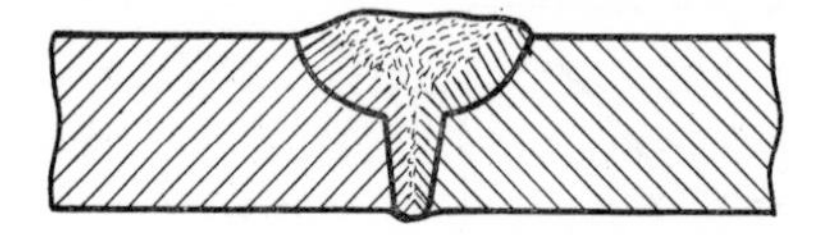

Fig. 2.70. Wineglass weld shape.

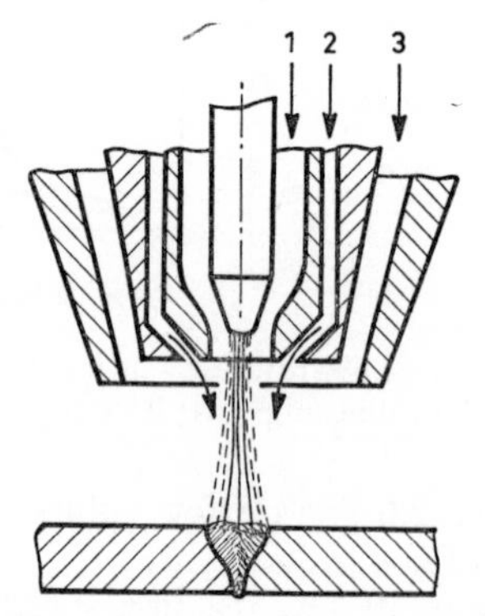

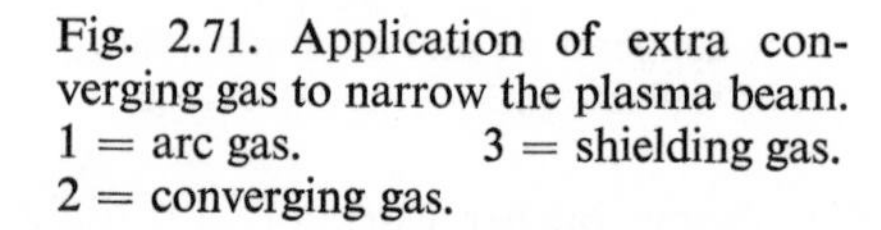

Fig. 2.71. Application of extra converging gas to narrow the plasma beam.
1 = arc gas. 3 = shielding gas.
2 = converging gas.

Application

Plasma welding is a further development of argon arc welding. A closer bundled beam of energy, resulting in a lower sensitivity to welding variables, a better weld quality and higher welding speeds, opens up the possibilities of welding high melting and chemically-sensitive metals. Furthermore, this includes all metals welded according to the argon arc method. It is still not possible to get a plasma arc with a.c. with the purifying effect of the negative period half. Thus aluminium, for example, cannot be welded by plasma arc.

2.2.9.3. Electron beam welding[78,79]

Although the thermal character of the electron beam has been known for over 100 years, it was not until 1952 that the first patent on electron beam welding was applied for, after a great deal of research work towards an intensive bundling of weld energy. Since that time, various countries have worked hard towards further development, and in about 1960 the electron beam welding machine found its way into industry. It is particularly in the field of space research that the electron beam technique opens the way to new possibilities.

Welding process

Under the influence of a high anode voltage, a cathode connected to a low voltage emits electrons in a vacuum. These electrons are focused by magnetic lens on to the workpiece (Fig. 2.72)[82].

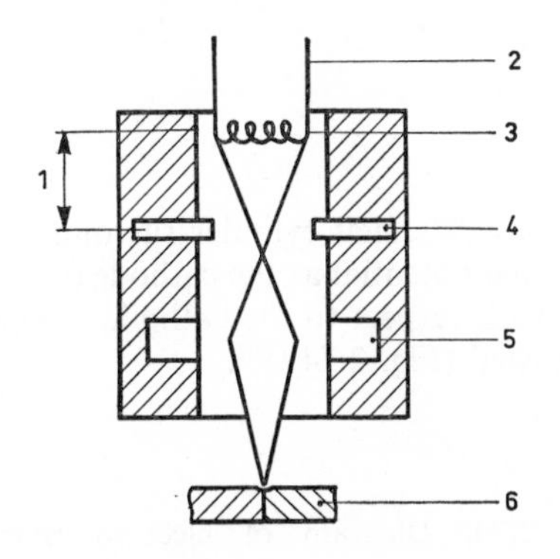

Fig. 2.72. Simplified diagram of an electron gun.
1 = high-voltage supply.
2 = low-voltage supply.
3 = wire cathode.
4 = anode.
5 = magnetic lens.
6 = workpiece.

Due to the enormous speed at which the electrons enter the workpiece material, the kinetic energy is converted into thermal energy whose density can assume very high values in the focus of the beam, depending on the anode voltage and vacuum used. With a voltage of 150 kV and a vacuum of 10^{-4} Torr, for instance, it is now possible to bundle an electron beam of 900W into one spot of 10 μm to yield a power density of 10^9 W/cm^2. For arc welding, the energy flux is 10^4–10^5 W/cm^2.

Only with the laser beam (see Laser welding 2.2.9.4) can a greater density be achieved during 2 msec; that is, up to 10^{16} W/cm^2.

Owing to the enormous heat concentration, a very narrow welding zone is produced, 1/20 to 1/25 of that occurring in argon arc welding (Fig. 2.73).

The heat consumption per seam length is also 1/20 to 1/25 that of the conventional fuse welding method.

With a 150 kV welding apparatus and a maximum current of 167 mA, the welding penetration in chrome nickel steel is 100 mm, the weld width 5 mm

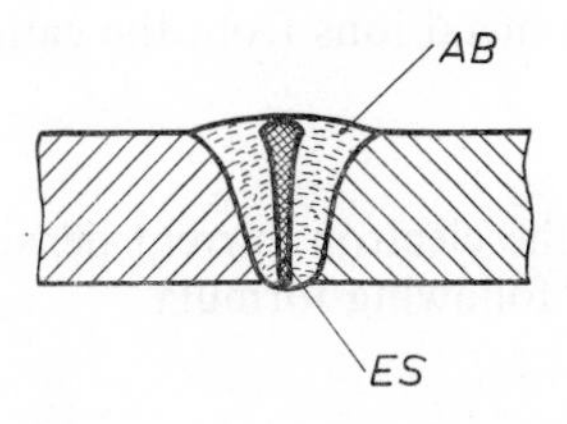

Fig. 2.73. Comparison between the welding zone argon arc welding (AB) and electron beam welding (ES).

and the welding speed 180 mm/min. One might ask how an electron beam, greatly diffracting in the atmosphere, can penetrate to such an extent in steel. If the electron beam is focused on the steel surface, a hole of about 0·3 mm cross-section is produced, in which a highly ionized gas vapour is developed, which keeps the beam sharply bundled. The melting zone of about 3 mm is located round the focused beam. Depending on the electron density and speed, the beam forces its way further into the material, the burning point remaining stationary. The gas beam is therefore constantly surrounded by highly-ionized metal vapours.

On the other hand, the penetration depth can be restricted to the surface, to give a component a special surface treatment, by adjusting the electron density.

Welding equipment

(a) *The electron optical part*

The widely-used electron gun consists of a Wehnelt cylinder surrounding the cathode. This cylinder has either the same potential as the cathode (diode system), or it receives a negative voltage with respect to the cathode (triode system). This negative voltage can be adjusted (Fig. 2.74)[79].

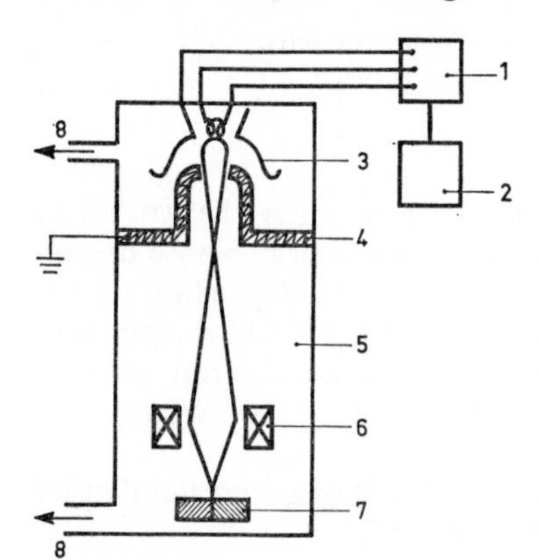

Fig. 2.74. Diagram of electron beam apparatus.
1 = generator for Wehnelt cylinder voltage and low cathode voltage.
2 = high-voltage generator.
3 = Wehnelt cylinder.
4 = anode (earthed).
5 = vacuum.
6 = magnetic lens.
7 = workpiece.
8 = to vacuum pump.

With the triode system, it is possible to control the current intensity, and consequently the electron emission, independently of the operating voltage, by varying the control voltage at the Wehnelt cylinder.

When welding aluminium and magnesium alloys, there is the danger of metal vapour ions reaching the cathode, and interfering with the electron emission. By mounting the cathode obliquely and controlling the electron beam magnetically, it is possible to divert the metal ions from the cathode.

High and low tension equipment

The effect of the operating voltage U_B on the electron current density j_v occurring in the weld spot can be seen in the following formula

$$j_v = K . j_e . U_B^2$$

where K is a factor depending on the cathode temperature and the degree to which the beam is bundled by the magnetic lens system, and j_e is the emission current density.

This means that, for the same j_v and $U_B = 125$ kV, a value of $j_e = 1$ A/cm^2 is required, whereas $j_e = 25$ A/cm^2 is needed at $U_B = 25$ V.

A high tension apparatus with $U_B \geqq 100$ kV has the following characteristics:

- A favourable ratio between seam depth and width; the value for steel, for instance, is 25:1.
- High penetration power and a narrow melting zone.
- Low heat effect on the surroundings.
- A rapid cooling of the weld, resulting in fine-grain recrystallization without undesirable segregations.
- A low voltage apparatus with $U_B = 25$–50 kV has the following advantages:
- Simple construction, no need for expensive screening against undesirable X-radiation, and less expensive installation equipment.
- Below 50 kV, the air is a good insulator.

(b) *Vacuum equipment*

The electron beam technique entails the use of a vacuum. The pressure of the vacuum depends on the requirements made. A pressure of 10^{-2} Torr is already sufficient for the welding of machine steel. Generally, a pressure of 10^{-4} Torr is used, the free path of the electrons being about 4 m.

Highly active chemical materials are welded in a vacuum up to 10^{-9} Torr. If a lot of products are to be welded, a special vacuum system operating with gates is preferable. The pressure difference of the workroom and the final gate chamber can then be substantially reduced.

(c) *Electron beam welding in the atmosphere*[73]

It would be a great advantage if there were no need to lead the products into the vacuum; attempts have therefore been made to lead the electron beam into the atmosphere by means of a special plate permeable to the beam, or by means of vacuum jets. But, when the electron beam enters the free air, the electrons are scattered as soon as they collide with the air molecules. After having covered a distance of 10 mm through the ambient air, a greatly accelerated, well-bundled electron beam of a 7 kW apparatus penetrates no more than 10 mm into steel. The position of the workpiece in the direction of the beam is then very critical. The non-vacuum welding method is used in aircraft industry.

Application

Electron beam welding is a method employed in reactor engineering, rocket manufacture and the aircraft industry for welding:

High melting metals such as Mo, W, Va, Ta, Ti.

Alloyed chromium steels used in gas turbines and missiles.

Shrink-proof nickel alloys of the nimonic and inconel series. When other thermal welding methods are used, cobalt and molybdenum alloys tend to fracture.

It is expected that this welding method will be particularly useful in future space stations where vacuum is free.

As it stands now, due to the need for vacuum and a critical positioning of the workpiece, electron beam welding is reserved for single-piece work of high quality products.

2.2.9.4 Laser welding[86,87,89,90]

One of the latest physical developments appealing to the imagination is denoted by the abbreviation, "laser", for "light amplification by stimulated emission of radiation". In contrast with ordinary white light, laser radiation is monochromatic and coherent, that is, of equal phase, wavelength and amplitude, so that optimum bundling is obtained. Apart from many other applications, the laser beam can be used in micro-welding engineering, with a few advantages over other welding methods. Vacuum is not required, and no other material (in contrast with soldering) need be added, heat development is highly concentrated, electrodes are not used and the workpiece itself is not even touched.

Ruby laser process

By a chain reaction of light quanta taking place in a rod of ruby with reflecting ends, the light flash produced by a gas discharge lamp is converted into energy of coherent parallel light waves. If the intensity (that is, the amplitude) of the light energy exceeds a certain value, it penetrates through the end face with low permeability in the longitudinal direction of the ruby rod (Fig. 2.75)[89].

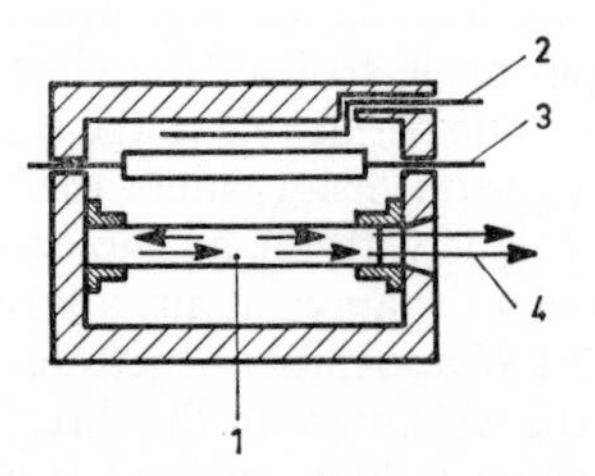

Fig. 2.75. Diagrammatic representation of the laser welding machine.
1 = ruby rod.
2 = sparking plug.
3 = gas discharge lamp.
4 = laser beam.

The latter (Al_2O_3 with 0·05% chromium) must be of a most homogeneous structure with accurate plan-parallel end planes, to ensure light resonance in the longitudinal direction. The ruby can be replaced by a column of gas, such as a mixture of helium and neon. As the gas is free from inhomogeneity, the coherent characteristics of the emitted light are better than those of a ruby laser. The output power is, however, lower.

Laser beam

If the initial radiation is parallel throughout the ruby area, the light energy can be bundled to the wavelength of green light, which is 0·6943 μm, but the inhomogeneity of the crystal sets a limit.

In 1963 an initial energy of 0·4 kW was bundled to a point of 0·1 mm diameter, corresponding to an energy current density of 5×10^6 W/cm^2. By means of an f/1 lens, sunlight can be condensed to 500 W/cm^2.

Energy emission takes place in 2 msec pulses, corresponding with the transition time of excited Cr atoms of 1·83 eV to basic condition. A modern welding laser can emit 1–10 joule per pulse at a frequency of up to 10 pulses per sec. The efficiency of the laser is very low: 0·5–0·1 %. To obtain a pulse of 10 joules, the discharge lamp must emit 2 000–10 000 joules.

Heat absorption of the workpiece is limited to the surface. The penetration, which is still very low for laser beams, depends on the thermal diffusion.

Application

Apart from the use of laser techniques in communications, navigation, measuring engineering, photography, chemistry and micro-surgery, there are many possibilities in micro-welding engineering:

- For welding integrated circuits, such as 10 μm nickel foil on to a substrate carrying a layer of Cr-Cu-Au, nickel foil on to a vapour-deposited aluminium film, Au on Al, Au and Cu foil on etched copper of a print-plate.
- Wire connections. For instance, two 50 μm nickel wires, as shown in Fig. 2.76[88].
- Electronic components. For instance, the edge of the lead-in at the housing of a thermistor (Fig. 2.77)[89].

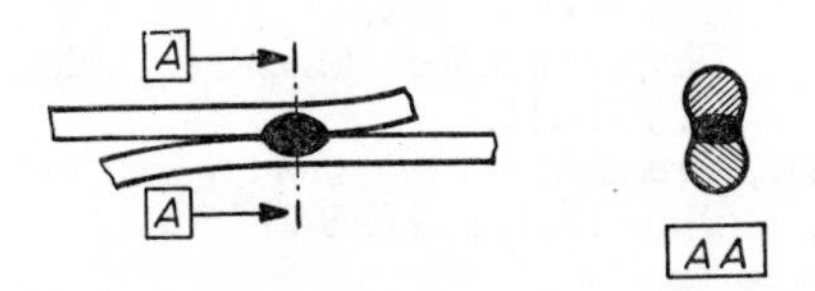

Fig. 2.76. Welded nickel wires of 50 μm.

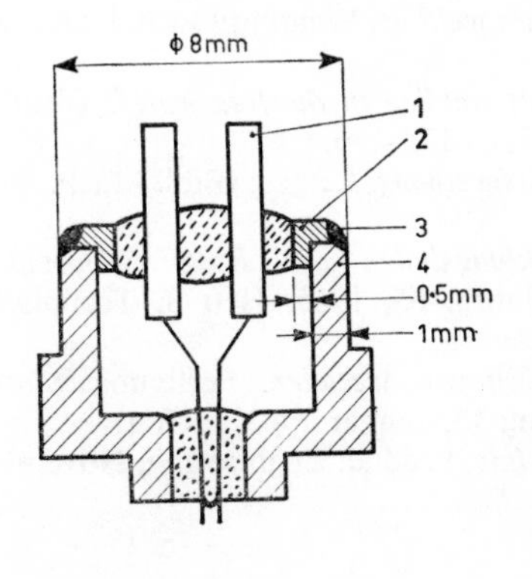

Fig. 2.77. Laser weld of steel sprout on housing of a thermistor.
1 = lead-in pins.
2 = steel.
3 = laser weld.
4 = chromium–nickel steel.

The welds shown in the Figs. 2.76 and 2.77 were made with an energy of 6 joules and one pulse per sec.

The possibility of welding glass offers certain advantages. The method is used for correcting resistors deposited on glass[92]. One of the difficulties is the very accurate aiming of the beam and the possibility of reaching the welding spot.

REFERENCES

[1] A. L. PHILIPS — *Welding Handbook*,
Section I : *Fundamentals of welding*, 1968.
Section II : *Welding processes*, 1969.
Section III: *Welding, Cutting and related processes*, 1965.
Section IV: *Metals and their weldability*, 1966.
Section V : *Applications of welding*, 1967.
American Welding Society, 1962.

[2] G. HERDEN — *Schweiß- und Schneidtechnik*, VEB Carl Marhold Verlag, Halle (Saale), 1960.

[3] E. SUDASCH, — *Schweißtechnik*, Carl Hanser Verlag, München, 1959.

[4] J. C. MERRIAM — *The new welding processes*, Materials in Design Engineering, January 1960, p. 105–119.

[5] J. F. PARR — *Welding developments . . . new tools for production*, The Tool Engineer, January 1960, p. 115–124.

[6] P. L. J. LEDER — *A review of some recent developments in welding processes*, Journal of the Institute of Metals, 1964–1965, Vol. **93**, p. 375–381.

[7] S. A. FRANCIS — *Techniques, equipment and procedures for production welding of electronics*, IEEE Transactions on Product Engineering and Production, January 1963.

[8] — *Booglassen*, Philips Bedrijfsapparatuur Nederland N.V., Eindhoven, 1964.

[9] B. C. MOTL — *Button welding*, Machine Design, November 5, 1964, p. 150–155.

[10] W. N. CANULETTE, S. A. AGNEW, N. E. ANDERSON — *Precision gas tungsten-arc spot welding*, Welding Journal, June 1965, p. 270.S-275.S.

[11] B. KOCH — *CO_2 welding for shipbuilding*, Welding Journal, December 1965, p. 1005–1015.

[12] P. E. MASTERS, R. S. ZUCHOWSKI — *Vertical submerged-arc welding*, Welding Journal, October 1965, p. 829–837.

[13] J. E. NORCROSS — *Electroslag/Electrogas welding in the free world*, Welding Journal, March 1965, p. 177–186.

[14] H. REINHARDT — *Die Elektro Slacke Schweißung*, Leipzig (Bibl. Mach. fabr. 5616).

[15] W. HOFMANN — *Stand und Entwicklungslinien der Kaltpreßschweißen*, Werkstattstechnik, Jahrg. **55**, 1965, Heft 3, Technische Hochschule Braunschweig.

[16] M. OLSZEWSKI — *Das Kaltpreßschweißen von Metallen*, Fertigungstechnik und Betrieb, Jahrgang **13**, Januar 1963, Heft 1.

[17] D. S. PRESTON, T. K. HENDERSON — *Wirewrapping saves time*, Product Engineering, November 13, 1961, p. 86–92.

[18] J. WODARA *Explosionsschweißen—ein Kaltpreßschweißen mit extrem kurzen Schweißzeiten*. Schweißtechnik, Berlin, Oktober 1963, Jahrgang **13**, Heft 10.

[19] M. D. CHADWICK, D. HOWD, G. WILDSMITH, J. H. CAIRNS *Explosive welding of tubes and tube plates*, British Welding Journal, Oct. 1968, p. 480–492.

[20] M. J. ALBOM *Solid state bonding*, Mechanical Engineering, 44, March 1965.

[21] M. J. ALBOM *Diffusion welding*, Machine Design, September 16, 1965, p. 179–182.

[22] P. M. BARTLE *Introduction to diffusion bonding*, Metal Construction and British Welding Journal, May 1969, p. 291–299.

[23] T. GIEBEN *Ervaringen met thermocompressie*, rapport 77–1962–JV, Ontwikkelingslab. Halfgeleiders N. V. Philips' Gloeilampenfabrieken, Nijmegen.

[24] J. EIGEMAN, A. L. M. GIEBEN *Scissorshechten* III–1965, Ontwikkelingslab. Halfgeleiders N. V. Philips' Gloeilampenfabrieken, Nijmegen.

[25] H. CHRISTENSEN *Electrical contact with thermo-compression bonds*, Bell Laboratories Record, April 1958, p. 127–130.

[26] E. T. Staff report, *Bonding and welding technique for microcircuits*, Electro-Technology, March 1963, p. 89–94.

[27] E. S. HODGE *Friction joining*, Batelle Technical Review 14, 1965, no. 8–9, Aug–Sept. p. 10–13.

[28] HOLLANDER, CHENG, QUIMBY *Friction welding*. Machine Design, June 18, 1964, p. 212–218.

[29] E. D. NICHOLAS, C. R. G. ELLIS *Friction welding of mild steel*, The Welding Institute, Research Bulletin, vol. **11,** no. 1, Jan. 1970, p. 3–10.

[30] D. PRESTON *Welded connections*, Product Engineering, August 5, 1963, p. 78–87.

[31] S. FLAX *Modern applications of resistance welding*, Welding and Metal Fabrication, Aug. 1971, p. 289–293.

[32] O. GENGENBACH *Übersicht über den Stand des Widerstandsschweißens*, Teil I: TZ f. prakt. Metallbearbeitung, Jahrgang **56**, 1962, Heft 10, p. 572–578.
Teil II: TZ f. prakt. Metallbearbeitung. Jahrgang **57**, 1963, Heft. 3, p. 135–139.

[33] O. BECKEN *Punktschweißen—eine Leistungsreserve*, Zeitschrift für Wirtschaftliche Fertigung, **57**, April 1962, Heft 4, p. 144–149.

[34] P. B. BONNER *The choice of resistance of fusion welding for sheet metal fabrications*, Welding and Metal Fabrication, May 1965, p. 202–207.

[35] W. EVANS *Resistance welding of dissimilar metals*, IEEE Transactions on Component Parts, June 1964, no. 2.

[36] P. LOWENSTEIN, W. B. TUFFIN *Bonding dissimilar metals by coextrusion*, Machine Design May 21, 1964, p. 228–234.

[37] D. PECKNER *Joining dissimilar metals*, Materials in Design Engineering, August 1962, p. 115–112.

[38] T. BALDER *De invloed van het Peltier-effect*, De ingenieur, jaargang **74**, no. 48, 30 november 1962, p. E 153-E 156.

[39] T. R. RUSSELL *High-frequency welding*.

[40] W. A. OWCZARSKI *Getting the most from projection welding*, Machinery, vol. **69**, no. 2, October 1962, p. 97–106.

[41] W. A. OWCZARSKI *Utilisation optimale du soudage par bossage*, Machine Moderne, Mai 1963, p. 37–44.

[42] W. H. MITSCH *Projection welding of light gauge galvanized steel*, Welding Journal, July 1965, p. 553–561.

[43] H. ALBERTI, F. FRÜNGEL *Das Widerstandsschweißen mittels transformierter Kondensatorenthladung*, Blech, no. 9, September 1961.

[44] J. WODARA *Widerstandsschweißen von Kleinteilen*, Schweißtechnik. January 1962.

[45] H. LEESER *Making minute parts faster is industry's big problem*, Welding Design and Fabrication, October 1961.

[46] H. EDER, W. P. UHDEN *Feinwerktechnische Verbindungen durch Schweißen*, Feinwerktechnik, Jahrgang **67**, April 1963, Heft 4, p. 110–131.

[47] R. D. ENGQUIST *Resistance welding of electronic assemblies*, The Tool and Manufacturing Engineer, October 1962, p. 89–94.

[48] G. MARONNA *Maschinen und Einrichtungen zum Widerstandsschweißen von Kleinteilen*, Schwei ßtechnik **14**, 1964, Heft 4, p. 166–169.

[49] H. BÖHME *Die Widerstandsschweißtechnik in der Feingeräteindustrie*, Feingerätetechnik, Jahrgang **12**, 1963, Heft 10, 0. 456–460.

[50] C. W. J. VERNON *Micro welding development*, Welding and Metal Fabrication, September 1964, p. 326–334.

[51] M. J. WATTS *Miniature Micro-miniature resistance welding and equipment*,
Part I: *Assembly & Fastener Methods*, October 1964, p. 36–38.
Part II: *Assembly & Fastener Methods*, November 1964, p. 49–51.

[52] W. T. ILLINGWORTH *Parallel gap resistance welding*, Electronic Packaging and Production, May 1964, p. 126–132.

[53] J. L. LAUB *Thin film welding*, Electronic Packaging and Production, March 1964, p. 20–27.

[54] C. R. SARGEANT, P. S. CLEMENTE *Bonding thin film and small foils*, Electronic Packaging and Production, March 1964, p. 28–33.

[55] L. D. ARMSTRONG *Adoption of welded electronic circuits for missile and space vehicles*, Fourth symposium on welded Electronic Packaging. Long Island, New York, 24 March 1961. From the space Technology Laboratories Inc., Los Angeles, California, U.S.A.

[56] G. ALLEN, J. WETTSTEIN *Welding of electronic modules for the polaris missile*, IEEE Transactions on product engineering and production, September 1963, p. 1–7.

[57] F. H. HOMAN *Percussielassen*, Metaal en Techniek, **9** (1964), no. 2, Jan. 23, p. 63–66.

[58] P. P. KING, J. E. SCHNEPF *Arc/percussive buttwelding of fine wireconductors*. Welding Journal, February 1965, p. 100–105.

[59] J. S. GALLATLY, K. F. JOHNSON, A. L. QUINLAN *Low-voltage percussion welding*, The Western Electric Engineer, July 1959.

[60] J. C. COYNE *Monitoring the percussive welding. Process for attaching wires to terminals*, The Bell System Technical Journal, January 1963, p. 55–78.

[61] L. SCHMID — *Percussive/arc electronic welder for butt weld applications*, Electronic Packaging and Production, May 1959.

[62] P. ANTONISSE — *De toepassing en ontwikkeling van het boutlassen*, Polytechnisch Tijdschrift, 6 maart 1964, p. 207–216a.

[63] *Philips stud-welding process*, Philips Welding News, **14**, no. 4, April 1963.

[64] R. C. SINGLETON — *Arc welded fasteners*, Machine Design, **21** March 1963 p. 73 ev.

[65] G. MARONNA, B. WEISS — *Das Ultraschallschweißen—ein Überblick*. Schweißtechnik, **15** (1965), Heft 4, p. 167–171.

[66] E. LEHFELDT — *Möglichkeiten in der Schweißtechnik durch Ultraschallschweißen*. Schweißtechnik, **15** (1965), Heft 5.

[67] P. MARVIN, D. C. TAO, L. SIMBECK, V. A. NOLAN, W. ROSSNAGEL — *Ultrasonic welding*, Machine Design, April 9, 1964, p. 129–138.

[68] T. VARGA — *Schweißen mit Ultraschall*,
Teil I: Europaischer Maschinen-Markt, 11/1963, p. d27–d29.
Teil II: Europaischer Maschinen-Markt, 12/1963, p. d34–d35.

[69] H. P. C. DANIËLS — *Introduction to Ultrasonic welding*, Nat. Lab. Verslag no. 3851, Philips research laboratories.

[70] H. P. C. DANIËLS — *Ultrasonic Welding*, Ultrasonics, October–December 1965, p. 190–196.

[71] J. WODARA — *Widerstandsschweißen und Ultraschallschweißen von NE-Metallen*. Schweißtechnik, **14** (1964), Heft 4, p. 149–154.

[72] C. ZGLENICKI — *Ultrasonic welding*, Product Engineering, February 15, 1965, p. 87–91.

[73] Committee AOC-G12, *Ultrasonic welding* (*of plastics*), OCOS (A) Plastics Technology, May 20th, 1965.

[74] FR. HALL — *Ultrasonics is newest way to join plastic to metal*, Product Engineering, May 23, 1966, p. 101–103.

[75] F. WENDLER — *Der lichtbogen-Plasmabrenner, eine neuartige Hochtemperaturwärmequelle*, ETZ-A, Band **83**, Heft 23, 5 November 1962, p. 773–776.

[76] G. COOPER, J. PALERMO, J. A. BROWNING — *Recent developments in plasma arc welding*, Welding Journal, April 1965, p. 268–276.

[77] E. F. GORMAN — *New developments and applications in manual plasma arc welding*, Welding Journal, July 1969, p. 547–555.

[78] L. DORN, H. ZÜRN — *Die anwendung des Elektronen des Strahlschweißens in der Industrie*, Schweißen und Schneiden, February 1965, **17**. Jahrgang, Heft 2, p. 41–57.

[79] J. G. SIEKMAN — *Electron beam machining at the Nat. Lab. Philips Research Laboratories*, Nat. Lab. report no. 4019.

[80] M. BRICKA, H. BRUCK — *Sur un nouveau canon électronique pour tubes à haute tension*, Ann. Radioelec. 1948, 3.

[81] J. W. MEIER — *New developments in electron beam technology*, Welding Journal, November 1964, p. 925–931.

[82] K. J. MILLER, L. M. REESE — *Electron beam welding*, Part I: Machine Design, April 23, 1964, p. 217–222. Part II: Machine Design, May 7, 1964, p. 165–168.
[83] R. E. TRILLWOOD — *Electron-beam welding of small components*, The Welding Institute Conference 'Advances in Welding Processes', April 1970, paper 15.
[84] J. KING — *Which way to weld? Resistance, TIG or electron beam. . . .*, Product Engineering, October 12, 1964.
[85] R. F. DUHAMEL — *Non vacuum electron beam welding technique development and progress*, Welding Journal June 1965, p. 465–474.
[86] G. ANDERS — *Der Laser*, Bild und Ton, Heft 10/1963, **16**. Jahrgang. p. 290–295.
[87] M. KRONENBERG — *Laser-ein neues Werkzeug der nähen Zukunft*. TZ für praktische Metallbearbeitung, **57** Jahrgang, March 1963, Heft 3, p. 129–135.
[88] A. R. PFLUGER, P. M. MAAS — *Laser beam welding electronic component leads*, Welding Journal, June 1965, p. 264.S–269.S.
[89] K. J. MILLER, J. D. NUNNIKHOVEN — *Production laser welding for specialized applications*, Welding Journal, June 1965, p. 480–485.
[90] J. R. SHACKLETON — *Laser microwelding*, SCP and Solid State Technology, May 1965, p. 15–19.
[91] J. E. ANDERSON, J. E. JACKSON — *Theory and application of pulsed laser welding*, Welding Journal, December 1965, p. 1018–1026.
[92] T. M. COMELLA — *Lasers as production tools*, Automation, July 1964, p. 68–77.
F. EICHHORN, H. J. OPPE — *Das Verbindungsschweißen von Kleinstbauteilen*, Feinwerktechnik, October 1967, **71** jg. no. 10.

2.3. Weldability of metals

2.3.1 Introduction

By weldability is understood the possibility of welding a metal, or a combination of metals, under specific conditions to form a structure satisfying certain metallurgical, mechanical and sometimes electrical requirements.

The total electrical resistance of the circuit should not be affected by the weld. The mechanical requirements, such as tensile strength, bending strength, resistance against notching and fatigue, will depend on the purpose and function of the weld. The properties of non-welded materials can serve as a measure of the mechanical properties. In general, however, the weldability of two metals depends on how far they match metallurgically. It is thus possible to forecast to some degree the weldability, by knowing the metallurgical properties of the metals. To a certain extent this applies to welds in which a melting zone occurs, and even more so to welds where the welding process takes place in the solid state. As fuse welding is little used in precision engineering, the weldability of metals by the resistance welding process will receive most attention. During the discussion, several links with the other methods will become evident.

2.3.2 Weldability by the resistance welding process

A resistance weld is made under the influence of heat and pressure during a certain period of time, factors that can be adjusted according to the welding method applied. How the welded metal reacts to these preset values depends on the physical properties, particularly the specific resistance ρ, the melting temperature θ_s and the heat conductivity λ.

Weldability L can then be roughly expressed as:

$$L = \frac{\rho}{\theta_s . \lambda} \tag{4}$$

Table 2.7 is a compilation of common metals and alloys and the different physical properties decisive in their weldability. For metals with a high ρ and a low λ, a melting zone may occur in the weld during the welding process, thus increasing the reliability of the weld. The weldability of these metals is therefore classified higher than those with a low ρ and a high λ.

The weldability of nickel, for instance, is better than that of copper.

Otherwise, the weldability table does not show that the heavy oxide layer of aluminium hampers welding or that the zinc vapours from the brass cause porosity of the weld. The table is thus supplemented below by a number of characteristics having particular effects on the welding of two different metals.

(a) *Coatings*[5,6,7]

Tin-coated steel sheet is not so easy to weld as bare sheet. The tin layer increases the contact area and consequently reduces the transition resistance (Fig. 2.78). Tin oxide contaminates the weld and the electrode surfaces.

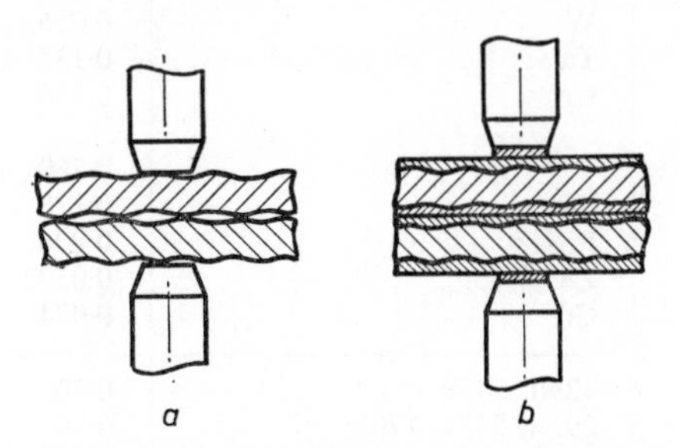

Fig. 2.78. The plate contact of *a.* bare and *b.* tinned steel plate.

After a number of welds have been made, the heat transmission between electrode and steel plate deteriorates. For the welding of tinned steel plate, therefore, the projection welding method is often preferred (see Section 2.2.8.2(e)(iii)). The projection increases the transition resistance and localizes the welding place: at the same time, the electrode area can be taken considerably larger, so the life of the electrodes is prolonged.

On the other hand, the weldability of copper plate is improved by the presence of a layer of tin. The transition resistance improves as compared with bare copper. But care must be taken that no low melting metals, such as tin, cadmium, lead and zinc, remain behind in the weld to cause embrittlement.

Lead in particular is disastrous for nickel and nickel alloys. Nickel-plated surfaces can give rise to higher transitional resistance and improve reproducibility of the weld, due to the uniformity of the layer.

TABLE 2.7

Metal	Composition	specific resistance $\times 10^{-6}\ \Omega$ m	melting temperature θ °C	heat conductivity
iron	Fe	0·0971	1537	0·01
cobalt	Co	0·0624	1492	0·01
nickel	Ni	0·0684	1453	0·02
copper	Cu	0·0167	1083	0·09
silver	Ag	0·0160	960	1·0
gold	Au	0·0222	1063	0·07
platinum	Pt	0·1060	1769	0·01
palladium	Pd	0·1075	1552	0·01
aluminium	Al	0·0269	660	0·05
magnesium	Mg	0·039	650	0·04
molybdenum	Mo	0·057	2600	0·03
tungsten	W	0·055	3380	0·03
tantalum	Ta	0·135	2980	0·01
chromium	Cr	0·130	1850	0·16
vanadium	V	0·260	1860	0·07
tin	Sn	0·128	232	0·01
lead	Pb	0·206	327	0·00
zinc	Zn	0·0592	420	0·02
cadmium	Cd	0·074	321	0·02
nickel-iron	42Ni, 58Fe	0·50	1425	0·03
fernico	29Ni, 54Fe, 17Co	0·70	1400	0·05
nickel-chromium	73Ni, 20Cr, Al+Fe	1·33	1400	0·03
nickel-copper	43Ni, 57Cu	0·49	1310	0·01
brass 58	58Cu, 42Zn, 0·2Sn, 2Pb	0·067	900–1080	0·02
brass 63	63Cu, 35Zn, 0·2Sn, 0·5Pb	0·070	905–920	0·02
brass 85	85Cu, 15Zn	0·047	1010–1025	0·03
phosphor bronze	93Cu, 7Sn	0·160	910–1036	0·03
steel	0·2C, 0·45Mn, 0·25Si	0·105	1500	0·01
chrome nickel steel	0·2C, 18Cr, 8Ni	0·276	1415	0·00

* The weldability L of the low melting metals is not approximated correctly in this colum

** 1 kcal/m.s °C = 41·868 W/m. deg

† 1m .²s/kcal = 0·23884 m.²s/J

‡ 1 kcal/kg = 4186·8 J/kg

¶ 1 kcal/kg °C = 4186·8 J/(kg. deg)

§ Average

$L = \frac{\rho}{\Theta_s \cdot \lambda}$ $\frac{m^2 \cdot s}{kcal}$ †	temp. coefficient electrical resistance °C × 10⁻⁶	boiling point °C	melting heat kcal/kg ‡	linear expansion coefficient 0–100°C × 10⁻⁶	specific heat kcal/kg °C ¶	crystal dimensions 10^{-10} m = Å	Crystal structure
3·72	6·51	3070	49	12·1	0·109	2·5 §	β body-centred cubic β face-centred cubic
2·54	6·04	2900	64	12·5	0·102	2·5 §	β hexagonal close packed β face-centred cubic
2·42	6·81	3000	73	12·3	0·100	2·49	β hexagonal close packed β face-centred cubic
0·164	4·3	2590	51	17	0·092	2·55	face-centred cubic
0·166	4·1	2210	25	19·1	0·054	2·88	face-centred cubic
0·31	3·9	2950	16	14·1	0·031	2·88	face-centred cubic
3·53	3·9	4240	27	9·0	0·032	2·77	face-centred cubic
4·09	3·8	3900	36	11·0	0·059	2·75	face-centred cubic
0·72	4·2	2480	85	23·5	0·219	2·86	face-centred cubic
1·50	4·2	1103	46	26	0·248	3·20	hexagonal close packed
3·65	4·6	5550	—	5·1	0·062	2·72	body-centred cubic
0·41	4·6	6700	60	4·5	0·033	2·74	body-centred cubic
3·49	3·8	—	—	6·5	0·034	2·85	body-centred cubic
0·43	5·9	2620	70	6·5	0·110	2·71	β body-centred cubic β hexagonal close packed
2·0	2·8	3350	—	8·0	0·190	2·63	body-centred cubic
—*	4·2	2450	14·2	23·5	0·054	3·10 §	body-centred tetragonal
—*	3·4	1740	5·7	29	0·031	3·49	face-centred cubic
—*	4·2	907	24·1	31	0·094	2·75 §	hexagonal close packed
—*	4·3	765	12·9	31	0·056	3·13	hexagonal close packed
9·94							
9·50							
0·15							
8·55							
2·62							
2·94							
1·25							
4·46							
5·38							
7·5							

The influence of the oxide layer on the weldability of aluminium is dealt with in Section 2.3.3.

(b) *Weldability of two different metals*

If a melting zone occurs in the transition area of the two metals, it can crystallize into a solid solution or into a mixture of solid matter.

Two metals form a solid solution, copper in nickel for instance, if the atom distances in the lattice of each of the solid metals do not differ too much ($<14\%$). The solid solution lends the weld toughness and strength. The formation of a mixture depends on the characteristic of atoms to take up electrons in the outer skin. Consult the condition diagram of the relevant alloy. If this connection is distributed in the weld lens as a very fine dispersion, the weld will be stronger; a close network of a solid mixture, on the other hand, causes hardness and embrittlement of the weld.[4].

If, owing to a low ρ and high λ, no nugget is formed, we are dealing with a pure pressure weld whose quality depends on the crystalline structure and crystal dimensions. The welding heat reduces the flow limit, so that, under the influence of the welding pressure, the two metals are deformed into a new orientation in the transition zone. If too little deformation takes place there will be voids weakening the weld (Fig. 2.79).

Too great differences in crystal structure and dimensions hamper adherence (Fig. 2.80). A difference between the expansion coefficients of the two metals can give rise to great stress during the cooling process[8], so this factor is given in Table 2.7, as are the boiling point, melting heat and specific heat. In general, all factors involved in the heat of a weld must be so handled that, during the welding process, the work temperature (and for a fuse weld, the

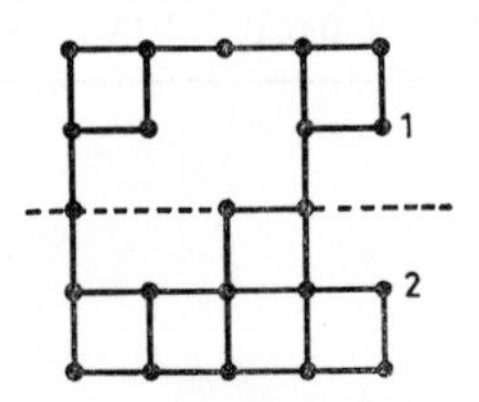

Fig. 2.79. Shows vacant areas in the transitional zone caused by a low degree of deformation of the metals 1 and 2.

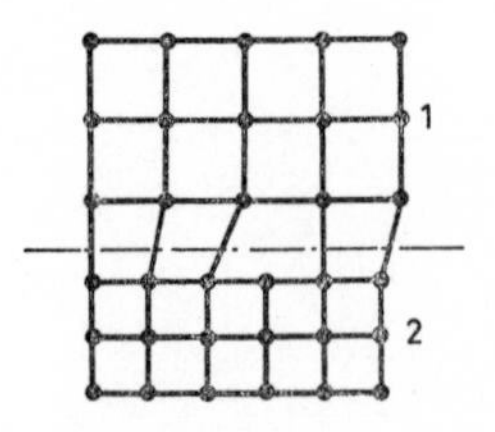

Fig. 2.80 Dislocations due to great differences in crystal structure and dimensions of the metals 1 and 2.

fuse temperature) of the two metals is reached simultaneously. The work temperature is the specific temperature at which a metal can be mechanically deformed without increasing its hardness, and without a deformation stress remaining behind.

The work temperature is about 80% of the melting temperature.

2.3.3 *A few metals difficult to weld*

(a) *Aluminium*[9,10,11]

The weldability of aluminium is greatly obstructed by the enormous heat conductivity and the presence of a thick layer of oxide. It is not possible to weld aluminium with a plasma arc, for an a.c. is needed to destroy the layer of oxide during the negative sine-half.

For fuse welding, the argon arc welding method is indicated, for the weld is then constantly surrounded by a gas protecting the weld against oxidation. The electron beam method is also suitable. Formerly, aluminium sheets were welded by Weibel's method (Section 2.2.4.2), but with the introduction of argon arc welding, that method became less important.

Where pressure welding for light components is concerned, the most important method is ultrasonic welding. The high-frequency vibrations break through the oxide film, and due to the low elasticity modulus of aluminium, ensure pure metallic contact over the whole welding area. Vapour-deposited layers of aluminium can be welded according to the thermo-compression method. Resistance welding of aluminium requires high power equipment, up to 6 000 KVA, and a welding pressure from 100 to 250 kg for the welding of sheet thicknesses up to 2 mm. Both the welding pressure and the welding current must be adjustable in three phases during one welding cycle: that is, the process must be programmed. With a pre-welding time and low power the electrode can make satisfactory contact throughout the whole area, so that the heat load during the welding time is lower. The surface must therefore be cleaned with a steel brush before welding.

(b) *Copper and copper alloys*[13]

Copper is a material difficult to weld, because of its great heat conductivity, strong inclination to oxidize, particularly above 400 °C, and great solubility for hydrogen. Fuse welding is only possible where sharply bundled energy can be administered in a neutral atmosphere, as with electron beam welding, for instance.

Resistance welding of pure copper, without special precautions, is not possible either. But, by applying a resistance-increasing coating, such as tin, there are certain possibilities. It should be noted that resistance welding is the most suitable method for alloys containing less than 80% copper. The welding conditions are about the same as those for welding soft steel of equal thickness, in which case, however, the welding pressure can be lower, as a certain contact resistance is maintained. Too low a pressure causes the electrode tip to adhere to the material. Projection welding (Section 2.2.8.2 (e)(iii)), is also one of the indicated methods. Tombac (85Cu, 15Zn) requires a great welding power, which increases the danger of sticking electrodes. Zinc vapours make brass with a larger zinc content porous, especially after longer welding times.

Manganese in brass increases the electrical resistance and reduces the heat conductivity, thus improving the resistance welding process. Silicon also largely improves the welding process. Silicon bronze is quite weldable, takes low welding energy and the electrodes do not stick.

Phosphor bronze has a reasonable welding characteristic, but the weld itself is rather brittle and the electrode surfaces require regular cleaning.

Copper nickel alloys are suitable for welding. For the welding of various kinds of wire to copper-zinc alloys, the percussion welding method (Section 2.2.8.3) is quite satisfactory. The short welding time, and the low penetration of the welding heat, prevent embrittlement of the weld.

(c) *High melting metals, tungsten, molybdenum and tantalum*[14]

Originally, high melting metals were prepared by the powder-sintering process. The sintering process, however, entails a large amount of W and Mo contaminations and renders these metals unsuitable for welding. The welding quality is improved by melting the sintered material in a vacuum or neutral gas with an electric arc.

Also, with the electron beam, a high melting material with a high degree of purity is obtained.

Mo and Ta are obtained in thin sheets: tungsten is obtainable in the form of wire.

All three metals have a cubic body-centred structure, which causes them to become very brittle below a certain temperature. The transitional temperature from brittle to flexible for Mo ranges from $-45°$ to 40 °C, and for tungsten from 175 °C to 455 °C (see Fig. 2.81). Pure tantalum remains flexible to the temperature of liquid hydrogen.

The brittleness of the metals depends to a large extent on contaminations. Tantalum is sensitive to hydrogen. The oxidation resistance of molybdenum is very low. At 450 °C Mo begins to smoke. Consequently oxygen forms no

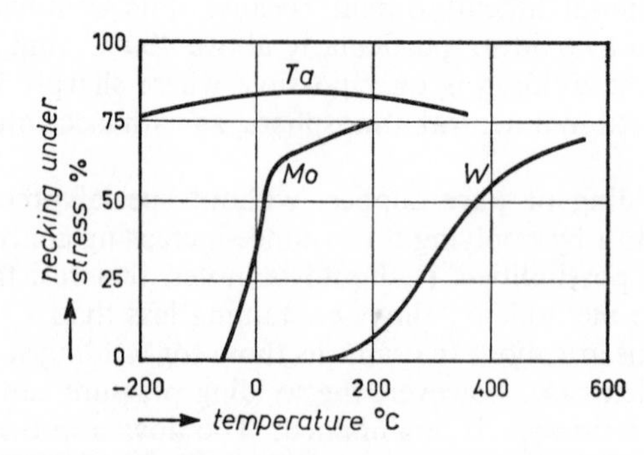

Fig. 2.81. Toughness as a function of temperature, for tantalum (Ta), molybdenum (Mo) and tungsten (W).

protection. Below the melting temperature of Mo, the MoO_3 evaporates so that the surface becomes porous.

Tungsten is more susceptible to contamination than molybdenum. It is possible to make molybdenum welds that are tough at room temperature; this is not possible with tungsten.

At room temperature, Mo and W have a low solubility for oxygen, nitrogen and carbon. Both metals are insensitive to hydrogen. Tantalum can be welded reasonably well by the argon arc method, using a direct current with the electrode connected as the cathode; the shield gas can be argon, with 0·1 % nitrogen and 0·1 % oxygen. The electron beam welding method is very suitable for tantalum. Very good results have been obtained in the resistance welding of Ta wire and Ta sheet.

Molybdenum is a material extremely difficult to weld. Argon arc welding gave no satisfactory results, but the electron beam welding method offers better possibilities, because of the highly concentrated heat supply in vacuum.

Certain results can be obtained by adopting the resistance welding method, but then high current intensities and short welding times are required. Overheating of the metal surface must be avoided by means of vaseline, distillated water or alcohol. Filler material, such as nickel, platinum and chromium, reduces the welding temperature and the dimensions of the welding zone. The percussion welding method also offers good possibilities for Mo-wire welding.

Tungsten is the most difficult metal. Argon arc welding gives no favourable results, and the same applies to tungsten alloys: at high welding temperatures the favourable effect of the alloying metal disappears.

Electron beam welding gives slightly better results, but the weld remains brittle and weak.

The resistance welding method requires a high current intensity, short welding times and low pressure, with iron, nickel or platinum as a sandwich material.

Electrode sticking can be reduced by using a high welding pressure or by adopting the sandwich method.

REFERENCES

[1] A. L. Philips — *Welding Handbook*, Section 1: *Fundamentals of welding*, Section IV: *Metals and their weldability*, American Welding Society, 1966.

[2] L. Jones — *The physics of electrical contacts*, Clarendon Press, Oxford 1957.

[3] J. K. Stanley — *Electrical and magnetic properties of metals*, American Society for Metals, Ohio.

[4] R. D. Engquist — *Evaluating weldability of various materials*, Welding Engineer, October 1964, p. 47–52.

[5] N. Pickel — *Der Einfluß der galvanischen Oberflächenbehandlung bei Punktschweißverbindungen*, Schweißen und Schneiden, Jahrgang **16**, 1964, Heft 4, p. 131–135.

[6] L. Schade — *Die Auswirkungen eines Zinnüberzuges auf den Schweißprozess beim Widerstandsschweißen von dünnen Stahlblechen.* Blech, no. 8, 1963, p. 518–524.

[7] J. B. WILLIAMSON *Characteristics of contact surfaces to make reliable electrical connections*, Machine Design, May 7, 1964, p. 172–179.

[8] D. PECKNER, *Joining dissimilar metals*, Materials in Design Engineering, August 1962, p. 115–122.

[9] M. BECKERT, J. RITTER *Widerstandsschweißen von Aluminium und einigen Aluminiumlegierungen*, Schweißtechnik, April 1964, Jahrgang **14**, Heft 4, p. 145–148.

[10] H. H. REINSCH *Die Durchführung des Widerstands-Schweißens von Aluminium*, Schweißtechnik, November 1962, p. 149–151.

[11] J. A. DONELAN *Cold welding aluminium and copper strip*, Sheet Metal Industries, December 1963, p. 863–880.

[12] J. E. ROBERTS *Spot welding of light alloys*, British Welding Journal, May 1955, p. 193–199.

[13] C. L. BULOW *Resistance welding of copper alloys*, The Tool Manufacturing Engineer, May 1964, p. 91–92.

[14] M. H. SCOTT, P. M. KNOWLSON *The welding and brazing of the refractory metals niobium tantalum, molybdenum and tungsten—A review*, Journal of the less-common metals, **5**, 1963, p. 205–244.

[15] H. RISCHALL *Laserwelding for microelectronic interconnections*, IEEE transactions on component parts, June 1964, no. 2.

[16] R. A. GESHNER *Welded electrical joints and criteria for process acceptance*, Report of the Proceedings; Electronic Components Conference, 1963, p. 61–66.

2.4. Quality control

2.4.1 Introduction

The quality of non-automatic fusion welds depends mainly on the skill of the welder. Any subsequent inspection consists principally of visual assessment. If more information is to be obtained, inspection becomes rather expensive.

The automatic welding process depends less on variables, and is therefore more reliable.

With pressure welding, and especially resistance welding, the results depend far less on the skill of the welder. The weld quality then depends mainly on the welding machine and how it has been adjusted. As many factors play a part in producing a pressure weld, one would like to follow the welding process in every detail, so that if anything goes wrong, the faulty weld can be removed via a feedback system. This monitoring during the welding process becomes more difficult as the dimensions of the weld become smaller. In precision and microwelding techniques, therefore, one is limited to visual inspection and taking random tests after welding.

Nevertheless, in this survey, some attention is devoted to non-destructive weld investigation, since it has various aspects important to precision welding.

2.4.2 Non-destructive inspection

2.4.2.1 Inspection of fusion welds

(a) *Visual inspection*

The quality of each weld can, to a certain extent, be visually assessed. In the case of an electric arc weld, for instance, faults such as irregular slag layer, too-deeply-burnt weld edges, cracks, pores and too little filler material, are indications of the quality of the weld. The quality inspector should be an expert and an experienced welder too.

(b) *Electromagnetic investigation*[1]

If magnetic poles are placed on each side of the weld, lines of force will run through the ferromagnetic metal. If the weld shows a crack or crevice, the force lines will spread out locally (Fig. 2.82).

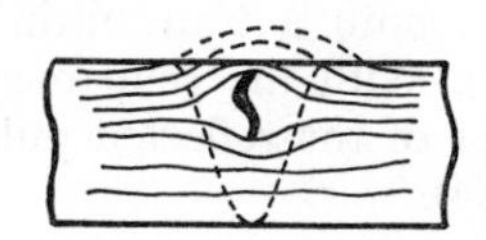

Fig. 2.82. Magnetic lines of force travel round a cavity.

The local deviation of the lines can be made visible with a thin flowing oil film containing iron filings. The penetration depth is 2–4 mm.

(c) *Capillary suction method*[1]

To localize superficial cracks, a pigment of low surface tension is applied. If after 1 min the surface is washed clean, an emulsion of chalk powder and methanol will then show-off the cracks in red lines against a white background.

(d) *X-ray test*[1]

Thanks to their short-wave length (10^{-6}–10^{-8} mm), X-rays penetrate through solid materials, but they are more or less attenuated, depending on the chemical composition of the material and the thickness of the workpiece. The X-rays can be made visible on a photo-sensitive plate. Faults in the weld can thus be photographed (Fig. 2.83). Thorough experience is a first requirement for the correct assessment of the photograph. For safety reasons, a screening against X-rays is required.

By employing radioactive isotopes, use is now made also of gamma rays which, because of their shorter wavelengths, have greater penetration.

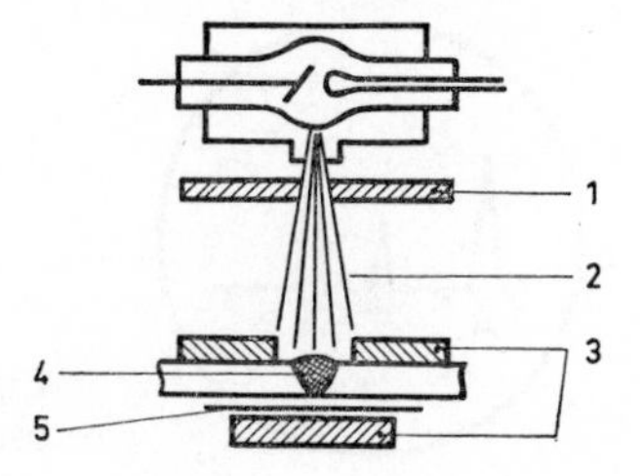

Fig. 2.83. X-ray weld examination. Lead plates and a lead diaphragm prevent scatter of the rays.
1 = diaphragm.
2 = X-ray.
3 = lead layer.
4 = weld.
5 = photosensitive plate.

(e) *Examination with ultrasonic waves*[2]

Ultrasonic waves, with a frequency of more than 20 kHz, take a straight path through homogeneous materials; irregularities, and the transition area between two different materials, produce reflections. A coarse-grained structure, porosity and the presence of fine dispersions impede the investigation of discontinuities. Generally, longitudinal waves with a frequency ranging from 1 to 25 MHz are used. The transverse waves are more liable to stray.

Two systems are used:

the through transmission method and the pulse-echo method.

The first method requires both a transmitting and receiving converter (see also Section 2.4.2.2.(c)). With the echo method, the converter serves as a transmitter and receiver of the reflected pulse. This method is widely used at present (see Fig. 2.84).

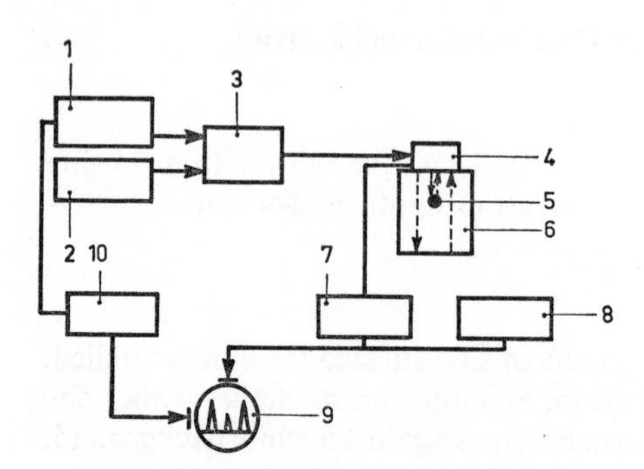

Fig. 2.84. Diagram of detector in the pulse-echo method.
1 = pulse generator.
2 = high-frequency generator.
3 = modulator.
4 = converter.
5 = discontinuity.
6 = workpiece.
7 = receiver amplifier.
8 = matching generator.
9 = screen of cathode-ray tube.
10 = time base.

The converter, a piezo-electric crystal of quartz, lithium sulphate, barium titanate or Rochelle salt, converts the high frequency voltage from the modulator into ultrasonic vibrations.

Between the converter and the workpiece, there is a film of water, motor oil or glycerine, to ensure good contact between converter and workpiece.

The time lag between the original pulse and reception of the echo is made visible on the screen of the oscilloscope via an amplifier (see Fig. 2.85). The distance A-A corresponds with the thickness of the section under investigation: B is the echo of a micro-crack or cavity in the weld. The distance A-B is then a measure of the position of the discontinuity. The pulse time-base and marker generator must operate synchronously.

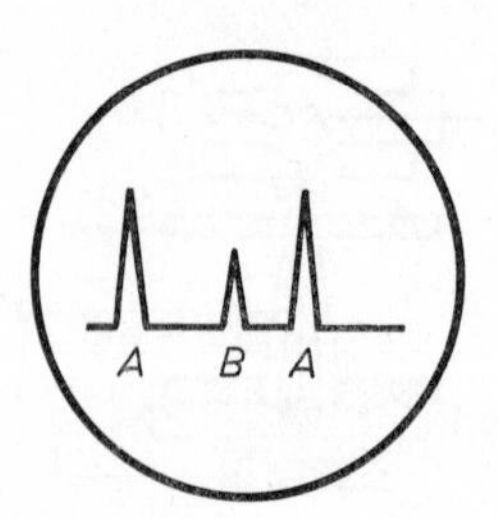

Fig. 2.85. Reflection echoes on the screen of an oscilloscope.

A correct interpretation of the image requires expert knowledge of the structure of the weld, the temperature and the shape of the part under investigation, as well as the surface condition of the workpiece. Calibres are often used for inspection purposes.

2.4.2.2 *Examining resistance welds*

(a) *Visual assessment*

A visual inspection of the resistance weld is often insufficient for quality assessment. In many cases, the weld is not even visible. Wire connections are usually easier to assess (Figs. 2.86 and 2.87).

After all, only random tests can be made. A few examples of wire connections, with their faults and causes, are shown in Figs. 2.86 and 2.87.

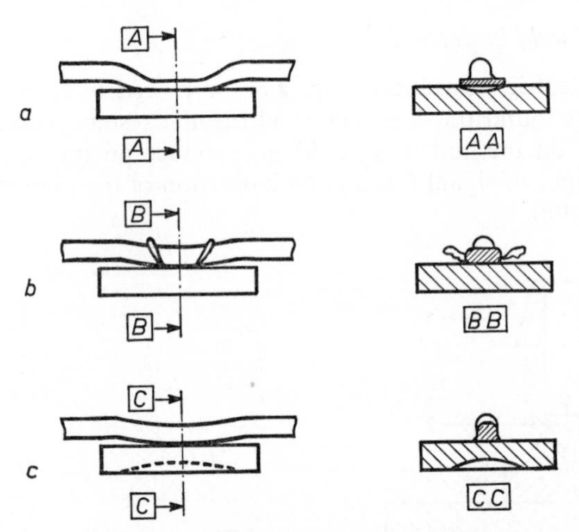

Fig. 2.86. Wire flat on the plate [3].
a. Weak weld: wire flattened too much owing to excessive welding pressure.
b. Sputtered weld: electrode does not follow quickly enough.
c. Plate material burns in: another electrode material must be selected.

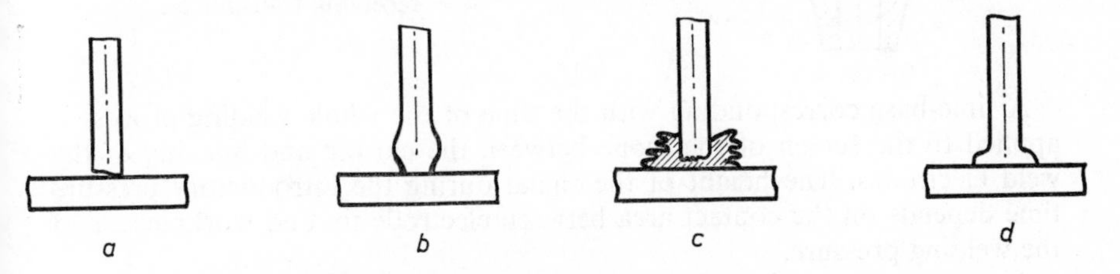

Fig. 2.87. Tip of wire on plate.
a. Weak weld: too little power.
b. Weak weld: too heavy pressure.
c. Brittle weld: too much power.
d. Correct weld: correct quantity of power and pressure.

(b) *Measuring the welding current*

Measuring the principal variables during the welding process is far more efficient than post-inspection.

For this purpose, welding current meters take up a foremost position. From an induction coil round the current conductor, a current signal is delayed and fed to an indicating instrument giving direct reading of the rms or maximum value of the current pulse. The meter can also be extended with a monitor which gives a warning as soon as the weld current deviates from a certain preset value. In principle, it is possible to use this deviating current value to remove the defective weld. The weld current meter is suitable mainly for the transformer welding method.

A lot of microwelding apparatus is fitted with an ohmmeter for measuring the initial weld resistance, that is, before welding, (Section 2.2.8.2(a)).

(c) *Ultrasonic weld inspection*[4]

If a piezo-electric transducer (Fig. 2.88) is placed in each of the welding electrodes, it is found possible, via an ultrasonic system based on the transmission-reception method (Fig. 2.89), to produce an image on the oscilloscope. The height of signal P is a good indication of the diameter of the weld nugget (Fig. 2.90).

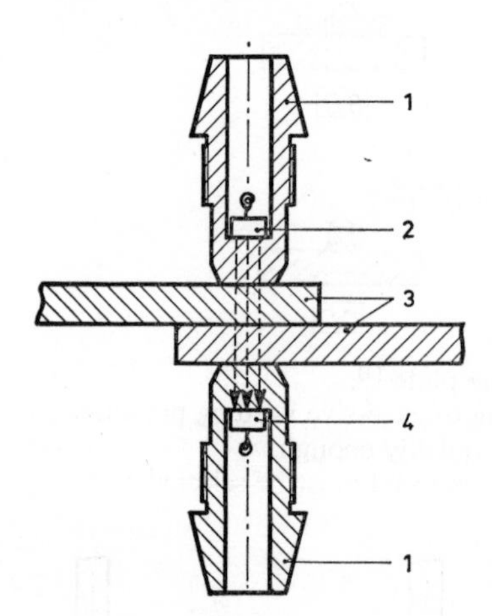

Fig. 2.88. Arrangement of converters in the electrodes.
1 = electrode.
2 = transmitting converter.
3 = workpiece.
4 = receiving transmitter.

A time-base corresponding with the time of the whole welding process is applied to the screen of the scope between the closing and opening of the weld electrodes. The height of the signal during the introductory pressure time depends on the contact area between electrode tip and workpiece, and the welding pressure.

At current conduction, the ultrasonic signal falls off to the reference line. As soon as a weld nugget is formed, the signal rises: it stops when the nugget has assumed its maximum value P, after which the signal falls off again. Switching off the weld current is accompanied by a steep rise, followed by a drop to the height of the post-pressure time. If the measured weld nugget diameter is plotted as a function of signal height P at various current intensities, the result is curve 1 in Fig. 2.91. At the same time, the corresponding strength of the weld is plotted in curve 2. The value of the welding current

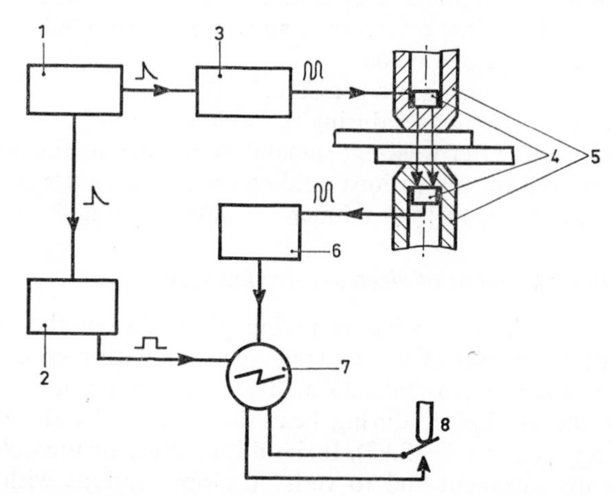

Fig. 2.89. Diagram of transmission-reception method for the ultrasonic investigation of resistance welds.

1 = pulse generator. 4 = converter. 7 = oscilloscope.
2 = square-wave generator. 5 = electrodes. 8 = trigger time-base.
3 = transmitter. 6 = amplifier.

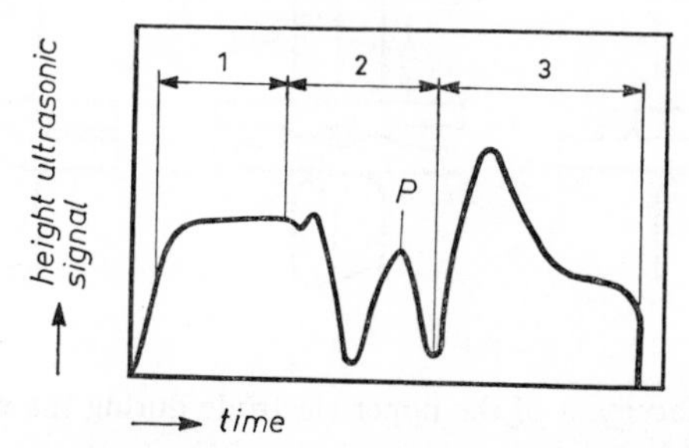

Fig. 2.90. Ultrasonic image obtained with the transmission-reception method.
1 = pre-pressure time. 2 = welding time. 3 = post-pressure time.

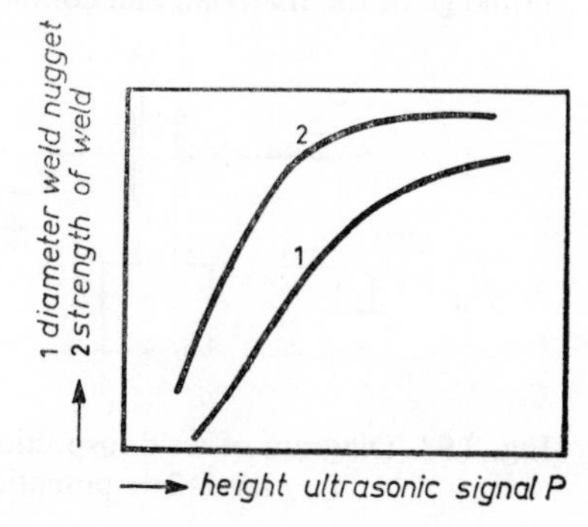

Fig. 2.91. Diameter (1) and strength (2) of weld, as a function of the height of ultrasonic signal.

has no direct effect on the ultrasonic signal, so the signal height depends only on the welding pressure, contact area between electrode tip and workpiece, and the metallurgical condition of the weld. At a certain current value, the weld nugget diameter remains constant. It is possible, therefore, using the ultrasonic method of inspecting resistance welds, to read the following information from the scope image:

Electrode pressure, during pre-pressure.

Diameter of the weld nugget during the welding period.

Material shrinkage after welding; signal decrease during the post-pressure.

The time required for the various welding operations, such as pre-pressure and post-pressure time, melting moment, welding time and cooling time.

(d) *Examination by means of electrode displacement*[5]

To gain an insight into what is taking place during the spot welding process, the displacement of the moving electrode with respect to the fixed electrode can serve as an indication. The starting point is the thermal expansion of the workpiece during heat supply and the shrinkage of the material during cooling (Fig. 2.92). It should therefore be possible to observe the electrode displacement and to verify the observations with the quality of the weld. Verification will become more difficult with smaller welds. A method of displaying the displacement on the screen of an oscilloscope is shown in Fig. 2.93. Through a lever, the electrode arm operates a precision

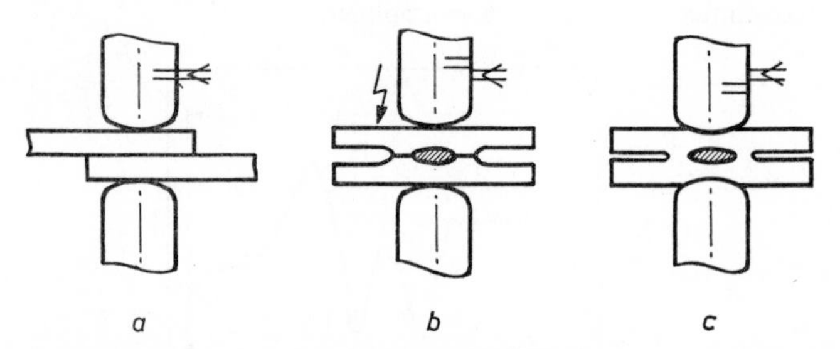

Fig. 2.92. Displacement of the upper electrode during the welding process.
a. Before welding begins.
b. Expansion of the material during welding.
c. Shrinkage of the material, and compression of the electrode, during cooling.

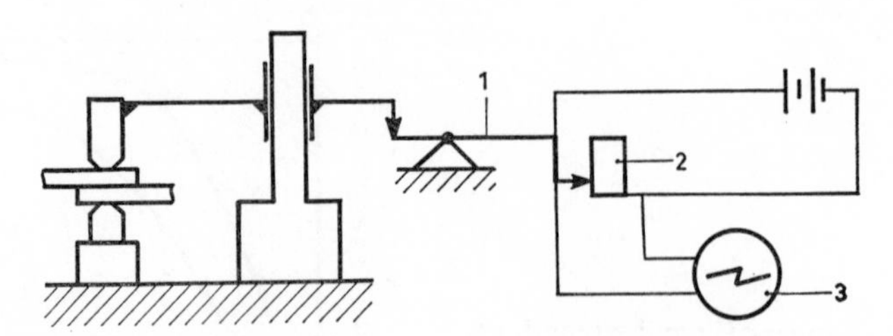

Fig. 2.93. Diagram of weld inspection by means of electrode displacement.
1 = level. 2 = potentiometer. 3 = oscilloscope.

potentiometer whose voltage gives an electrical indication of the electrode displacement.

Fig. 2.94 is a diagram of the electrode displacements when welding 1·5 mm mild steel with a welding time of 9 cycles, 180 ms.

Moreover, the variation of the welding current and the welding pressure have been included. If the welding current is switched on, the material expands immediately and moves the electrode upwards according to BC ($\approx$ 0·14 mm) until the current is switched off again. The electrodes are forced into the softened material so that at C the curve is slightly flattened. During period CD the weld cools off and the material shrinks, after which the electrodes move upwards again.

Angle α is decisive for the quality of the weld. If α is too great the result will be a sputter weld; if α is too small, it indicates that too little welding power has been supplied, to give a weak weld (Fig. 2.95).

Tan α can be fed to a monitor by measuring the time of electrode displacement A (Fig. 2.95) from the moment the welding current is switched on. The monitor is so adjusted that for a time $t_2 < t < t_1$, for instance, a red

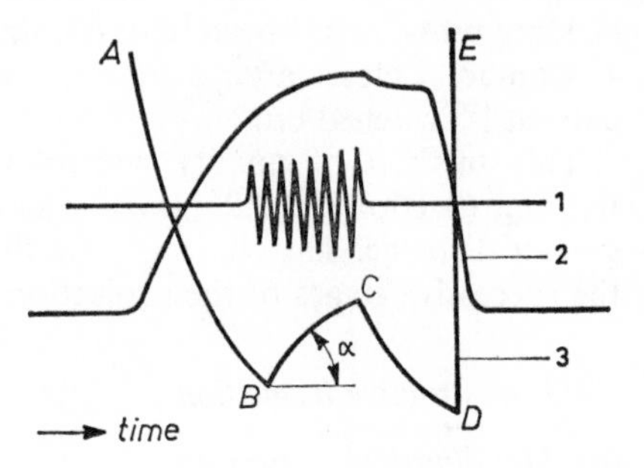

Fig. 2.94. Diagram of electrode displacement, current and pressure of a mild steel spot weld.
1 = current.
2 = pressure.
3 = electrode displacement.
AB = closing of electrodes.
BC = material expands.
CD = cooling of weld.
DE = opening of electrodes.

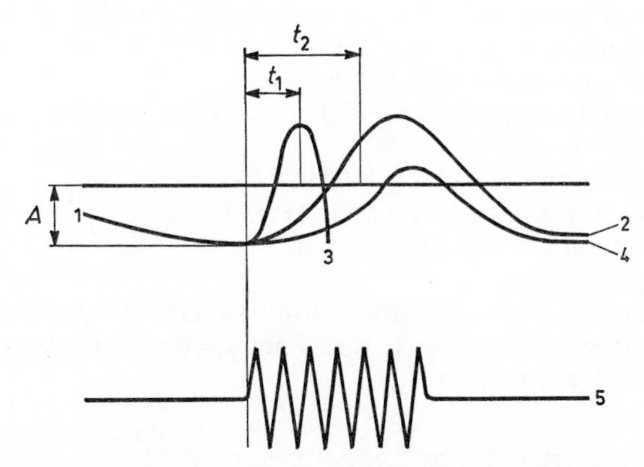

Fig. 2.95. Electrode displacements depending on the quality of the weld.
1 = electrode displacement.
2 = normal weld.
3 = sputter weld.
4 = weak weld.
5 = current curve.
A = time of electrode displacement

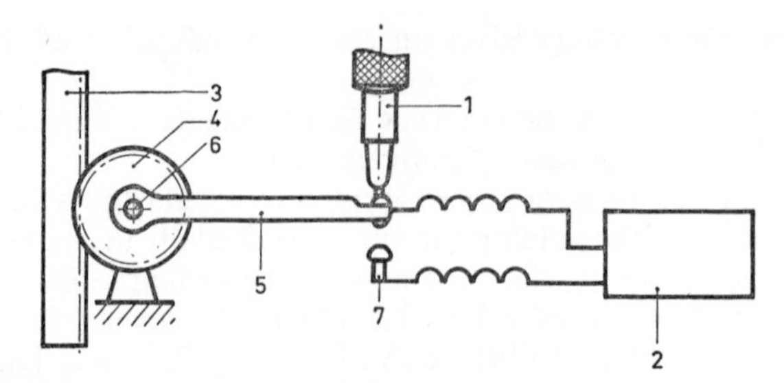

Fig. 2.96. Diagram of inspection system of the electrode displacements during a time *t*.

1 = microscrew. 3 = tooth rack 5 = level. 7 = contact.
2 = monitor. 4 = pinion. 6 = axis.

lamp indicates that the weld is rejected. The apparatus feeding time t to the monitor is schematically shown in Fig. 2.96.

By means of a pinion 4, tooth rack 3, connected to the moving electrode arm, drives lever 5 which is connected to shaft 6 via a clutch. As soon as the welding pressure has been built up, the end of 5 is placed against microscrew 1. Contact 7 closes after a time t, calculated from the moment the welding current is switched on.

This method of quality control is particularly suitable for projection welding (Section 2.2.8.2(e)(iii)), where the electrodes are displaced over a greater distance. The characteristic thus obtained gives a reliable picture of the successive stages of the projection welding process[6].

2.4.3 *Destructive inspection*

(a) *Metallurgical inspection*

If nothing can be learned from the appearance of a weld, a slice (that is, a polished and etched cross-section) of the weld might supply more information. In case of a resistance weld this involves:

The presence of a weld nugget and its dimensions, position and structure (Fig. 2.97).

Recrystallization of the metal (Fig. 2.98).

The presence of gas cavities or cracks (Fig. 2.99).

Contamination in the weld (Fig. 2.100).

Shrinkage (Fig. 2.101).

Otherwise, a slice shows the situation of one cross-sectional area of the weld, for one specific case only. It is still not known how far the welds not investigated are in accordance.

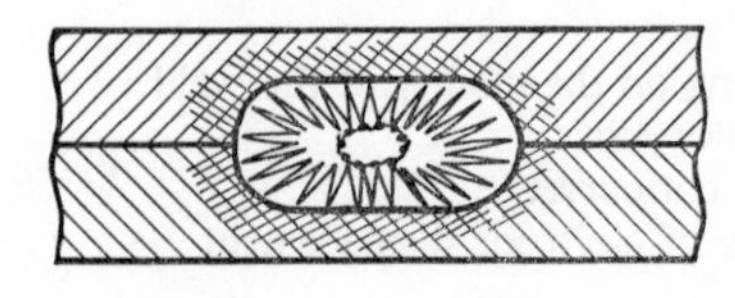

Fig. 2.97. Normal weld nugget. The melting zone shows a dendritic cast structure [7].

Fig. 2.98. Recrystallization.
a. Resistance weld of gold-coated Ni-wire and Ni-strip; with melt zone.
b. Resistance weld of tin-coated Cu-wire and Ni-wire; no melt zone.

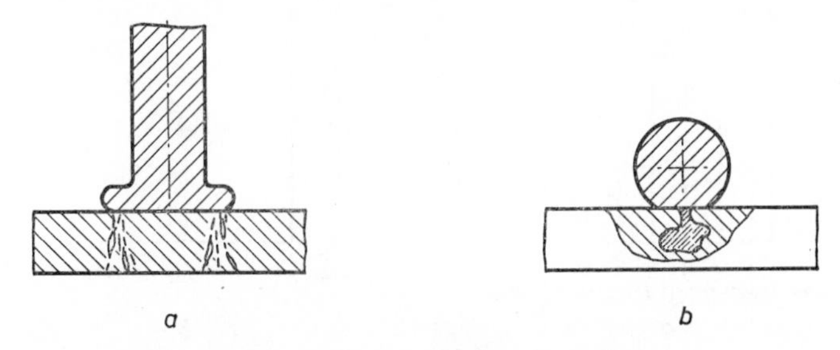

Fig. 2.99. Gas cavities and cracks.
a. Cu-wire and NiFe-strip: occurrence of microcracks.
b. Tin-coated Cu-wire and Ni-strip: cavitation in Ni-strip.

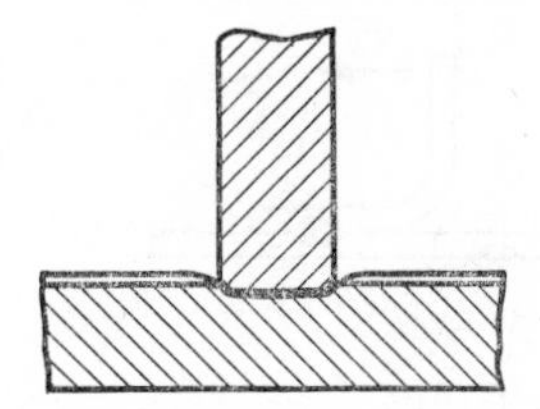

Fig. 2.100. Steel pin on tinned plate: presence of burnt tin in the weld.

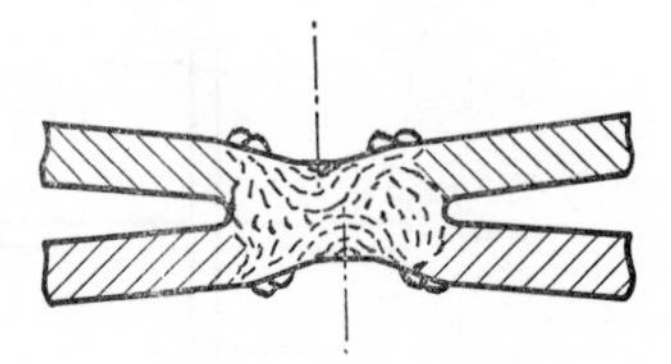

Fig. 2.101. Shrink effect: welded strips move apart. Causes: too small an electrode tip area, electrodes not properly aligned, too great a current intensity [7].

(b) *Strength test*

The quality of fusion welds has to meet certain standards[1] as regards mechanical tests in regard to: tensile strength, malleability, notching, fatigue and hardness.

The strength of projection welded products depends also on the variable shape and dimensions of the welded components. Moreover, function determines the requirements of the weld. For instance, a compression weld between gold wire and a vaporized metal film requires a different approach to strength assessment than a spot weld in coach building.

Consequently, there is no standardization of strength requirements in pressure welding.

Test standards can, of course, be made for certain categories of weld, and the results can be brought together in a histogram or iso-strength diagram (Fig. 2.102) to find a correct adjustment of the welding machine. For the testing of wire and ribbon connections, for instance, the breaking force *F* is measured (Fig. 2.103).

For a strength test of the weld between plate and wire, the bending-tensile test is applicable (Fig. 2.104).

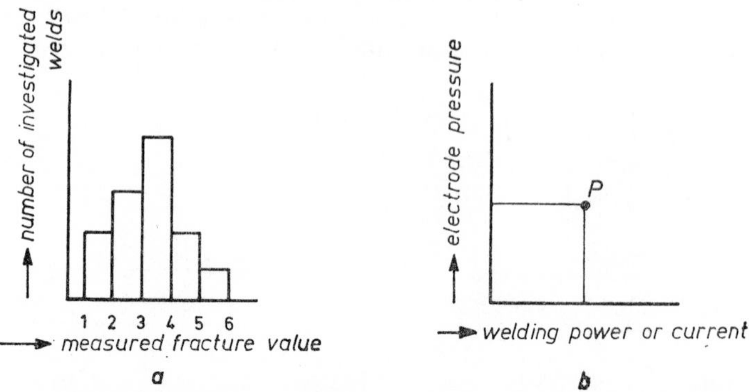

Fig. 2.102. *a.* Histogram. Number of investigated welds as a function of the measured fracture value.
b. Iso-strength diagram. The measured value is located in *P*.

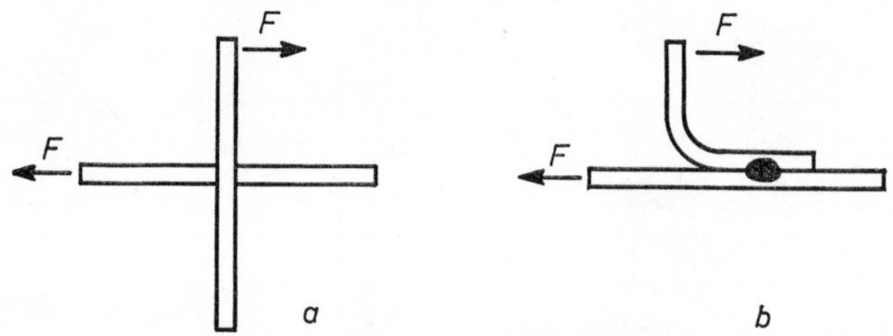

Fig. 2.103. Strength investigation of wire *a.* and strip *b.* connections.

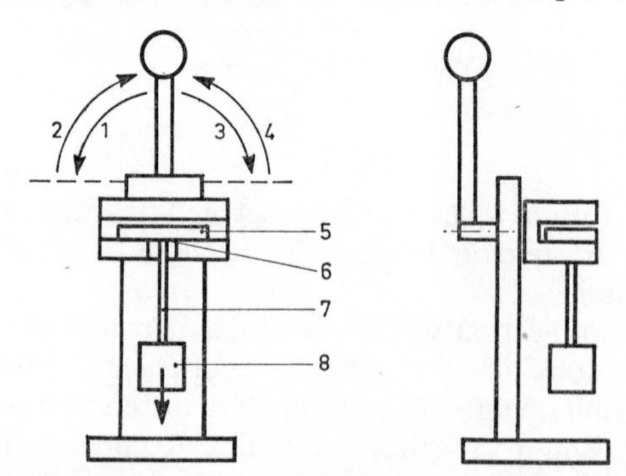

Fig. 2.104. Bending-tensile tests for a plate-wire connection.
1, 2, 3 and 4 = sequence of movement; 6 = weld. 8 = weight.
5 = plate. 7 = wire.

The weld carrying a weight via the wire is placed in the apparatus, after which the lever bends the weld through 90°. The number of quarter-cycles made before weld fractures is a measure of the weld strength.

REFERENCES

[1] E. SUDASCH — *Schweißtechnik*, Carl Hanser Verlag, München, 1959.

[2] B. OSTROFSKY — *Ultrasonic inspection of welds*, Welding Journal, March 1965, p. 97.S–106.S.

[3] J. C. NEEDHAM, D. B. BENTON, M. D. HANNAH, R. G. NEWLING — *Automatic quality control in resistance spot welding of mild steel*, The Welding Institute, Research report P/35/69, July 1969.

[4] G. E. BURBANK, W. D. TAYLOR — *Ultrasonic in-process inspection of resistance spotwelds*, Welding Journal, May 1965, p. 193.S–198.S.

[5] D. N. WALLER, P. M. KNOWLSON — *Electrode separation applied to quality control in resistance welding*, Welding Journal, April 1965, p. 168.S–174.S.

[6] O. E. WEISS, L. F. KREISLE, M. L. BEGEMAN — *Projection weld quality determination based on dynamic displacement of the electrode*, Welding Journal, September 1965, p. 417, S-423.S.

[7] A. L. PHILIPS — *Welding Handbook*, Section I. *Fundamentals of Welding*, American Welding Society, 1968.

Chapter 3

Soldering and brazing

F. van der Hoek

3.1 Introduction

(a) *Definition*

Soldering is a method of joining metal components (or at least components with a metal skin) nearly always by means of a molten metal (alloy), the solder, with a melting point below that of the components to be joined and a composition different from that of the components. The components are wetted without melting.

A distinction is made between soft and hard soldering: for **soft soldering** a solder is used with a melting point below 450 °C: for **hard soldering** the melting point lies above 450 °C. Some authors use 600 °C as the lower limit, so that solders on an aluminium base, having a melting point between 450 °C and 600 °C, could not be classified. As many kinds of these solders can no longer be worked up with a soldering bit, one sometimes feels inclined to classify them under hard solders, notwithstanding the limit of 600 °C. However, in accordance with the definition of the International Institute of Welding* the limit of 450 °C will be maintained here for both soft and hard

* *International definition of soldering:*
Soldering is the process of joining metal components by means of a molten additional metal (the solder), whose melting point lies below that of the components to be joined. The latter are wetted without melting. For a definition and explanation, see, for example, DIN 8505.

solders. Below 450 °C we speak of soft soldering, above that temperature the process is called brazing*, hard soldering, silver soldering, silver brazing, and high temperature brazing. In this review, the word "soldering" embraces soft soldering and hard soldering or brazing.

(b) *Aim of soldering*

Soft soldering is employed for joining components with at least a metal coating, in order to centre or fix these components mechanically, to fill the seam between them, to establish electrical and/or thermal contact between the components and to build-up the surface to obtain a special shape. The strength of soft solder is very low.

Hard soldering is applied to obtain rapidly a moderately cheap, lightweight, simple, detachable connection that can take static and dynamic loads, is corrosion-proof, vacuum-tight and can be used at higher temperatures. Distortion is generally low.

Soldering or brazing is used in general engineering (vessels and boilers, pipe lines, cars, coolers, inter-connection of blades, bicycle frames, honeycomb constructions, filling cavities in castings), instruments, tools, vacuum and gas treatment equipment, laboratory set-ups, precision engineering (connecting miniscule components or extremely thin wires), electrical engineering (standing wave tubes, lamps, radio valves, radio and TV sets, conductors, fuses), connecting metal components of unequal thickness and components which, due to local heating (welding), might fracture, and last, but not least, for joining metals to non-metals (such as glass and ceramics) and ferrous metals to non-ferrous metals.

3.2 Principles, physical and chemical aspects

Soldering and welding have this in common: they both serve to establish a stress-free joint between metal components by means of liquid metal.

Soldering differs from welding in that:

The components to be joined do not liquidize.

The solder is of another composition, and consequently has a different and lower melting point than the components to be joined. This leaves open the possibility of separating these components, without causing damage to them, by melting the solder again. This lower melting point also implies: less oxidation, less distortion of the seam and the product, less need for annealing and less risk of damaging heat-sensitive materials;

Thin and thick material can be joined together, as can very thin material with itself.

A flux and/or a certain atmosphere is required.

Different materials can be joined.

Because of the difference in composition between base metal and solder, corrosion is more likely to occur.

The capillary effect is an actual contribution.

Brazing and heat treating can sometimes be combined.

* "Brazing" is derived from "brass", formerly the most important filler metal.

Non-physical points of difference are: the possibility of making a large number of joints together and to suffice, sometimes, with less need for thorough schooling and experience. The appearance is better, and less finishing is required, but pretreatment is more elaborate, and the solder can flow into inaccessible places. The smoother seam scarcely holds dirt, so the product is suitable in contact with foodstuffs, etc.

The mechanism of soldering depends, in the first place, on *wetting* the components to be joined with the liquid solder (a process greatly facilitated by the capillary effect). This implies that the liquid solder has a lower surface tension than the solid material and spreads widely over a horizontal plane. It can be explained by introducing imaginary forces on the boundary line: **force** *a* on the boundary plane base metal-air (the so-called surface tension of the metal), **force** *b* on the boundary plane solder-air (the surface tension of the solder) and **force** *c* on the boundary plane solder-base metal (the boundary plane tension solder-metal).

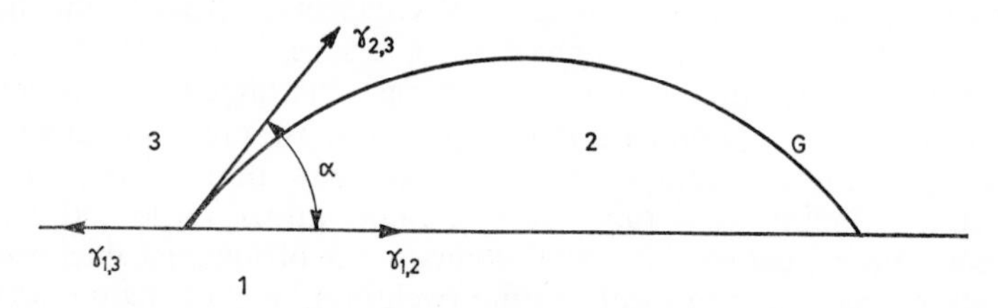

Fig. 3.1. The boundary angle α as a measure of the boundary plane tension and wetting.

1 = solid. 2 = liquid (solder). 3 = air, protective gas or flux.

For spreading, $\gamma_{1,3} > \gamma_{1,2} + \gamma_{2,3} \cos \alpha$. When, due to the roughness of the surface, the real area is n times the area of a "smooth" surface, the equation becomes: $n.\gamma_{1,3} > n.\gamma_{1,2} + \gamma_{2,3}.\cos \alpha$.

Each boundary plane has an amount of energy proportional to the surface of the boundary plane. As every system has a tendency towards a minimum of free energy, these boundary planes, too, try to shrink as much as possible. But they counteract one another; reduction of one boundary plane increases another.

Ultimately the boundary line G is so shaped between the three phases that the whole system has a minimum boundary plane energy (Fig. 3.1).

When the boundary angle is called α, tension $\gamma_{2,3}$ can be divided in a tension ($\gamma_{2,3}.\cos \alpha$) along the surface and a tension ($\gamma_{2,3}.\sin \alpha$) perpendicular to the surface.

When the forces acting on the edge of the solder drop, and those along the surface of the component to be soldered are balanced, no further spreading (or unspreading) of the solder occurs. Then $\gamma_{1,3} = \gamma_{2,3}.\cos \alpha + \gamma_{1,2}$ and $\cos \alpha = (\gamma_{1,3} + \gamma_{1,2})/\gamma_{2,3}$. If $\cos \alpha$ is large, α is small and there is much spreading. So, it is necessary to obtain a small α.

Spreading is thus greatly influenced by (changes of) the forces *a*, *b* and *c*. Therefore soldering is a process in which three components play a part: the solder, the base metal and the environment (air, gas, flux and/or vacuum).

The wetting and adherence of the solder to the wetted plane depends on the *adhesive forces*. As these operate only on an atomic scale, it is necessary to establish atomic contact. This is the reason why so much value is attached to a pure surface of solid components and the droplet of solder. Oxidation reduces the surface tension of the metal to be joined, which is in itself an unfavourable effect. But it does the same for the solder, so that it flows nicely. (Perhaps that is the reason why oxidizing agents like B, Li and P have such a favourable effect on the flow capacity*). Oxidized base metal is the reason why $\gamma_{1,3}-\gamma_{1,2} < 0$, so $\cos\alpha = \gamma_{1,3}-\gamma_{1,2})/\gamma_{2,3} < 0$, hence $\alpha > 90°$. The solder drop has the shape depicted in Fig. 3.2; it does not spread, notwithstanding the direction of $\gamma_{2,3}$, which is favourable to spreading.

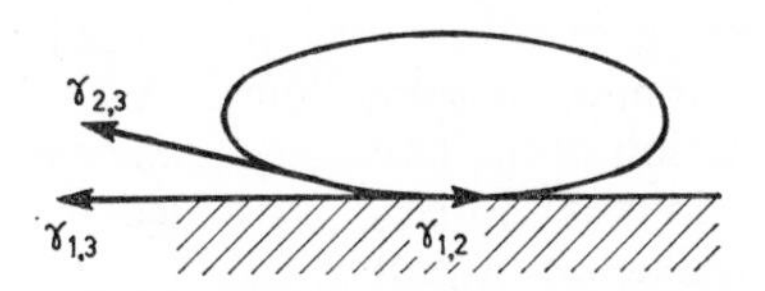

Fig. 3.2. The solder drop does not spread in oxidized base metal.

The diffusion, dissolving, *alloying* and the formation of chemical compounds of the metals concerned, generally play a part in joining the components. By mutual dissolving of base metal and solder $\gamma_{1\cdot 2}$ decreases, so wetting is improved. Roughly then, it may be stated that more adhesion takes place in soft soldering, whilst more alloying takes place in hard soldering[14].

The wetting amounts to the formation of chemical compounds (for instance Cu_6Sn_5 in soldering brass, and Fe_2Sn_2 in soldering steel with a solder containing tin), or mutual dissolving of the elements. The alloying rate determines the spreading rate and the contact angle. When solder and base metal are mutually insoluble at soldering temperature, the contact angle is sometimes > 90°. By increasing the brazing temperature, however, a satisfactory joint is usually formed, e.g. between Mo and Cu.

Too much alloying prevents flowing over the surface by the solder; it penetrates the base metal instead. The best solution to this problem might be to employ a filler metal which is insoluble with the base metal, and to add a controlled amount of material that is soluble in both the filler and the base metal. No wetting can take place without dissolving and alloying, which can be seen from the combinations: Fe-Ag, Fe-Pb, Fe-Cd, Fe-Bi, Cu-Pb, Cu-Bi, Cu-Mo, Cu-W, Al-Pb and Al-Cd†.

The degree of wetting, the surface tension, the gravity, the free energy of the solder and the energy of the components to be joined, as well as the surface roughness (the grooves caused by machining act as capillaries and also increase the contact area and so the boundary layer energy), are all

* The boundary plane energy of solid oxide-free metals is 1 000–2 000 J/m², of liquid oxide-free metals 1·06–1·1J/m², and of solid oxidized metals 0·06–0·1J/m².

† To predict dissolving, see [22].

decisive for the flow of solder required to obtain a good joint. The heat produced in the mutual alloying of solder and base metal seems to provide the energy for the spreading of the solder on the substrate. Two parallel plates, with a very small gap between them, show how their respective expansion energies reinforce each other, thereby pressing the solder into the gap: the capillary fill pressure. This pressure, inversely proportional to the gap width, is quite different from the capillary action by the surface roughness with respect to the liquid solder. Anyhow, the shape and dimensions of the joint affect its strength.

The hydrostatic pressure of the solder plays its role, too. In a horizontal plane, it provides a force beneficial to spreading. This force decreases with increased spreading. With a different relative position of solder and base metal, e.g. in dip soldering, this force acts differently, and not always favourably.

In designing a soldering system, spreading should not be based on the amount of solder to be applied. A small boundary angle should be obtained only by a good combination of the boundary plane energies, more in particular by aiming at a minimum force *c*. Provided *c* is small, spreading is adequate, despite any influence of gravity.

In good solder joints, the boundary angle is smaller than 30° and the components are wetted so well that their shape can be traced through the solder. This is impossible when too much solder has been used; here the result is like a joint with insufficient wetting. We conclude that too much solder conceals faulty joints. For the influence of absorbed gas, see[18]; the wetting effect and the surface tension decrease. Pure metal generally forms a thinner liquid than a compound solder.

By dissolving base metal in the solid or liquid solder, the composition and flowing capacity of the latter changes; the gap is sometimes filled with the earlier melting phase of the solder, or even incompletely filled, so that at a stronger interaction between the base metal and the solder (at a higher temperature for instance) the gap must be chosen wider. The change in the composition of the solder can even give rise to solidification at the same temperature.

This dissolving of the base material often takes place along the grain boundaries, especially if the workpiece is subjected to tensile stress, or if it has undergone cold deformation or uneven heating. Such load becomes manifest especially at the grain boundaries, where the structure changes direction and composition, and where, as a result, the tension is higher. At these grain boundaries an eutectic is formed, with a structure often enriched with elements extracted from the grains. The molten solder widens the cracks into which it is absorbed, and its danger lies in its behaviour as a corrosive medium, with an effect strongly resembling stress corrosion. This may drastically reduce the strain and the tensile, bending and fatigue strengths.

Particularly are stainless chrome nickel steel and monel, with their low heat conduction and consequently high heat stress, susceptible to this brittleness fracture, which is usually more pronounced with coarser grains. Nickel and its alloys, e.g. Fernico, are sensitive to solders containing silver, while among others zinc content in solder can give rise to strong corrosion of iron and its alloys.

Means of avoiding this are: the tempering of cold-deformed metal, annealing, reducing the rigidity of the joint, minimizing the heat stress by slow heating, dip soldering or furnace soldering; in general, by aiming at an equilibrium of structure and phases. Another possibility is to give the components a coat of metal not susceptible to brittle fracture. Sometimes, electroless nickel is used, because it is corrosion-proof and works as a solder when heated, so that a corrosion-proof joint can be made.

Due to the high temperature at which soldering takes place (often above 650 °C), crystal growth, recrystallization and annealing of hardened steel may occur. On the other hand, air-hardening steels, for instance, may harden, so they must be annealed after soldering. Beryllium copper and chromium copper must be soldered after heat treatment, in view of the high temperature used. Nevertheless, it is interesting that, because of this phenomenon, there is a possibility of combining soldering and hardening into one operation.

A final phenomenon worth mentioning is that hydrogen, due to its varying solubility at varying temperature, as well as its change from atom to molecule, is practically insoluble. The heating that takes place in soldering can lead to destruction along the grain boundaries.

REFERENCES

[1] K. Bailey — *The meaning of solderability*, British Non-ferrous Metals Research Association, Report A1104, October 1953.

[2] G. L. J. Bailey, J. C. Watkins — *The flow of liquid metals on solid metal surfaces and its relation to soldering, brazing and hot-dip-coating*, Journal of the Institute of Metals, vol. **80**, 1951, p. 57–81, and p. 692–694.

[3] A. Bondi — *The spreading of liquid metals on solid surfaces.*

[4] R. D. Wasserman, J. Quaas — *Dilution and diffusion aspects of nonfusion welding*, The Welding Journal, December 1951, p. 1098–1101.

[5] E. R. Funk, H. Udin — *Surface tension and the joining of metals*, Welding Research Supplement, May 1952, p. 247–252.

[6] I. Milner — *A survey of the scientific principles related to wetting and spreading*, British Welding Journal, March 1958.

[7] F. B. Bowden, D. Tabor — *Mechanism of adhesion between solids*, Proceedings of the second international congress on surface activity, 1957.

[8] N. F. Lashko, S. V. Lashko-Avakyan — *Brazing and soldering of metals*, London 1961.

[9] H. H. Manko — *Solders and soldering*, McGraw-Hill, New York.

[10] J. Colbus — *Versuche zum Deutung der Bindevorgänge beim Löten*, Schweissen und Schneiden, 10, 1958, p. 50.

[11] H. H. Manko — *Soldered Connections*, Machine Design, May 21, 1964.

[12] L. Pessel — *Assured reliability in soldered connections*, IEEE Transactions on product engineering and production, January 1963, p. 28.

[13] W. B. Harding — *Solderability testing*, Plating, October 1965, p. 971.

[14] W. H. Kohl — *Soldering and brazing*, Vacuum, vol. **14**, p. 175–198.

[15] J. O. Outwater — *Composite materials, particle materials*, Mechanical Engineering, February 1966, p. 32–35.

[16] H. Zürn, Th. Nesse — *Die metallurgischen Vorgänge beim Weichlöten von Kupfer und Kupferlegierungen und das Festigkeitsverhalten der Lötverbindungen*, Metall, November 1966.

[17] *Löten*, Vorträge der Sonder-Tagung Augsburg 1964 des Deutschen Verbandes für Schweisstechnik e.V., Arbeitsgruppe 26 "Loten".

[18] H. PARTHEY *Grenzschichtprobleme beim Löten*, Mitt. Forschungsgesellschaft Blechverarbeitung, Nr. 23/24, p. 367, 1965.

[19] H. Lange, H. PARTHEY, I. N. STRANSKI *Grenzschichtprobleme beim Löten*, Mitt. der Forschungsgesellschaft Blechverarbeitung e.V., Nr. 19, p. 209, October 1, 1958.

[20] R. J. KLEIN WASSINK *Wetting of solid-metal surfaces by molten metals*, Journal of the Institute of Metals, vol. **95**, p. 38, February 1967.

[21] A. L. MAVERICK, R. FICHTER *Untersuchungen über Diffusionserscheinungen an Lötverbindungen mit der Mikrosonde*, Schweizer Archiv, January 1967, p. 22.

[22] B. W. MOTT *Immiscibility in liquid metal systems*, Journal of Materials Science, **3**, 1968, p. 424–435.

[23] Y. V. NAIDICH *Interfacial surface energies and contact angles of wetting of solids by liquid in equilibrium and non-equilibrium systems*, Russian Journal of Physical Chemistry, **42**(8), 1968.

3.3 Fluxes

Soldering is a combination of three components: the solder, the base metals and the environment. The environment can be air, another gas, vacuum or a flux. In practically all cases, use is made of a flux applied before or during soldering: its function is to accelerate the spreading of the solder over a larger area. This is achieved because a good flux at soldering temperature covers both base metal and flux, and thereby:

Removes (dissolves, breaks down, binds) contaminations (oxides, sulphides, etc.) of the base metal, so that its surface energy increases, not only beside the drop of solder, but also between solder and base metal. These phenomena combined have the effect that $\gamma_{1,3}-\gamma_{1,2}$ increases, so the solder spreads.

Removes all contamination from the surface of the liquid solder, so that $\gamma_{2,3}$ decreases. Provided $\alpha < 90°$, a smaller value of $\gamma_{2,3} \cos \alpha$ advances spreading.

Protects all metals concerned against the setting-in of a new oxidation process, etc.

Reduces the surface tension of the molten solder.

Sometimes tins the base metal electrochemically.*

* With copper as the base material, the following reactions take place when using $ZnCl_2$:

$ZnCl_2+H_2O \rightarrow Zn\,OHCl+HCl$

$2\,HCl+CuO \rightarrow CuCl_2+H_2O$

$2\,HCl+Sn$ (from the flux) $\rightarrow SnCl_2+H_2$

$SnCl_2+Cu \rightarrow CuCl_2+Sn$.

Together with the base metals copper and iron, this tin forms respectively Cu_6Sn_5 and $FeSn_2$ [26]. These reactions require an intensive exchange of ions; hence the preference for acids, bases, salts and water. Colophony misses this exchange, so does not promote tinning. Apart from that, the boundary plane energy drops from about 1·3 J/m² to about 0·54 J/m² for tin. Nevertheless, spreading is highly accelerated.

Furthermore a soldering flux must satisfy the following requirements:

- It should not react with the solder, unless metals useful to the soldering process are secreted.
- The specific mass must be lower than that of the solder, in order to shield it and be replaced by it.
- It must be liquid at soldering temperature.
- The melting point must be lower than that of the solder, to ensure a timely decontamination of the base metal.
- The melting point must be lower than that of the base metal.
- It must not react with the base metal.
- It must not react with the PVC insulation of wire, components of printed wire panels, etc.
- It must not decompose, carbonize, evaporate, etc. during storage or at soldering heat, the more so if the vapour condenses elsewhere and is detrimental.
- It must spread easily, viscosity and surface tension being of importance. The viscosity should be high enough for the flux to adhere to vertical planes, and low enough for the liquid solder to drive away the flux.
- The residue must not be corrosive and electrically non-conducting, unless it can easily be removed, or evaporates rapidly.
- The residue should not be sticky, otherwise dirt will adhere. This will attract water, which can lead to short-circuiting.
- The residue, when not removed, should be transparent, to enable checking of the underlying joint.

A few further points that are decisive in choosing a flux will now be discussed.

(a) *Corrosivity*

The value of a flux is assessed by its aggressiveness to decontamination. An active flux will also attack the metal itself (and perhaps even the PVC insulation of wire) and a flux can never be administered without there being harmful remnants after soldering. These remnants must be carefully removed: due to their often hygroscopic character, they attract water, thus forming electrolytes, which sometimes cause serious corrosion. Consequently, the electrical and mechanical characteristics of, for instance thin wires, as well as their appearance, can be drastically changed, even due to inter-crystalline penetration. The corrosive effect depends less on the acidity than on the halogen compounds.

Unfortunately, the flux remnants cannot always be removed (for instance, in radio sets). Hence, an effort will always be made at first to apply the least reactive flux, even if it works slower and is less effective.

There are soldering fluxes which:

- Do little more than protect, such as colophony.
- Are hardly corrosive at all, such as mineral oils, stearin, aniline phosphate or aniline chloride.

- Are corrosive to a limited degree: for instance, organic acids and some chlorides.
- Are suitable for removing unstable oxides: for instance zinc chlorides (copper, low-carbon and low-alloy steels).
- Are suitable for removing stable oxides (of stainless steel, aluminium, monel) for instance, stannous chloride and hydrochloric acid.
- Are suitable for removing very stable oxides (Al, Nb, Ti, Mo), such as triethanolamide, ethylglycol, metallic borides, fluorides and chlorides, e.g. HF, NaF, NaCl, NH_4Cl, KCl, $ZnCl_2$, $CdCl_2$ and LiCl.
- Are suitable for high temperatures: for instance, fluxes containing borium.

In order to obtain the properties required, fluxes are often:

- Mixed with one another, e.g. 75% zinc chloride (melting point 262°C, covering well) and 25% ammonium chloride (covering badly), to obtain a lower melting point (170°C, covering excellently).
- Provided with a humidifier, such as glycerin, manitol, triethanolamine.
- Diluted, with vaseline for instance (if the mixture is not homogeneous, other problems arise).
- Intensified by means of activators in colophony, such as glycollic acid, urea, ammonium chloride, halogenides of amines and dimethyl amine hydrochloride.
- Provided with inhibitors, to decelerate aggressiveness.
- So selected that the aggressive constituent evaporates at the soldering temperature: for instance, hydrazine. Usually it condenses elsewhere with all the consequences involved; hydrazine is sometimes a constituent in resin core solder.*
- So selected that the aggressive constituent disintegrates at the soldering temperature (but all the flux does not always reach this temperature).
- So selected that the reaction speed at room temperature is very low.
- Provided with a means of preventing spattering, e.g. glycerine or glycol.

It will be clear that to assess the corrosivity of a flux practically calls for a specialist. Methods of assessment are given in Section 3.8.

The fluxes are roughly divided as follows:

(1) Fluxes on a base of resin, whether or not activated, (activators with or without halogens).
(2) Other organic fluxes, such as acids (citric, phthalic, glutamic, lactic, oleic, stearic), bases (diethylamine), amines and halogen compounds (for instance aniline-hydrochloride, cetylpyridinium bromide, hydrazine hydrochloride and bromide, glutamine hydrochloride, palmityl bromides) all with a working temperature of about 400°C); generally salts are more corrosive than acids.

* The round- or star-shaped resin core is located concentrically in the round soldering wire, or a ring of 2–5 thin resin cores runs through the solder. This is done so that nowhere in its length is the solder wire devoid of flux. Or the wire consists of a core of solder, a ring of resin, the whole being surrounded by a jacket of solder.

(3) Inorganic fluxes: hydrochloric acid, zinc and ammonium chloride, etc., borax, boric acid, hydrogen, bromic acid, phosphoric acid, etc.

(1) and (2) have a small temperature traject, within which they are active; they work slower and less intensely than the fluxes in (3).

(2) and (3) are highly corrosive and therefore to be avoided in precision engineering[21].

There is but one flux that is absolutely harmless: natural rosin or colophony, which is obtained from certain coniferous trees. Rosin consists mainly of abietic acid ($C_{20}H_{30}O_2$, melting point 173 °C, working temperature up to 300–350 °C), dissociating at the soldering temperature and aggressive only to weak oxides (of Cu, for instance). It does not dissolve in water, does not dissociate at room temperature, and consequently forms no electrolyte and has no corrosive effect. Moreover, it does not attract dirt. It can be removed from metal surfaces with alcohol, petrol, turpentine, etc. Sometimes "activators" (organic bases and acids, more in special halogenic compounds of both) are added to the resin (for instance, dimethylamine hydrochloride); these activators are nothing but powerful fluxes increasing the reactivity of the mixture. Sometimes rosin also contains aggressive chemicals that have remained behind in the preparation process; resin should therefore be obtained only from reliable suppliers. As colophony is a natural product, its properties are not always constant per batch.

Table 3.1 gives a few fluxes on a colophony basis. Some amines are good inhibitors for iron, but accelerate the corrosion of copper, which can be misleading in corrosion tests [23]. Borax ($Na_2B_4O_7 . 10H_2O$, melting point 760 °C, obtainable as a solution, a paste or in powder form) and boric acid (H_3BO_3, melting point 580 °C) remove the oxides of Fe, Cu, Sn, Zn, Cd, Si, Ni and Ag and those of Al, Be and Cr less well, but owing to their low solubility in water, these fluxes are difficult to remove. The combination gives a better wetting than each of the constituents alone.

Boric acid works quickly enough only at about 900 °C; if this temperature is maintained for more than five minutes, the metal oxidizes again.

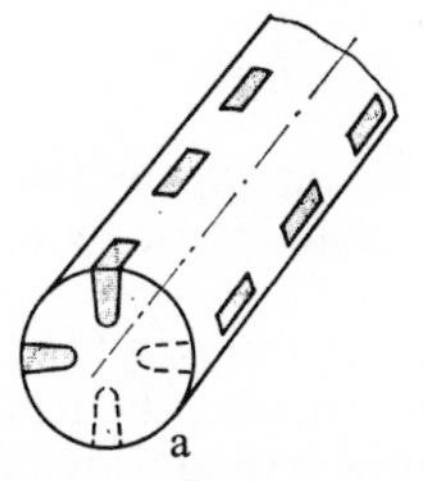

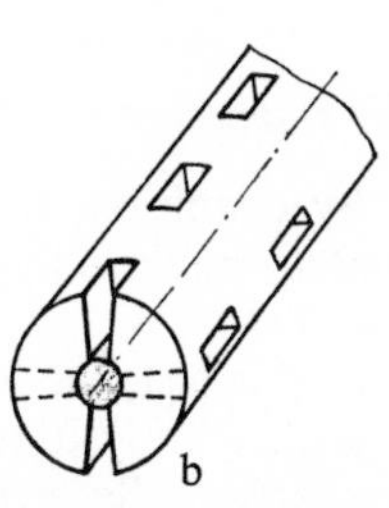

Fig. 3.3. Some less-common types of rosin-cored solder wire.

a. resin in the slots only.
b. resin in the central core only.

The rosin lies as close as possible to the outside of the wire, so that in soldering it will rapidly melt, run out and spread. For the same reason, massive wire is provided with circular grooves containing rosin. This wire is said to be easy to manufacture, and gives little rise to spattering during soldering, but it sticks to the hands.

Other types are marketed by Gardiner, Chicago, see Fig. 3.3.

As to the advantages of rosin cored solder wire, see the first footnote of Section 3.6.3.

TABLE 3.1

Some fluxes on a colophony basis

Colophony (%)	Activator		Solvent	
	Name	Content (%)	Name	Content (%)
65	—	—	ethanol, isopropanol	35
55	—	—	monoethylglycolether	45
59	dimethylamine hydrochloride	1·2	isopropanol	40
52	urea	5·3	isopropanol	42
44	aniline 9, plus lactic acid 36	45	ethanol	11
40	—	—	ethanol	60
40	urea	2	monoethylglycol	58
±40	DMA	0·8–0·16	isopropanol	±60
35	—	—	ethanol, isopropanol	65
35	DMA 4, plus triethanolamine 2	6	ethanol	60
25	DMA	0·5	diacetone alcohol	75
25	urea	2·5	monoethylglycolether	72·5
25	—	—	isopropanol	75
24	zinc chloride	1	ethanol	75
24	diethylamine or methaphenylenediamine 4, triethanolamine 2	6	ethanol	70
21	aniline hydrochloride 7, triethanolamine 2	9	ethanol	70
20	aniline hydrochloride	1	ethanol	79
20–8	—	—	ethanol or isopropanol	80–92
5	phosphoric acid	28	methanol or ethylene glycol	17
2	—	—	ethanol	98

(b) *Improving the flow capacity*

By definition, the flow capacity is the area (mm^2) of horizontal surface coated, under standard conditions, by a certain quantity of heated solder and flux.

In this test, the following can vary: the flux, the solder, the base metal as regards composition and surface, the temperature and the heating rate and atmosphere. So, even after several dozen tests, no well-founded opinion can be expressed on the flow capacity of this flux. The maximum flow is important when soldering a particularly difficult metal; otherwise the main interest is focused on the *activity*, which is the same as the flow speed. Flow capacity and activity depend largely on the temperature.

To prevent the flux from reaching undesirable places (e.g. electrical connections), a moderate flow will sometimes be preferred. Or use can be made of a stop-off stick, a sort of tallow-dip applied as a pencil to mark the product area to which flow must be restricted.

(c) *Composition of the solder*

The lower the flow of the solder, the more active will the flux have to be. The flow of the solder is partly determined by the temperature (see paragraph (g)) and the surface of the base metal (see Section 3.5.1). Quickly-oxidizing solder materials, such as aluminium, zinc and cadmium, require active fluxes. Solder with a large melting traject must be heated rapidly, otherwise it alloys too much with the base metal.

(d) *Composition of the base metal* (see also Section 3.5)

Brass and bronze often have a layer of zinc or tin oxide (respectively) that cannot be removed by a weak flux like colophony. These, and other parts difficult to solder, can be pre-soldered individually with strong fluxes, which are washed-off afterwards. The actual soldering then takes place on the previously prepared base, and now a non-corrosive flux is used.

Apart from the phenomenon that the metal surface is not wetted by the solder, it sometimes occurs that the solder retracts from a surface originally wetted, the so-called balling-up effect.

An extreme case in this respect occurs with a thin silver coating on ceramic material. Silver is not only wetted well by tin, but it dissolves well into tin, too. So, a thin layer of silver dissolves rapidly in the molten solder, so that the latter comes in contact with the ceramic base material. The solder does not wet it and will retract from this surface. This is the reason why, in soldering silver-coated ceramic material, use is made of silver-containing solders, which counteract a fast dissolving of the silver layer.

The same phenomenon occurs in layers (solder, silver, gold) galvanically applied to a dirty substrate; moreover they dissolve easily in the solder. The liquid solder then comes in contact with the contaminated base material, which has a low boundary plane energy. This involves reduction of tension $\gamma_{1,3}$ and a strong increase of tension $\gamma_{1,2}$, leading to balling-up of the solder on the substrate, and sometimes even total de-spreading.

The same applies to copper and brass, when they are not properly cleaned and contain contaminations like oxides in the surface layer. During soldering, the flux will clean the base metal superficially, wet it and alloy with it. Due to this alloying and dissolving of copper in the solder, the boundary plane solder–base metal will gradually move deeper into the base metal, so that the liquid solder comes in contact with the contaminations originally embedded in the surface layer. This results in a reduction of tension $\gamma_{1,3}$ and a strengthening of tension $\gamma_{1,2}$, and hence balling-up of the solder.

In actual practice, these phenomena also occur simultaneously.

(e) *Soldering method, including soldering time and heating rate* (see also Section 3.7)

In surface soldering (see Section 3.7.2(f)), it is a long time before the soldering temperature is reached. To reduce oxidation of the workpiece, flux is often applied before it has become warm. During the long heating-up period, the flux should neither dissociate nor attack the base metal. The long heating time gives the flux sufficient time to evaporate its solvent and to spread out uniformly.

In quick soldering, however, dip soldering for instance, (see Section 3.7.2(e)) or h.f. soldering (see Section 3.7.2(i)), the flux must take effect within a fraction of a second. In h.f. soldering, the flux is not heated by the eddy current, and so can remain too cold. Too rapid heating, in combination with a flow restricting seam, could cause the boiling flux to blow away the solder.

In soldering printed circuit panels use is sometimes made of foam fluxes, with the following items of interest: repulsion of the foam by the panel, temperature of the jigs, pressure of compressed air applied, (very important) influence of flux temperature, type and quantity of contaminations and water content of the foam, increasing with the water content of the compressed air. This water, raising the boiling temperature, results in spattering; water-soluble fluxes must therefore be processed free from water. The solvent isopropanol prevents spattering almost completely.

(f) *State of the flux* (see also Section 3.7.2(f))

Fluxes occur in any of the following states: (i) solid, (ii) paste, (iii) liquid and (iv) gaseous.

(i) *Solid flux*

Borax and boric acid are employed above 600 °C. Ammonium chloride powder is often used (also as a paste or a liquid), sometimes in combination with zinc chloride, as a layer on tinning baths, but especially in resin core soldering wire and foil core solder.

Pure powder does not adhere to the cold workpiece, does not protect it during heating (oxidation) and can be blown away by a torch flame.

(ii) *Paste flux*

The paste is usually resin, glycerine, lanolin, etc. This is mixed with chlorides. The corrosivity and bad removability of the mixture restricts its application. Paste flux does not drip like liquid flux. It can thus be accurately positioned, can be mixed with solder powder (in favour of furnace soldering, resistance soldering, high frequency soldering, flame soldering, etc.) and is used in resin core soldering wire.

Solder powder combined with paste flux is applied by brush, spatula or drop-bottle, and with air-driven sprayers, transfer rollers, printing devices or syringe applicators. On heating, the flux spreads as well as the solder powder, thereby limiting its application.

(iii) *Liquid flux*

Liquid fluxes are liquid by themselves or are solutions of solid fluxes. The latter is the type most used in electrical engineering.

In mass soldering, liquid fluxes are more suitable than solid or paste fluxes, since large areas and many solder spots can simultaneously be covered with flux by dipping, smearing with a brush, spraying, roll brushing, or by horizontally moving a printed wiring panel over a wave of flux (similar to the technique described in Section 3.7.2(e)) or over a vessel from which rises a mass of foamed flux. Afterwards, the flux is dried to reduce its volume. Otherwise, during soldering, too many gas bubbles might form, and prevent the access of solder.

The most common solvents are water and organic solvents. The flux has a maximum activity when most of the solvent has been evaporated. (A solvent vaporizing quickly in soldering will, however, do the same when stored in open vessels.)

Another drawback of liquid fluxes is that they may move (by capillary action or by gravity) to places where they are undesirable, e.g. to switches, and so contaminate them. Besides, the surface to be cleaned can run short of flux.

Fluxes on a basis of inorganic salts are practically only soluble in water and are, therefore, nearly all dissolved in that solvent. When water has not been sufficiently well removed from the flux, it will cause violent spattering during soldering. To prevent this, polyethylene glycol or glycerine is sometimes added.

The remnants of this type of flux are highly corrosive; due to their good solubility in water, they can easily be removed, in principle, by water, to which additions are made if necessary.

The reaction products formed in soldering this group of fluxes are less soluble in water, however. Moreover, capillaries in the product suck in flux, which can scarcely be washed away.

Table 3.2 gives some chloride-containing fluxes. Other types are acids and salts on a fluorine basis (poisonous), chlorides of Na, K and Sn and bromides, H_3PO_4 etc.

TABLE 3.2

Some chloride-containing fluxes

Composition by weight %				
$ZnCl_2$	NH_4Cl	HCl	$SnCl_2$	Water
72	7	—	—	21
60	—	—	—	40
48	—	22	—	35
48	—	10	—	42
48	12	—	—	40
48	—	—	—	52
48	5	—	—	48
40	20	—	—	40
39	4	2	5	50
19	6	—	—	75
16	3	—	—	81
10	—	35	—	55
6	4	5	—	85
3	16	—	—	81

Fluxes on a basis of organic compounds are generally dissolved in organic solvents. The best known organic flux is colophony, a non-corrosive compound, whose activity is, however, less than for many inorganic fluxes. To cope with this problem, activators are often added to the resin.

Colophony in itself is soluble in practically any organic solvent; since the activators must be soluble, too, the suitable solvents are restricted to the alcohol type, in particular ethanol and isopropanol (which is cheaper and has more dissolving power than ethanol), sometimes toluene or polyethylene glycol. When special requirements are to be met, such as slow evaporation from open flux vessels, other solvents are used, like ethyl glycol and diacetone alcohol.

These technically preferred alcohols are not objectionable from the medical point of view.

Corrosion never occurs from the remnants of pure colophony, but it does when using certain activated types. The need to remove them depends on the flux and the type of product. Removal sometimes presents difficulties and demands specialist advice.

Colophony is a natural product, whose properties can vary to some extent. For unknown reasons, many colophony resins are precipitated from their alcoholic solution after a certain period of time. This raises the activator content of the solution, thereby increasing the risk of corrosion by the flux residues and reducing, sometimes, the insulation value.

(iv) *Gaseous flux*

The advantages of gaseous (including vacuum) flux over other types are: the amount of flux at the solder joint is always correct; the distribution of the flux along the joint is quite homogeneous; there is no risk of entrapment of flux; the flux is suitable for inaccessible places; there is no restriction on the maximum soldering time; and flux remnants need not be removed.

In many cases, the liquid flux is replaced by gas, or gas is merely added. The gas will be inert, or protective against corrosion, or active against corrosion products. Sometimes a vacuum is preferred.

The rate at which oxides dissociate increases with the temperature and as the partial pressure of the oxygen in the atmosphere decreases. As soon as this partial pressure becomes lower than the vapour pressure of an oxide, the oxide will dissociate (and the other way round). [1] gives the dissociation pressure of the most important oxides in relationship to temperature.

The partial oxygen pressure of the atmosphere is reduced by removing the atmosphere (vacuum) or replacing it by an inert gas, for instance.

(1) *Soldering in inert gas*

Very low oxygen pressures are obtained by simultaneously evacuating and supplying argon, also during the soldering process. For less stable oxides, those of copper and nickel, for instance, dry and pure nitrogen will suffice. Nitrogen, however, always contains some oxygen, and besides, cannot bind oxygen. When unavoidable, the expensive gases argon and helium are used.

(2) *Soldering in a reducing gas*

The gas used in practice is sometimes partially-burnt* town gas, but usually H_2 or N_2, obtained from the dissociation of NH_3. Often, CO, CO_2, CH_4 and NH_3 are subsidiary constituents. Pure hydrogen is seldom used: it is expensive and involves an explosion risk when brought in contact with oxygen.

When copper-containing components are soldered, sulphur must be removed, whatever its form or shape. CO, in principle, acts carburizing; CH_4 even deposits soot.

The hydrogen oxidizes into water; wet gas can, however, give rise to oxidation, and hence the gas must be kept dry to achieve the equilibrium H_2O–H_2 in a condition under which the metal oxides are dissociated. The humidity of the gas is indicated by its dew point. For steels containing Cr, Si, Mg, Ti, etc. it must lie at a temperature as low as −60 °C[7],[8]. If this requirement is not satisfied, flux (KF, for instance) is added. More information, from which Fig. 3.4 is derived, is given in [5] and [6]; a line forms the boundary between oxidation (left) and reduction (right). In general, oxides with a formation heat lower than that of water vapour are easily dissociated by that vapour, and need no flux.

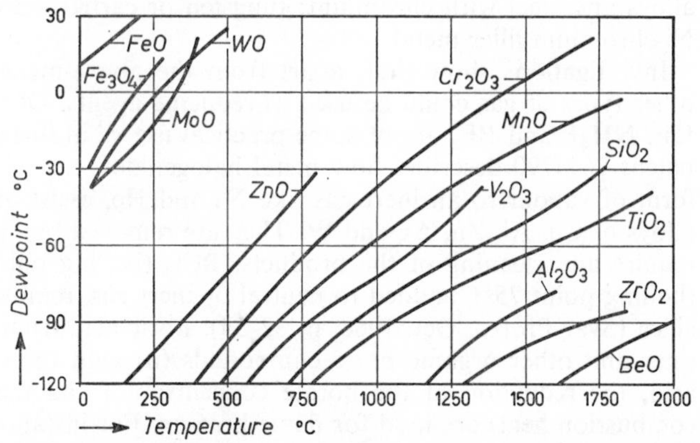

Fig. 3.4. The dew point of H_2 is decisive for oxidation or reduction of the metal oxides.

Hydrogen gives hydrides in Pb, Cd, Sn, Ti, Ta, Zr and Nb, and should therefore not be used. The solid solutions of hydrogen in Al, Co, Fe, Mn, Mo and Pt are undesirable, too. In this respect, steel is notorious; a low soldering temperature and relative humidity reduce the risk of brittle fracture. [9] and [10] give more information on the subject.

In contrast with oxygen-containing copper, hydrogen can cause hydrogen brittleness; that is why oxygen-free copper is generally used when employing hydrogen.

Moreover, hydrogen leads to decarbonization ($Fe_3C+2H_2 \rightarrow 3Fe+CH_4$), just as moisture and CO_2. In view of the reaction $CO_2+H_2 \rightarrow CO+H_2O$ it is advisable to remove CO_2 if the atmosphere must remain dry. The first-mentioned reaction can be suppressed by adding some CH_4, and completely avoided by copper-plating or nickel-plating the components involved.

* The gas is burnt to increase its quantity and to reduce the risk of explosion. No more gas is burnt than is necessary to maintain a sufficient reducing capacity of the gas.

The reducing gas often contains nitrogen, which, together with steel, gives iron nitrides, and greatly increases the hardness of the metal. As in practically all chemical processes, this phenomenon can be suppressed by keeping the temperature low and the reaction time short.

Moisture is removed by drying the gas with, for instance, silicagel, followed by contact with P_2O_5 or intensive cooling. The gas can also be dried with caustic soda or potassium[1]. Oxygen, for example, combines with hydrogen and is removed in the form of water. CO can decompose into soot and O_2, which gas reacts with CO and H_2, resulting in CO_2 and H_2O respectively.

Soldering in a reducing atmosphere is applied in the manufacture of contacts, conductors, lamps, radio valves, cutting-bits for tools, medical equipment, turbine blades, bicycle frames, motor cars, aircraft, etc.

Usually, preheating takes place at a temperature where the oxides dissociate; the solder is then made to melt at a higher temperature. This process is followed by cooling at about 200 °C in a protective gas or hardening in a tank. Usually, tunnel ovens are preferred. To avoid gas leakage, box furnaces are used, sealed by welding for each batch of very stable oxides.

Copper and its alloys, including monel, nickel, carbon steel and alloyed steel are brazed with copper, copper-phosphorus, brass or silver; chromium alloys and steel with chromium; tungsten or carbides are also brazed with Ni-chromium filler metal.

Investigations show that, apart from the above-mentioned gases, some other types of gas could be used as reducing agents. Of these Cl_2, HCl, F_2, HF, NH_4F and BF_3 are of some practical use [1] in furnace soldering. U.S. patent 2674790 describes how metal halogenides, e.g. $MoCl_5$, added in the form of vapour to an inert gas like N_2 and He, assist in brazing steel and alloys of Cu, Al, Zn, Ag and W. They are removed by inert gas and do not require any cleaning of the products. BCl_3 (boiling point 17 °C) and PCl_3 (boiling point 75 °C) added to neutral or inert gas, remove all oxides except silica (Svar Proizo, Oct. 1968, p. 32–34). Boric acid trimethyl(ethyl)(propyl) ester and other organic boric compounds (to fight the hydrolysis of boric acid, the reduction of the borate concentration and the reduction of the combustion heat) are used for flame brazing. For instance, according to[31], boric anhydride is deposited on the workpiece when the acetylene to be used in the oxygen-acetylene burner is led over an element soaked in flux. Advantages claimed for this method of flux supply are: quantity of flux proportional to that for the flame; no separate flux supply; no need for post-treatment.

(3) *Vacuum soldering*

Vacuum soldering causes more evaporation of the base metal than of the oxides, but does indeed eliminate many oxides[2]. Soldering takes place in an evacuated furnace (low heat conduction due to the vacuum) or in an evacuated vessel, placed in the furnace (high working rate as the vessels are exchanged while the furnace keeps warm)[2,3]. Heating takes place by means of resistance or h.f. induction, the latter method being technically the more difficult of the two ([1] gives many useful details on the subject). A better method of heating is by radiation, despite its non-homogeneous heating of the product.

The depth of the vacuum depends on the oxide to be removed: for copper and nickel about 0·07 mm Hg, for stainless steel about 10^{-4} mm Hg, for aluminium about 10^{-6} mm Hg. The pressure used is generally much higher than the equilibrium pressure between the metal and its oxide, a fact that makes it none the easier to gain an insight into the effect of the vacuum. An explanation might be found in the carbon produced by the metal, binding the oxygen of the oxides, evaporation of the oxide, or base metal (Ti) absorbing oxygen from the oxide.

Where necessary, a flux supports the effect of the underpressure, or the latter is partly replaced by inert gas. Vacuum-soldered joints are often better and stronger than those made in an atmosphere of hydrogen or with the aid of a flux. This may be due, for instance, to the absence of enclosed gas or flux, so that there will be no corrosion afterwards. Besides, solder flows better in vacuum than in a normal atmosphere, so that the joint produced seldom needs mechanical after-treatment.

Vacuum soldering is constantly finding more applications in the rocket industries (engine design, sandwich panels, liquid lines), in the atomic energy industry (heat exchangers, seals), in electronics and other precision applications such as wave tubes. Vacuum soldering is time consuming, unsuitable for rapidly-evaporating metals, and requires expensive equipment.

The metals to be soldered are copper and nickel and their alloys, armco iron, alloyed and stainless steels, super-alloys, aluminium, molybdenum, niobium, tantalum, titanium, beryllium, zirconium, tungsten carbide, graphite, glass and ceramics[4]. Using hydrides of titanium, zirconium or niobium, a layer of metal is applied to glass and ceramics. For information on base and soldering metals see [25] and [29].

(g) *Soldering temperature*

This temperature is decisive for the flow capacity and the soldering time. The maximum flow capacity occurs at a certain temperature. As the temperature increases, the alloying and dissolving of oxides sometimes sets-in earlier and the solder begins to flow sooner, so, to obtain optimum soldering speed, a temperature is sometimes selected at which the flow capacity is no longer a maximum. The temperature is limited by increasing wear of the soldering tools; roughness of the surface because of oxidation; burning of the solder; recrystallization; more risk of intergranular corrosion; annealing or hardening of the base metal and a change in its mechanical properties in general (tensile strength, yield strength and strain); distortion of the product; too intensive alloying of the solder with the base metal (so that too many brittle inter-metallic compounds are formed, and, besides, holes sometimes appear in thin metal sheet) and with its coating (so that the melting temperature of the (altered) solder composition can change[28]). Other factors involved are the longer heating and cooling times of the product; the evaporation or decomposition of the flux; the evaporation of elements from the substrate; the reaction of the flux with the insulation of the wire; and the temperature stability of components like transistors and materials such as resin-bonded paper and, in general, non-coated metals. By the way, soldering at high temperature over a short period of time is often less detrimental than soldering at a lower temperature over a longer time.

For soldering at lower temperatures corrosive inorganic soldering fluxes are available only.

(h) *Hardness of the flux residue*

Too great a hardness results in chipping or hair fractures. The chips that jumped away can give rise to malfunctioning of contacts, etc., and the hair fractures attract moisture. The resulting stickiness can cause soiling, as can be seen from the behaviour of resin in the tropics.

(i) *Colour and transparency of the flux residue*

A certain transparency is required for visual inspection. Some colours give the false impression of corrosion: pigments are added to remove this impression, to distinguish resins, to check whether unauthorized after-soldering has taken place, to identify the functions in an intricate circuit, to identify what work was done by what person (each with his own colour) or group of people all working on the same apparatus, to find out in an easier way whether a joint was made and to identify different solders used in one workpiece (as applied in stepped soldering).

On the other hand, the absence of colour is seldom a guarantee that no corrosion occurs. Very instructive is [32]. Any misinterpretation is avoided by removing all flux remnants.

(k) *Specific resistance of the flux residue*

If the specific resistance is to meet certain requirements, resins are given preference. Activators seem to reduce the specific resistance.

(m) *Removability of flux residue*

This is sometimes required to avoid corrosion (after soldering the flux residue must be removed when the flux was corrosive). Rinsing with warm and cold water is suitable for some kinds of flux* and for products without slots or crevices, which are not sensitive to moisture and temperature. Whereas resin dissolves in tri, ethanol, ethylacetate and butylacetate, acetone, toluene and propanol-2, the far-more-dangerous activators sometimes remain behind in a higher concentration (pure colophony offers protection against corrosion). Nitric acid, warm sulphuric acid and hydrofluoric acid remove active fluxes. The usually brittle remnants can be removed by brushing, tapping and sandblasting, but there is then a danger of the joint getting damaged. A thermal shock (dipping non-ferrous products in cold water) and ultrasonic vibration are two other possibilities. Remnants of fluorine-based fluxes are removed by brushing in (warm) water; and the remains of borax by pickling in a sulphuric acid solution, followed by rinsing.

(n) *How the flux affects health*[30]

Zinc chloride, ammonium chloride and hydrochloric acid can irritate the mucous membranes and the eyes. Hydrazine compounds, such as hydrazine

* e.g. guanidine carbonate or DMA-glycerin.

hydrobromide, for instance, can cause dermatitis on face and hands, especially when there has been direct contact. Aliphatic amines and low-molecular diamines and polyamines can bring out eczema or cause considerable irritation to the skin. Benzidine and β-naphthylamine hydrochloride are powerful cancer generators. Due care should be taken when using aniline; fortunately, the safety limit is rather high, the Maximum Allowable Concentration (MAC) being 19 mg/m^3. This MAC applies to healthy persons inhaling it for 8 h per day without suffering any permanent harm. Low-boiling mono-basic unsaturated carbonic acids and acid anhydrides, such as maleic acid anhydride, have an irritating effect. Certain halogenized carbon acids are poisonous. Improper handling of flux can create fluorine compounds, which irritate the mucous membranes. So, great care should be exercised in handling hydrofluoric acid, the more so as it causes painful burns on the skin.

Nitric acid decomposes in contact with organic materials and produces highly toxic gases.

Sodium chromate can promote dermatitis under certain conditions.

Of all the flux solvents, methyl alcohol (MAC = 260 mg/m^3) is the most poisonous. Formaldehyde and allylaldehyde, produced by the decomposition of organic solvents, are most irritative.

Caustic soda solutions can cause respiratory and eye irritation and, besides, react with Al to evolve hydrogen, which is explosive under certain conditions.

It is a wise policy to analyse unfamiliar matter and generally to reduce the vapour concentration of less-desirable components by means of an exhaust system, by ventilating or by stirring the air. Having an exhaust is the best solution of the three.

(p) *The position of the workpiece and its shape, especially of the solder seams.* (See Sections 3.6 and 3.7)

REFERENCES

[1] N. F. Lashko, S. V. Lashko-Avakyan — *Brazing and soldering of metals*, London 1961.

[2] Jonson — *Brazing progress*, Journal of the American Ceramic Society, Ceramic Abstracts, **33**, No. 5, 1950.

[3] — *The vacuum furnaces for brazing*, Engineering, **183**, No. 4757, p. 600, 1957.

[4] Anon. — *Metal to non-metal brazing*. Metal Treatment and Drop Forging, **21**, No. 101, p. 66, 1954.

[5] L. S. Darken, R. W. Gurry — *Physical Chemistry of Metals*, New York, 1953.

[6] W. H. Chang — *A dew point-temperature diagram for metal-metal oxide equilibria on hydrogen atmospheres*, Welding Journal, **35**, No. 12, p. 6629, p. 6645, 1956.

[7] D. M. Dovey, K. C. Randle — *High temperature furnace brazing*, Metal Treatment and Drop Forging, **22**, November 123, p. 501, 1955.

[8] R. L. Peaslee, W. M. Boam — *Design properties of brazed joints for high temperature applications*, Welding Journal, **31**, No. 8, p. 651, 1952.

[9] J. Stanley — *The embrittlement of pure iron in wet and dry hydrogen*, Journal American Society for Metals, **44**, 1952.

[10] G. SIMON *Hartlöten in Schutzgas-Atmosphäre im elektrischen Widerstandsofen*, Zeitschrift für Metallkunde, Band **117**, Heft 7, 1957.

[11] H. H. KELLOG *Metallurgical reactions of fluorides*, Journal of Metals, vol. **191**, no. 2, p. 137–141, 1951, no. 6, p. 864–872.

[12] E. LÜDER *Handbuch der Löttechnik*, Berlin 1952.

[13] M. F. JORDAN, D. R. MILNER *The removal of oxide from aluminium by brazing fluxes*, Journal Institute of Metals, **85**, No. 2, p. 33, 1956.

[14] H. R. CLAUSER *How to select brazing and soldering materials*, Materials and Methods, **35**, No. 3, p. 105, 1952.

[15] W. I. GABE *Improve your soldering with non-corrosive flux*, Iron Age, **172**, No. 10, p. 115, 1953.

[16] S. FREEDMAN *Fluxless aluminium joining avoids joint corrosion*, Iron Age, **177**, No. 9, p. 71, 1956.

[17] H. KÜNZLER, H. BOHREN *Untersuchungen an Feinlötstellen*, Technische Mitteilungen P.T.T. (Bern), No. 9, p. 329, 1954.

[18] A. Z. MAMPLE *Soldering fluxes and flux principles*, Western Union Technical Review, **35**, January 1957.

[19] C. G. KEEL, G. B. BRUBACHER *Flussmittel zum Weichlöten*, Journal de la Soudure, No. 5, p. 95, 1952.

20] H. C. SOHL *Non-corrosive fluxes, evaluation of spread and corrosion properties*, Symposium on solder, A.S.T.M.-S.T.P., No. 189, p. 81, 1956.

[21] H. H. MANKO *How to choose the right solder flux*, Product Engineering, 13 June, 1960, p. 43.

[22] W. R. BJORKLUND *Non-spattering solder flux*, Tin and Its Uses, No. 70, p. 12, 1966.

[23] B. M. ALLEN *Some factors involved in soft soldering*, Electronic Components, July 1966.

[24] H. H. MANKO *Solders and Soldering*, McGraw-Hill Book Company.

[25] M. M. SCHWARTZ *Vacuum brazing—from aluminium alloys up through the refractory metals*, Metals Engineering Quarterly, ASM, November 1966, p. 47–51.

[26] W. R. LEWIS *The action of fluxes that assist tinning and soldering*, Tin and Its Uses, No. 72, 1966.

[27] DIN 8511: *Flußmittel zum Löten metallischer Werkstoffe*, Blatt 1: *Hartlöten von Schwermetallen*, Blatt 2: *Weichlöten von Schwermetallen*, Blatt 3: *Weich- und Hartlöten von Leichtmetallen*, Deutscher Normenausschuβ, August 1966.

[28] G. CIRIAK *Werkstoffeinflüsse auf die Löttemperatur*, Metall, **21**. Jahrgang, Heft 6, p. 590, June 1967.

[29] M. M. SCHWARZ *Clean, fast vacuum brazing joins diverse materials*, Materials Engineering, December 1967, p. 76.

[30] W. WUICH *Einwirkung schädigender Gase auf den menschlichen Organismus beim Schweissen und Löten*, Draht-Welt, vol. **53**, No. 4, p. 273, 1967.

[31] B. LIEBESMAN *Développement de l'emploi du flux gazeux dans l'assemblage des métaux*, lecture to La Société des Ingénieurs Soudeurs, February 24, 1955.

[32] H. H. MANKO *Color, corrosion and fluxes*, Electronic Packaging and Production, February 1969, p. 26.

[33] W. P. MCQUILLAN *Guide to Soldering*, Welding Engineer, April 1965, p. 112.

3.4 Solder alloys

A very important point in choosing a solder is the *melting point or melting traject*. It varies from −39·2 °C (Hg+0·1% Bi), via 29·8 °C (gallium)[42], bismuth solders (41 to 227 °C), eutectics of Sn, Pb, Cd or Tl (143 to 183 °C), alloys of Sn(-Sb)-Pb (183–324 °C), Pb-Ag alloys and cadmium alloys (150–400 °C), zinc alloys (300–500 °C), aluminium alloys (470–624 °C), silver alloys (600–980 °C), copper alloys (625–1 240 °C), to nickel alloys (880–1 452 °C). With palladium, platinum and rhodium, the temperature rises to 1 970 °C.

Solder with a large melting traject has a greater viscosity and is therefore used for wide gaps, large areas, smoothing-out roughness, and on the whole for less accurate work. When heating is slow, as in furnace soldering, and the melting traject larger, there is more segregation between high- and low-melting constituents of the solder. In the beginning, the higher-melting components are not sucked-up by the seam and can give the impression that it is filled, so that no more heat is applied and the joint is considered to be ready. This is not true, however, and the solder remaining outside must be removed mechanically. When just enough solder is supplied to fill the gap completely, a residue of solder indicates a failure. See also under preforms in Section 3.6.3.

The soldering temperature is usually selected at least 50 °C below the melting point of the base material; the soldering temperature lies below or above the liquidus temperature of the filler metal. The melting temperature, and hence the solder composition, is determined by the strength requirements upon the filler metal at working temperature. In the case of a high working temperature, diffusion between filler and base metals can affect their mechanical behaviour.

The mechanical properties of the solder are decisive for its form as a semi-product. The solder is usually given its form by casting, extruding and the usual techniques (for the shape, see 3.6.3). With resin core soldering wire, the ratio between flux and solder is of very great importance; the flux weighs but a few percent of the total weight of the wire.

Apart from melting point and mechanical properties at the operating temperature, the choice also depends on the base metals*, the thermal and electrical† properties, the shape and dimensions of the seam, the soldering method, time and temperature, corrosion and oxidation risk, the behaviour in vacuum (lead and cadmium are poisonous and should not be used in hospitals and near food), the difference in colour between soldered seam and base metals (important for decorative products), the amount of post-treatment required, the skill of the labour involved and the total costs of the joint. Corrosion is fought by using solder lying close to the base metal in the

* As the quality of the joint depends on the wetting by the solder, its wetting power is of primary importance. When the base metals cannot be joined with the solder selected for other reasons—e.g. joining stainless steel and aluminium—a junction piece is sometimes used, consisting of a laminate of the two metals to be joined. This junction piece is soldered/brazed to the base metals by common means.

† In practice, the transition resistance across the joint is generally compensated for by its increased cross-sectional area, so that the resistance per unit length of the joint is less than that of the base metal. Hence, thermal overloading of the joint is very unlikely.

electromotive series. As, moreover, the surface area of the soldered seam is generally smaller than that of the components joined, it is (in view of the selective corrosion by the environment) advisable to use a filler metal that is cathodic to (nobler than) the least-noble component part. The two diffusion zones on both sides of the solder seam have a corrosion resistance differing from both the filler and the base metal.

Apart from these considerations, the filler metals must meet the following requirements:

The melting point should be below that of the base metal, even below the temperature at which the structure and properties of the base metal begin to change;

The solder must wet the base metal, and have a viscosity low enough to enter the seam and fill it completely by capillary action;

The solder must not, however, alloy too intensively with the base metal;

The alloying must not result into a brittle or fragile joint;

The solder should not contain elements which, when processed by the chosen technique, evaporate, thus altering the solder composition adversely.

(a) *Bismuth solders** expand during cooling and ageing. They are therefore used in some cases to compensate for shrinkage of alloys. Due to its high price, bismuth is added in small quantities only; it does not wet very well and its joints are very weak. Bismuth alloys are specially used as the safety element in fire safety devices.

The group of low melting solders (melting point below 180 °C, see Table 3.3) includes compositions of Bi, Pb, In, Cd and Sn; these are used for soldering close to already existing soldered joints, near heat-sensitive materials such as textiles and paints, plastics and wood, combustible matter, hardened material and for joining metal and glass, e.g. injection syringes.†

This group of low-melting solders has but very little strength, often oxidizes strongly and due to their low melting temperature cannot be worked up with non-corrosive fluxes. See [11], [16] and [31].

(b) *Indium solders* can have very low melting points. They give a stronger and harder seam with an attractive appearance, have a good wetting capacity (they wet glass, Si and Ge) and are sometimes highly oxidation resistant. Due to its very low vapour pressure, indium solder is used in vacuum equipment, e.g. 90In/10Sn. See [31] and [39]. Indium is expensive.

(c) *Tin-lead and tin-zinc solders* are the most-often-used soft solders (because tin alloys with a very wide range of metals), especially in the composition 50Sn/50Pb and 33Sn/67Pb (for resin core soldering wire 60Sn/40Pb is preferred). In tin-lead solders, tin improves the ductility, spreading, electrical conductivity and up to 63 % Sn the strength. Lead serves as a filler metal to

* In general, alloys containing more than 55 % Bi expand and those with less than 48 % contract during solidifying.

† Other applications of these fusible alloys are: lost-wax models, moulds for electroforming products, castings, temperature gauges, thermal fuses, hardening-bath liquid, chuck jaws for fragile components, casting-in of dies and metallographic samples, filling thin-walled pipes that need bending, and the manufacture of simple dies and seals, also of bolted connections.

TABLE 3.3

Some soft solders

Composition by weight (%)								Melting traject (°C)	Notes; application
Sn	Pb	Bi	In	Sb	Cd	Ag	Other elements		
—	—	—	—	—	—	—	100 Hg	−38·9	
—	—	—	—	—	—	—	100 Ga	29·8	
10·7	22·2	40·3	17·7	—	8·1	—	1 Tl	41·5	
10·8	22·4	40·6	18	—	8·2	—	—	46·5	
8·3	22·6	44·7	19·1	—	5·3	—	—	47·2	eutectic
12	18	49	21	—	—	—	—	58	expands when solidifying
12·8	25·6	48	4	—	9·6	—	—	64–61	
14·2	25	50	—	—	10·8	—	—	67–63	
13·1	27·3	49·5	—	—	10·1	—	—	70	Wood's metal, eutectic, bad wetting, weak joint, expands when solidifying
17	—	57	26	—	—	—	—	79	
41·4	—	—	44·2	—	13·6	—	0·8 Zn	90	
—	40·2	51·7	—	—	8·1	—	—	91·5	eutectic, requires active flux
18·7	31·3	50	—	—	—	—	—	95·0	eutectic, requires active flux, Newton's metal
25	25	50	—	—	—	—	—	98–96	D'Arcet's metal
19	36·5	35	—	—	9·5	—	—	100–70	
—	40	45	—	—	15	—	—	100–92	
26	—	53·5	—	—	20·5	—	—	103·0	eutectic, expands when solidifying, requires active flux; fuses
15	40	45	—	—	—	—	—	110–95	
22	28	58	—	—	—	—	—	110–96	Rose's metal
25	31·5	43	—	—	—	—	0·5 Cu	115–95	
48	—	—	52	—	—	—	—	117	eutectic; for glass and enamelling
—	—	—	27	—	73	—	—	123·0	eutectic
—	43·5	56·5	—	—	—	—	—	124	eutectic, expands when solidifying, requires active flux

TABLE 3.3—*continued*

Composition by weight (%)								Melting traject (°C)	Notes; application
Sn	Pb	Bi	In	Sb	Cd	Ag	Other elements		
—	43	55	—	—	—	—	2 Zn	124	
50	—	—	50	—	—	—	—	125–117	soldering glass (rubbing in)
11·4	—	—	65·4	—	23·2	—	—	126–105	
40	—	56	—	—	—	—	4 Zn	130	
10	—	—	90	—	—	—	—	135–130	
42	—	58	—	—	—	—	—	138	eutectic, expands when solidifying, any flux suitable
33·3	33·4	33·3	—	—	—	—	—	140–95	
49·35	32·56	—	—	—	18·1	—	—	143·3	eutectic, any flux suitable
—	—	60	—	—	40	—	—	144	eutectic, any flux suitable, expands when solidifying
—	—	—	97·5	—	—	2·5	—	150–140	Wets glass, silicon and germanium
43	43	14	—	—	—	—	—	155	
—	—	—	99·5	—	—	—	0·5 Ga	155	
53	30	—	—	—	—	17	0·5 Zn	156–122	reduced attack when soldering gold coatings
—	15	—	80	—	—	5	—	156	
—	—	—	100	—	—	—	—	156·4	metallized glass
68	—	—	—	—	27	—	5 Zn	158	good wetting and strength
50	25	—	—	—	25	—	—	160–145	pre-tinned copper wire to silver-coated glass
65	13·5	—	—	—	21·5	—	—	160–145	
66·5	—	—	—	—	31	—	2·5 Zn	163	
—	20	—	80	—	—	—	—	165–159	
50	40	10	—	—	—	—	—	167–142	
60	—	40	—	—	—	—	—	170–138	no expanding or shrinking when solidifying
57	—	—	—	—	—	—	43 Tl	170	eutectic
66	27·5	—	—	—	6·5	—	—	172–145	

TABLE 3.3—*continued*

Composition by weight (%)								Melting traject (°C)	Notes; application
Sn	Pb	Bi	In	Sb	Cd	Ag	Other elements		
37·5	37·5	—	25	—	—	—	—	175–130	
—	—	—	99·3	—	—	—	0·5 Ga 0·2 Al	175–155	
67·5	—	—	—	—	32·5	—	—	176	eutectic, any flux suitable
59·4	36·6	—	—	—	—	4	—	177–170	printed circuit panels
60	38	2	—	—	—	—	—	178	
62	36	—	—	—	—	2	—	179–170	for silver plated component parts
60	38	—	—	2	—	—	—	180	
60	38	—	—	—	—	—	2 As	180	
63	34	3	—	—	—	—	—	180	
59	35	—	—	—	—	6	—	180	
37·5	37·5	—	25	—	—	—	—	181–134	
62·3	37·7	—	—	—	—	—	—	183	eutectic; radio and instruments, quality work
55·8	37·2	—	—	—	—	7	—	185–180	
76	21	—	—	—	—	3	—	187–176	
60	40	—	—	—	—	—	—	187–183	tough, low viscosity, strongest Sn/Pb solder; radio and delicate work, electrical apparatus, copper tubing
45	45	—	10	—	—	—	—	192–159	
70	30	—	—	—	—	—	—	191–183	for zinc
91·1	—	—	—	—	—	—	8·9 Zn	198·6	eutectic; aluminium soldering
50	46	—	—	—	—	4	—	200–174	
55	45	—	—	—	—	—	—	200–183	
80	—	—	20	—	—	—	—	200–117	
80	20	—	—	—	—	—	—	201–183	

TABLE 3.3—*continued*

Composition by weight (%)								Melting traject (°C)	Notes; application
Sn	Pb	Bi	In	Sb	Cd	Ag	Other elements		
48·5	48·5	3	—	—	—	—	—	202–176	
50	47	—	—	3	—	—	—	204–183	
50	50	—	—	—	—	—	—	212–183	hand soldering; electrical apparatus, radio, telephone sets, copper tubing, up to 100°C, at lower temperature higher tensile strength and lower impact strength
90	10	—	—	—	—	—	—	214–183	
50	48·7	—	—	—	—	—	1·3 Cu	215–180	Sav bit No. 1, electrical apparatus, telephone sets, batteries, Cu compounds with Sn, reducing flow
—	85	—	—	—	—	—	15 Au	215	eutectic
—	50	—	50	—	—	—	—	216–183	alkali resistant
93	—	—	—	—	—	—	7 Au	220–217	silver plated component parts
96·5	—	—	—	—	—	3·5	—	221	creep resistant, strong, no discolouring. Very conductive, eutectic; stainless steel, delicate work, cryogenic and elevated temp., food-handling machinery and surgical implants
95	5	—	—	—	—	—	—	223–183	
9	65	—	—	—	26	—	—	225–145	
—	88	—	—	0·5	—	—	11·5 Au	225–215	
95	—	—	5	—	—	—	—	225–220	
96·4	—	0·1	—	—	—	3·5	—	225–220	
14·5	28·5	48	—	9	—	—	—	227–103	expands strongly when solidifying
90·5	—	—	—	5	2	—	2·5 Cu	227	
100	—	—	—	—	—	—	—	231·9	lead free; Cu to Pt, max. electrical conductivity
23	68	—	—	—	9	—	—	235–145	

TABLE 3.3—*continued*

Composition by weight (%)								Melting traject (°C)	Notes; application
Sn	Pb	Bi	In	Sb	Cd	Ag	Other elements		
40	60	—	—	—	—	—	—	235–183	hand soldering, electrical apparatus, mechanized can soldering
3	75	—	—	—	22	—	—	237–145	
30	—	—	—	—	70	—	—	240–170	no thermo-electric-motive force
95	—	—	—	5	—	—	—	240–222	good wetting and strength, also at higher and lower temp., high electrical conductivity, lead free; electric motors and sanitary copper tubing
50	32		—	—	18	—	—	245–145	
15	75	—	—	—	10	—	—	245–145	
33	65	2	—	—	—	—	—	245–181	
35	65	—	—	—	—	—	—	245–183	
95	—	—	—	—	—	5	—	245–221	lead free; for somewhat elevated temp., e.g. brass and copper, hot copper tubing
27	70	—	—	—	—	3	—	247–173	glass and silver coated ceramics
—	82·5	—	—	—	17·5	—	—	248·0	eutectic
30	rest	—	—	1–1·7	—	—	—	248–185	fuses, motors, cables, dipping and pre-tinning baths
88	—	—	—	—	10	2	—	250–205	
98	—	—	—	—	—	—	2 As	250–232	
—	70	—	30	—	—	—	—	250–240	
—	88·9	—	—	11·1	—	—	—	252	eutectic
10	80	—	—	—	10	—	—	253–145	
30	70	—	—	—	—	—	—	255–183	fuses, motors, dynamos, cables, car body repairs
6	80	—	—	14	—	—	—	256–239	
5	85	—	—	—	10	—	—	257–145	
—	80	—	—	—	20	—	—	259–248	

TABLE 3.3—*continued*

Composition by weight (%)								Melting traject (°C)	Notes; application
Sn	Pb	Bi	In	Sb	Cd	Ag	Other elements		
10	85	—	—	—	5	—	—	260–145	
—	70	—	—	—	30	—	—	260–248	
25	75	—	—	—	—	—	—	266–183	
—	—	—	—	—	82·6	—	17·4 Zn	266	eutectic for friction soldering of Al
—	90·8	—	—	—	7·8	—	1·4 Zn	267–237	
—	87	—	—	13	—	—	—	267–252	
80	—	—	—	—	—	—	20 Zn	270–199	soldering of Al
—	—	100	—	—	—	—	—	271·3	
—	90	—	—	—	10	—	—	274–248	
10	88	—	—	—	2	—	—	275–145	
18	81	—	—	1	—	—	—	275–185	lamps, motors, dipping baths
20	80	—	—	—	—	—	—	275–183	lamps, motors, dynamos
20	78·7	—	—	—	—	1·3	—	276–181	
—	91·5	—	—	—	8·5	—	—	276–248	
30	—	—	—	—	50	—	20 Zn	277–157	
—	85	—	15	—	—	—	—	280–275	
20	—	—	—	—	—	—	80 Au	280	eutectic; silver plated ceramics, Si, Cu/Mo
40	57	—	—	—	—	3	—	284–178	
33·5	61·5	—	—	—	—	5	—	285–183	
—	—	—	—	—	79	5	16 Zn	285–270	Cu to W
—	97·2	—	—	—	—	—	2·8 As	288	eutectic
15	85	—	—	—	—	—	—	290–223	
—	—	—	—	—	69	1	30 Zn	290–272	
90	—	—	5	—	—	5	—	290	

TABLE 3.3—*continued*

Composition by weight (%)								Melting traject (°C)	Notes; application
Sn	Pb	Bi	In	Sb	Cd	Ag	Other elements		
5	—	—	—	—	65	—	30 Zn	294–229	
90	—	—	—	—	—	10	—	295–221	Cu to W
—	—	—	99·5	—	—	—	0·5 Ga 0·5 Ge	300–155	
10	90	—	—	—	—	—	—	300–267·5	
5	93·5	—	—	—	—	1·5	—	301–296	automotive radiators, cryogenic and high temp.
—	97·5	—	—	—	—	2	0·5 Ag, As or Sb	304·0	fair wetting, corrosion sensitive, cryogenic temp.; flame soldering Cu-alloys, higher temp., Mo/Ag
8	92	—	—	—	—	—	—	305–270	
—	—	—	—	—	73	5	22 Zn	307–272	
—	98	—	—	—	—	2	—	308–304	
—	83	—	—	12	—	5	—	310–254	
1	97·5	—	—	—	—	1·5	—	310–304	some strength to 200°C: flame soldering, hot-water copper tubing
70	—	—	—	—	—	—	30 Zn	312–199	soldering of Al
5	95	—	—	—	—	—	—	315–300	suitable to —180°C, at cryogenic temp. tensile and impact strength somewhat less than 97·5 Pb/2·5 Ag
—	—	—	—	—	100	—	—	320·9	
2	98	—	—	—	—	—	—	324–318	
—	—	—	—	—	45	—	55 Zn	327–266	friction soldering of Al
—	100	—	—	—	—	—	—	327·4	
50	30	—	—	—	—	—	20 Zn	328–175	
10	—	—	—	—	60	—	30 Zn	332	
—	—	—	—	—	40	—	60 Zn	335–265	soldering of Al

TABLE 3.3—*continued*

Composition by weight (%)								Melting traject (°C)	Notes; application
Sn	Pb	Bi	In	Sb	Cd	Ag	Other elements		
60	—	—	—	—	—	—	40 Zn	340–199	soldering of Al
—	—	—	—	—	96	3	1 Zn	340–320	
84	—	—	—	—	—	16	—	345–221	
—	97·3	—	—	—	—	1·7	1 Zn	350–320	
—	95	—	—	—	—	5	—	360–304	
—	87·5	—	—	—	7·5	—	5 Zn	368–235	
19	1	—	—	—	—	—	80 Zn	375–198	
30	—	—	—	—	—	—	70 Zn	376–199	soldering of Al
—	—	—	—	—	67	10	23 Zn	380–271	
—	—	—	—	—	95	5	—	396–343	soldering of Al to itself or other metals, steel
—	—	—	—	—	10	—	90 Zn	398–265	soldering of Al
—	—	—	—	—	—	—	100 Zn	419·5	
70	—	—	—	—	—	30	—	420–221	

reduce the price of the alloy, and as a means of influencing the melting point and boundary plane energy (it moderates the alloying action of tin). Lead is toxic and should not be used in food-handling equipment.

Pure tin is seldom used, for instance, in joining Cu to Pt.

Tinpest (disintegration into powder at a low temperature) can practically always be traced back to oxidation, since the normal alloying elements and contaminants are most unlikely to cause tinpest. For the remainder, Sn/Pb solders containing at least 35% Sn are somewhat brittle at low temperature. See also [14], [41], [53] [57] and [68]. No whisker formation* takes place in

* *Whiskers.*

Definition, properties

Whiskers are hair-shaped crystals growing spontaneously from solids. Their diameter is expressed in microns; their length can be several mm's.

Whiskers also form on many metals and alloys, mainly on Ag, Sn, Cd, Sb and Zn, but to a lesser extent on Fe, Ni, Mg, Mo, W, Ta, Pa, Pd, Pt and Au. In general, the phenomenon appears with rather soft materials having a low melting point; lead and indium, however, seldom show whiskers.

In electrical engineering, especially in miniature designs, whiskers often cause sparking, short-circuits and noise at high frequency. In view of the materials used in soldering, tin whiskers are specially to be feared. These tin whiskers grow at a rate of 0·01 to several Ångström/sec. (1Å/s equals about 3 mm/year). Under favourable conditions, this rate can be 100 times more.

Growth mechanism

It is generally assumed that whiskers are caused by dislocations generated by compressive stress in the lattice of the surface zone. These stresses can be caused by mechanical and thermal pretreatment of the material and/or external factors such as pressure, temperature, etc.

Appearance, factors of influence

Shape, structure, size, amount, growth rate and direction, and the moment where whisker growth starts, cannot be predicted accurately; sometimes, however, some regularity is noticed and these properties also can be influenced.

Apart from the internal stresses that affect the formation and properties of whiskers we mention:

(1) *Base metal*

In the base metal, a role is played by its composition and by the mechanical and thermal history, affecting structure and residual stress. A steel substrate displays the fastest whisker growth, but the phenomenon is much less pronounced with brass, copper, nickel, silver and gold.

When these elements are applied as an intermediate layer on steel, the growth rate is reduced. But not always satisfactorily: the influence of intermediate layers is not always clear.

(2) *Top layer*

For the sake of solderability or otherwise, the materials used in electrical engineering are commonly provided with a galvanically- or thermally-applied layer.

continued on p. 142

The former often has high internal stresses, the second is stress-free, in principle. For this reason, and because of the relatively small thickness of the layer, whisker growth mainly occurs with layers that are applied electrochemically. The growth can be suppressed by post-melting this layer.

Growth rate also depends on the layer thickness: it is maximal at 1 to 2 μm (perhaps as the porosity of the layer is high, hence the corrosion through and the stress beneath it; thinner layers are insufficiently investigated in this respect).

The composition of the galvanically-applied layer is important, too. Some additions (e.g. for improving the lustre) to the tin bath promote whisker growth, others reduce it. Films of alloys show much less whiskers than one-component layers. Tin-lead is practically whisker-free: it can be easily applied, is cheap, has a fair appearance and good solderability. Tin-nickel is another top layer; it is, however, hard and brittle, more expensive and only moderately solderable. Whisker growth is also reduced by small additions of Cu, Co, Ge, Sb and Au to the bath, which additions are absorbed by the top layer.

(3) *Intermediate layer*

To suppress whisker growth, intermediate layers are sometimes applied; copper and gold are effective between tin and a steel base. But their influence is not always clear: gold galvanically-applied to nickel silver substantially reduces tin whisker growth, but electroless Atomex gold markedly increases the growth rate (a galvanically-applied intermediate layer should have a minimum thickness of 3 μm). Here, too, the ultimate effect largely depends on the stress pattern of the top layer. Moreover, as the price of gold prohibits its commercial application as an intermediate layer, copper and nickel are generally used.

(4) *Temperature and temperature gradient*

Temperature rise increases the mobility of the atoms and hence the growth rate of the whiskers. On the other hand, an increase in temperature has a stress-annealing effect, thereby reducing growth. At higher temperature, the latter influence dominates, so that any piece of metal has a maximum growth rate at a well-defined temperature. For tin, this temperature is as low as 60 to 70 °C, so the temperature must be well controlled.

Local difference in temperature, leading to unequal expansion and hence stresses, also play a role.

(5) *Atmosphere*

High pressure, humid and warm atmosphere, and a high oxygen content promote whisker growth.

(6) *Radiation*

Neutron radiation, for instance, has the same effect.

(7) *Mechanical stressing and residual stress by machining*

Whisker growth is nearly proportional to the pressure exerted on the base metal (4 mm/hour at 600 kgf/cm^2). Tin layers should, therefore, be avoided in stressed structures. Clamped joints (for instance, bolted constructions) are rather susceptible, even when the top layer has been applied thermally.

Tinned material can also display serious whisker growth when being processed. This applies both to plastically deformed material in general and to cut edges.

normal solder compositions. Tin-lead solder containing little tin, contracts a fair amount on solidifying and can even display cracks.

Impurities in Sn-Pb solder affect flow (Al, Zn, <0·005%) by producing oxides, and ductility (Fe <0·02%, Cu <0·5%, As <0·005% and Au—see Section 3.5.2(d)) by producing intermetallic compounds.

Antimony[39] is added to improve the strength (also the fatigue and creep strength at room and high temperature) and to reduce the cost of the alloy,* but it reduces wetting, adhesion, elongation, corrosion resistance and does not have such a good appearance. Antimony-zinc joints are brittle[48], so brass and zinc coatings should not be soldered with Sb-containing solder.

Bismuth makes the solder (e.g. 48 Sn/49 Pb/3 Sb) dull, which facilitates inspection of soldered prints.

Additions of silver reduce oxidation of liquid tin-lead, improve the mechanical characteristics (at increased temperature too) and prevent silver from dissolving from the base metal.

The addition of copper is very effective against the dissolving of copper from objects (of interest for very thin wire, for instance). Similarly, cadmium fights the dissolving of silver.

Because of its high price, tin is replaced by zinc (see Table 3.3). Alloys containing 10–30% zinc are used for the ultrasonic tinning of aluminium.

There are solders on a lead basis containing no tin. There are also tinless alloys of lead with bismuth, cadmium, silver, antimony, and a few ternary combinations (see Table 3.3 and [38], for instance).

At a slightly increased temperature, and a somewhat longer load duration, the strength of the already weak soft solders becomes much less (see Section 3.6.1): the brittle compound layer thickens at higher temperature.

Under prolonged heating at a high temperature, the concentration of lead vapour in the air can exceed the safe value of 0·2 mg/m^3 (see also [54]). Lead is not suitable for use in vacuum.

(d) *Solders on a cadmium basis* can produce poisonous[56] cadmium vapour,† all the more since Cd is very volatile. The safe value is 0·1 mg/m^3. Cd-Ag solders have a high creep strength and can easily be deformed. Due to the low melting point of cadmium solders, they are often used to make joints close to already existing joints which should not come loose. For the remainder, cadmium is used as a basis for heat-resistant solders used for radiators, etc.; cadmium-zinc solder is a heat-resistant solder for aluminium.

Table 3.4 gives a few examples.

(e) *Solders on a zinc basis.* Zinc oxidizes rapidly and therefore has a lower flow capacity. It also reacts strongly with aluminium and copper, so that the

* Tin is about 12 times as expensive as lead: antimony 4 times.

† Cadmium oxide fumes are insidious, cannot be smelt, and lethal doses need not be sufficiently irritating to cause discomfort, until the worker has absorbed sufficient quantities to be in immediate danger of his life. Symptoms of headache, fever, irritation of the throat, vomiting, nausea, chills, weakness, and diarrhoea, generally may not appear until some hours after exposure. The main injury is to the respiratory passages.

TABLE 3.4

Some aluminium solders

Composition by weight (%)									Melting traject (°C)	Techniques and properties
Al	Si	Zn	Sn	Cd	Cu	Fe	Mg	Other elements		
—	—	8	92	—	—	—	—	—	200	Bit, flame, dipping, heating plate or fluxless ultrasonically. For sheet etc. Corrosion-sensitive
—	—	10	90	—	—	—	—	—	210–200	
—	—	3	45	—	—	—	—	52 Pb	220–170	Cable ends: sprinkling with solder Corrosion-sensitive
—	—	15	85	—	—	—	—	—	230–200	Bit, flame, dipping, heating plate For sheet etc. Corrosion-sensitive
—	—	20	80	—	—	—	—	—	265–200	As 15Zn/85Sn
—	—	36	54	10	—	—	—	—	300–170	Fluxless smearing with solder stick Corrosion-sensitive. For casting repairs
—	—	30	70	—	—	—	—	—	310–200	Fluxless smearing with solder stick Corrosion-sensitive
—	—	55–65	—	45–35	—	—	—	—	338–266	Fair corrosion resistance
—	—	40	60	—	—	—	—	—	340–200	
5	—	90	—	—	3	—	—	—	370	Furnace and flame, or fluxless smearing with solder stick. For sheet. Corrosion-resistant
—	—	70	30	—	—	—	—	—	376–200	
5	—	95	—	—	—	—	—	—	395–380	Furnace and flame soldering. For sheet. Corrosion-resistant

TABLE 3.4—*continued*

Composition by weight (%)									Melting traject (°C)	Techniques and properties
Al	Si	Zn	Sn	Cd	Cu	Fe	Mg	Other elements		
—	—	100	—	—	—	—	—	—	419	
—	—	95	—	—	—	—	—	5 Ag	450–420	Flame soldering. For sheet and castings. Joining aluminium to copper and steel. Corrosion-resistant
30	—	70	—	—	—	—	—	0·1–0·5Mn	510–440	Flame soldering. Touching up castings. Corrosion-resistant
75·2	10	10	—	—	4	0·8	0·07	0·07Mn 0·07Cr	557–515	
86–83·25	9·3–10·7	0·2	—	—	3·3–4·7	0·8	0·15	0·15Mn 0·15Cr	567–510	
87·5–85·5	11–13	0·2	—	—	0·3	0·8	0·1	0·05Mn	567–556	
88	12	—	—	—	—	—	—	—	577	
86	10	—	—	—	4	—	—	—	584–522	
90	10	—	—	—	—	—	—	—	590–577	
91·25	7–8	0·2	—	—	0·25	0·8	0·1	—	610–575	
94·6–92·6	4–6	0·1	—	—	0·3	0·8	0·05	0·05Mn	626–575	

gap must be rather wide (0·2 mm). The strong evaporation of zinc (the seam becomes porous) makes it unsuitable for use in vacuum equipment: in general, it requires short soldering times at low temperatures. An oxidizing atmosphere forms zinc oxide, which suppresses evaporation; 95Zn–5Al can solder aluminium without flux.

Zinc delays the dissolving of silver from the product into the solder.

(f) *Solders on an aluminium basis* are practically only used to join aluminium components. Al-Cu-Zn, Al-Zn and Al-Si are the usual types. Cu, Zn and Si lower the melting point of the aluminium. The corrosion resistance of Al-Zn alloys is reduced considerably by contamination. (See also Section 3.6.3, first footnote.)

Table 3.4 gives a few aluminium solders. (An XYZ-solder is an alloy containing XYZ; in everyday speech, however, an aluminium solder is a solder that is used to solder aluminium-base metal components.)

(g) *Solders on a silver basis* are widely used (flame, induction, resistance, furnace, salt or flux bath soldering) because of their excellent spreading and capillary action, strength (little filler metal needed), plasticity, corrosion resistance, little need for after treatment and easier removal of flux residue, attractive appearance of the joint, good electrical and thermal conductivity, low melting point (hardly any risk of grain growth, short brazing times) and less energy cost. In short, a total joint cost that is quite acceptable notwithstanding the cost of the silver alloy.

Silver suppresses the dissolving of silver out of the product. Ag-Cu solders are the most popular in this group, and are suitable for vacuum applications, just as Ag-Cu-Sn. Additions of zinc, tin and cadmium increase the electrical resistance. Table 3.5 gives a few Ag-Cu solders.

They are used to join copper and copper alloys, steel, stainless steel and silver.

Nickel increases the strength and the liquidizing temperature of Ag-Cu solder considerably, improves the gloss, and has a better wetting effect.

Manganese decreases the liquidizing point of silver solders.

Palladium improves the wetting capacity of the solder[27].

Cu-P-Ag solders are used for copper and its alloys; silver increases the strength.

(h) *Solders on a copper basis* (see Table 3.5) are used for brazing copper alloys, steel, nickel, molybdenum and tungsten. Pure copper (melting point 1 083 °C) is used for steel components, flows excellently (requires a very narrow gap: about 15 μm), is vacuum tight and cheap, gives a strong joint, has a low vapour pressure and is little sensitive to the type of brazing atmosphere. OFHC copper is used for monel and fernico (for fernico, say, 85Cu/7Ag/8Sn, 985 °C). Due to its high melting-point, copper promotes crystal growth; besides, a high-temperature brazing furnace is expensive. To lower the melting point, Zn is added. Copper-zinc filler metal, used for brass, has a relatively low melting-point. Moreover, zinc dissolves in copper and brittleness increases with the zinc content. Hence, the maximum zinc content of 45%. Zinc evaporates during soldering, produces poisonous

TABLE 3.5

Some brazing filler metals, such as silver, copper and cadmium solders

Composition by weight (%)				Melting traject (°C)	Notes; application
Ag	Cu	Zn	Other elements		
40	16	18	26Cd	605–600	
40	19	20	21Cd	615–550	strong; Cu+steel alloys, incl. cast iron and cast steel
45	15	16	24Cd	620–605	Easy-flow 45, low visc., good flow, strong, corr. resist.; general purpose incl. Cu alloys, Ni alloys, austen. stainless steel
45	16	17	22Cd	630–610	Easy-flow, low visc., tough, salt water resist.; steel, Cu, Ni, Ag and their alloys
50	15·5	16·5	18Cd	635–627	(as above)
56	22	17	5Sn	650–620	little stress corr.; carbon and stainless steel, Cu, Ni and their alloys, food contact applic.
35	26	21	18Cd	670–610	Easy-flow 35, tending to segregation; not for furnace brazing and thick parts, for (stainless) steel, hard metal, Cu, Ni, and wide gaps
64	17	15	Cd+Sn=4	675–650	steel, alloys of Cu and Ag
49	16	23	7Mn, 5Ni	680–650	steel, incl. corr.-resist. and heat-resist. steel, hard metal
30	28	21	21Cd	690–600	cheap, matches brass colour; alloys of Cu and steel
50	15·5	15·5	16Cd, 3Ni	690–630	Easy-flow 3, tendency to segregation, corr.-resist.; corr.-resist. steel, steel to Al-bronze, alloys of Cu and Ni, hard metal, wide gaps
61·5	24	—	14·5In	705–625	alloys of Cu and Ni
15	80	—	5P	705–640	Sil-fos; not for alloys of Fe and Ni; for Cu alloys, fluxless for Cu and Ag
5	88·75	—	6·25P	705–640	Sil-fos 5; not for alloys of Fe and Ni; for Cu+alloys
44	30	26	—	715–670	tensile strength 20 kg/mm^2 at 400°C; steel, stainless to brass
64	21	15	—	718–693	Ag+alloys
60	30	—	10Sn	720–600	seawater-resist.; radio valves, non-ferrous and ferrous metals
30	32	24	14Cd	720–620	low viscos.; steel incl. stainless, hard metal
65	20	15	—	720–670	silver parts
60	25	15	—	720–675	silver parts, Ni-alloys incl. monel
75	—	25	—	720–705	applied where Cu is attacked
64	26	—	10In	723–698	vacuum, incl. radio valves
50	28	22	—	727–677	
57·5	32·5	—	3Mn, 7Sn	730–605	difficult to wet metals as W, Cr-C
20	38	26	16Cd	740–610	Cu+alloys, steel incl. stainless, hard metal, wide gaps, thick component parts
70	20	10	—	740–690	silver parts

TABLE 3.5—*continued*

Composition by weight (%)				Melting traject (°C)	Notes; application
Ag	Cu	Zn	Other elements		
18	39	28	15Cd	745–640	alloys of Fe and Cu
45	30	25	—	745–665	austen. corr.-resist. steel+Cu alloys
68	24	—	8Sn	746–672	Cu+alloys, stainless steel, vacuum
5	89	—	6P	750–645	low viscos.; Cu+alloys, fluxless for Cu and Ag
31·5	34	15·5	19Cd	753–628	general purpose
71·8	28	—	0·2Li	760–754	thin stainless sheet, furnace brazing monel; Cu to Ag, Cu, Ni, Be and Cr alloys
20	42	26	12Cd	765–605	high viscos., hence wide gaps
30	38	32	—	765–675	non-ferrous metals, nickel silver, radio electrics
2	91	—	7P	770–645	not for Fe and Ni, fluxless brazing Cu+alloys
40	36	24	—	770–670	wide gap, mild steel, non-ferrous metals, monel, furnace brazing Ni
75	20	5	—	774–732	alloys of Cu and Ag, small joints
25	37	22	16Cd	775–635	Cu+alloys
50	34	16	—	775–675	steel, incl. austen. stainless, non-ferrous metals, steam turbine vanes
64	26	—	Mn+Ni+ In = 10	777–725	stainless steel, Mo, W, vacuum
40	30	28	2Ni	780–666	mild and stainless steel, alloys of Cu and Ni, food contact
43	37	20	—	780–695	general purpose
25	40	35	—	780–700	cheap, high viscos., heat-resist.; steel, cast iron, alloys of Cu and Ni
15	43	30	12Cd	780–760	steel and Cu alloys (medium-thickness component parts)
72	28	—	—	780	vacuum, Fe, Ni, Cr, kovar, alloys of Cu and Ag, Be; Cu to Ni, monel, W and Pt
72	20	—	8Ti	780	ceramics to metals in vacuum
80	—	20	—	785–725	
65	28	—	5Mn, 2Ni	785–750	nickel iron
75	22	3	—	790–740	silver parts
71·5	28	—	0·5Ni	795–754	vacuum, silver coated ceramics to metal
63	28·5	—	2·5Ni, 6Sn	800–690	stainless steel
67	33	—	—	805–778	
80	16	4	—	810–725	silver alloys
20	45	30	5Cd	815–615	steel and non-ferrous metals, especially brass
20	45	35	—	815–690	non-ferrous metals, especially brass, combined brazing and heat treating
12	50	31	7Cd	825–620	high viscos., hence wide gaps; alloys of Fe and Cu
27	40	25	Mn+Ni=4	830–675	tough and heat-resist.; Fe alloys and high melting metals
15	52	33	—	835–810	Cu+alloys
—	42	58	—	845–835	alloys of Cu and Ni

TABLE 3.5—*continued*

Composition by weight (%)				Melting traject (°C)	Notes; application
Ag	Cu	Zn	Other elements		
65	28	—	5Mn, 2Ni	850–750	Kovar and invar to Cu for vacuum
9	53	38	—	850–765	Cu alloys, especially extruded ones
12	52	36	—	850–815	strong to 400°C; alloys of Fe and Cu and Ni, thick component parts
25	52·5	22·5	—	855–675	steel and non-ferrous metals
54	40	5	1Ni	855–725	strong to 450°C; stainless steel, jet engine parts
40	30	25	5Ni	860–660	stainless steel, carbides
5	55	40	—	870–820	Cu+alloys
10	52	38	—	871–821	Cu, steel, incl. stabilized stainless
50	50	—	—	875–780	Cu, Cu to Ni
5	58	37	—	880–840	Ni–Cr resistance wire, combined brazing and heat treating
90	—	—	10Sn	885–770	for vacuum
90	—	—	10In	887–850	for vacuum
92·5	7·3	—	0·2Li	890–760	stainless steel foil, fluxless brazing
—	60	40	(0·2Si)	900–885	very cheap but not very strong; wide gaps, cast iron, steel incl. austen. stainless, alloys of Cu and Ni
—	95	—	5P	910–714	P attacks base metal copper: braze quickly
—	63	37	—	925–890	
—	48	42	10Ni	935–920	silver coloured, high fatigue strength; steel incl. austen. stainless, Ni alloys
35	65	—	—	940–779	
100	—	—	—	960	OFHC Cu to nickel plated W
85	—	—	15Mn	970–960	heat- and ammonia-resist.; steel incl. stainless, stellite, Inconel, Cr- and Ti carbide
7	85	—	8Sn	985–665	steel incl. stainless. Fluxless furnace brazing
25	75	—	—	1000–780	
—	90	—	10Sn	1000–910	brass in hydrogen, Cu, low-alloy steel
—	—	—	32Ni, 68Mn	1015	
—	87	—	10Mn, 3Co	1030–980	thermally and mechanically highly loaded hard metal joints
—	—	—	72Au, 22Ni 6Cr	1040–975	

ZnO and causes hydrogen embrittlement (moreover, zinc combines with boron and fluorine from the flux[59]). Silicon and an oxidizing atmosphere suppress these phenomena. Silver and silicon increase the strength, elongation, corrosion resistance and flow of brass solders; silver lowers the melting-point.

Cu(-Ag)-P solder (4–8 % P) is very often used to make copper and copper alloy joints (reduces copper oxide); the seam is brittle when no Ag is used. Prolonged influence of a sulphur-containing atmosphere at high temperature can cause corrosion. In brazing pure copper this filler metal can be applied without a flux. Under certain conditions Cu-P might be used to solder steel and nickel alloys, although the resulting iron phosphides are considered to be very brittle. The phosphides oxidize preferentially, thereby reducing the working temperature to about 200 °C.

Sometimes a flux, generally consisting of borate compounds, is required.

The melting point is lowered by alloying, for instance, 64Cu/5P/1Sn/ about 30Zn (melting point 670 °C). Cu-Zn-Ni and Cu-Zn-Mn (for instance, 64Cu/11Zn/25Ni; 48Cu/42Zn/10Ni) are also found. Here, Mn adds to the strength and plasticity, nickel increases the liquidizing temperature. 53Cu/15Mn/30Ni/1-2Si is very suitable as a solder without volatile constituents.

(j) *Solders containing gold* are used in dentistry and (ultra high) vacuum engineering (e.g. 37·5 Au/Cu), and sometimes to improve strength and corrosion resistance at high temperatures (e.g. 82·5 Au/Ni, 950 °C). Gold solders are little sensistive to variation in gap width, and can easily be worked into strip, wire, foil, etc. They contain up to 94 % gold and for the remainder practically always copper, often silver, nickel, palladium, chromium, cadmium and zinc[28], and sometimes Ge, Sn, Sb, Si, In and Te. Cadmium and zinc are not used in vacuum, owing to their high vapour pressure. Copper can be brazed to fernico with 42Au/3Ag/55Cu (960 °C). For many solder compositions used in vacuum engineering, see [32]. For the Au-Pb-Sn system see [43].

(k) *Solders on a nickel basis*. Pure Ni (1 453 °C) is used for brazing heat-resistant Ni-, Co- and Fe-base alloys, Mo and W, Cr and ceramics. Apart from that, Ni is used as Ni-Cr (hard, heat resistant), and in the group of Nicrobrazes, Ni-B*, Ni-Mo, Ni-Mn (lower melting point) and Ni-P.

Nickel solder is highly corrosion-resistant, hard but not tough, and very strong even at high temperatures. See also [29,30,32,33,34] and Table 3.6. The brazing atmosphere is pure dry hydrogen, inert gas, vacuum and sometimes dissociated NH_3 (dew point −60 °C). Furnace brazing is the most-applied technique.

(m) *Solders containing palladium* have a very good wetting capacity[50], are strong and very tough, corrosion resistant, heat resistant (up to 800 °C), reduce any crystalline attack, and cause hardly any erosion of the base metal. They can therefore be used for extremely thin sheet metal. Pd is often combined with Ag and Cu, Mn, Ni, Au and Co.

Applications: for joining Cu, Ni, Co, Be, Au, Mo, W, Ti, Zr, niobium, Kovar, Nimonic, stainless steel, high-strength steels, unequal metals, metal to ceramics, very thin sheet metal, products for very high operating temperatures and vacuum (in this case 48Ni/31Mn/21Pd or 60Pd/40Ni will be satisfactory; for joining graphite to Mo and W, 60Pd/35Ni/5Cr is suitable).

* In joining corrosion-resistant steels, the boron from nickel-base filler metals forms Cr-B compounds at the expense of the base metal, thereby making it susceptible to intercrystalline corrosion in wet environment. Homogenizing at 1 060°C for an hour largely remedies the evil.

TABLE 3.6
Some nickel solders

Composition by weight (%)									Melting traject (°C)	Notes; application
Ni	Cr	Fe	Mn	Cu	B	C	Si	Other elements		
27–29	—	bal.	1	—	—	0·05	0·2	22–22Co	—	metal, glass, metal-ceramics
bal.	13–16	1	1	—	—	0·2	1	1Ti, 4·5Al 5Mo 18–22Co	—	gas turbine vanes
>70	14–17	9–11	2	0·5	—	0·15	0·5	—	—	insensible to stress corrosion
>52	14–16	4–7	1	0·2	—	0·1	1	14–18Mo 5·5W	—	highly corrosion-resistant
62·5	11·5	3·75	—	—	2·5	0·55	3·25	16W	975–1110	very tough up to 380°C, oxidation-resistant up to 1200°C; mechanically and thermally highly stressed parts containing Co, W and Mo
72·5	10	2·5	—	—	2	<0·45	2·5	—	970–1160	very tough up to 380°C, oxidation-resistant up to 960°C, less hard, better machinability, for wide gaps
81	15	—	—	—	3·5	<0·15	—	—	1000–1000	tough up to 390°C, good spreading, little erosion
75	—	—	17	—	—	<0·15	8	—	1000–1025	very tough up to 400°C, oxidation-resistant up to 950°C, very little erosion
77	13	—	—	—	—	<0·15	—	10P	890–890	oxidation-resistant up to 860°C, very little erosion of Fe and Ni alloys
94·6	—	—	—	—	1·9	—	3·5	—	—	relatively soft, corrosion-resistant and machinable; stainless steel, Mo and W
92·6	—	—	—	—	2·9	<0·06	4·5	—	985–1040	low viscosity, oxidation-resistant up to 980°C; stainless steel, Réné 41, Inconel X, 17–7 PH, etc.

TABLE 3.6—*continued*

Composition by weight (%)									Melting traject (°C)	Notes; application
Ni	Cr	Fe	Mn	Cu	B	C	Si	Other elements		
83	6·5	2·5	—	—	3	<0·15	4·1	—	970–1000	low viscosity at liquidus temperature. As standard Nicrobraz; long seams etc.
72·9	16	3·5	—	—	3·5	—	4·1	—	—	high viscosity, warm-hard, very corrosion-resistant, non-machinable; stainless steel
78	11·5	3·5	—	—	3	<0·15	3·5	—	970–1090	oxidation-resistant up to 1000°C; wide capillaries
93·7	2·2	—	—	—	1·5	—	2·6	—	—	wide capillaries, heat-resistant steels
88·8	5·6	—	—	—	2·4	—	3·2	—	—	short brazing time!; wide capillaries
76·8	15·2	—	—	—	—	—	8	—	—	very little erosion; stainless steel
73·6	17·1	—	—	—	0·1	—	9·2	—	—	very little erosion
61	19	—	10	—	—	—	10	—	—	very little erosion; stainless steel in H_2
25	—	—	—	75	—	—	—	—	1200	Mo and W, Fe-kovar
21	21	—	—	—	0·8	0·4	8	45Co, 3–5W	—	cobalt alloys
73	13·3	—	—	—	0·23	—	7·6	—	—	wide capillaries of heat-resistant steels
73	13·5	4·5	—	—	3·5	0·8	4·5	—	980–1030	very tough, oxidation-resistant up to 1100°C, standard Nicrobraz; high thermal and mechanical loading
89	—	—	—	—	—	<0·15	—	11P	880–880	little tough, very low viscosity; oxidation resistant up to 770°C; extremely little erosion of Fe- and Ni alloys
71	19	—	—	—	—	<0·15	10	—	1080–1150	very tough up to 400°C, oxidation-resistant up to 1100°C, very little erosion; high thermal and mechanical loading
53·5	—	—	—	—	—	—	—	46·5Mo	1320	Fe-Fe

The width of the soldering gap is not very critical and can be as much as 0·5 mm. Any technique can be adopted, but induction and furnace brazing are mostly used. See [27,35,36,40] and Table 3.7.

(n) *Solders used without flux* must have a greater affinity for oxygen than the metals to be joined. They contain elements that reduce and dissolve the oxides of the base metal, or combine with them to form a slag floating on the liquid solder and reducing its surface tension. The use of an inert gas improves this effect.

Lithium is the best deoxidizing agent; unfortunately, the melting point of its oxide is as high as 1 430 °C. Moist atmosphere here gives a better result than a dry one. Boron strengthens the effect of lithium, but increases its viscosity, an effect that is counteracted by adding silicon, in itself a deoxidizing agent.

12% Si added to aluminium as a filler metal is self-fluxing in vacuum (10^{-6} Torr); brazing temperature 590 °C.

In an oxidizing atmosphere, Cu-P and Cu-Ag-P are suitable for copper and copper-rich components; the joint is strong but not very ductile. In special atmospheres, mild steel can be furnace-brazed with pure copper without flux. Most solders are on a basis of Ag-Cu(-Zn). Heating must take place rapidly (to prevent dissociation of the solder), preferably in hydrogen, argon or vacuum. When possible, Zn (and Cd) are completely omitted in filler metal as they vaporize quickly, thereby breaking the protective layer of oxides floating on the liquid solder.

Techniques applied are: furnace, induction and resistance soldering, as described in [55].

(p) *Solders for use in vacuum*

The requirement these solders must meet is that the vapour pressure of their constituents must be extremely low at the operating temperature of the product (and at the soldering temperature, so that the change in composition remains within acceptable limits).

For this reason zinc, cadmium and lead are out of the question, as are materials such as molybdenum, carbon and phosphorus, which give volatile reaction products when reacting with oxygen in the vacuum. It goes without saying that these solders contain no gas and, especially, no oxygen. For the remainder, the vapour pressure of the elements in alloys, etc. lies below that of the element proper, all the more so as the affinity between these elements is greater.

The remaining suitable solder metals are then Ag, Au, Cu, In, Ni, Pd, Pt, Rh and Sn (see also Section 3.4 (g), (h), (j) and (m)). Table 3.8 gives the vapour pressure of several elements at different temperatures.

Table 3.9 gives a few solders for use in vacuum, arranged in order of liquidizing temperature.

Mn and Si improve the wetting capacity (protect Mn against oxidation), but Si forms brittle intermetallic compounds with Fe, Ni, Mo and W. A coating of galvanically-applied and fused nickel often improves solderability

TABLE 3.7
Some palladium solders

Composition by weight (%)						Melting traject (°C)	$\sigma_B \times 10^5$ (N/m²)	Elongation (%)	Application
Pd	Ag	Cu	Mn	Ni	Other elements				
5	68·4	26·6	—	—	—	810–805	45·7	23	silver coated ceramics
5	95	—	—	—	—	1010–970	22	28	
8	70	—	—	22	—	1037–1005			
10	58·5	31·5	—	—	—	850–825	48·8	22	
10	90	—	—	—	—	1055–1000			
13	—	—	—	—	87 Au	1306–1260			vacuum solder
15	65	20	—	—	—	900–850	47·2	20	
18	—	—	—	40·6	32Cr, 4·8Si, 2·8Ti, 1·8Al				
18	—	82	—	—	—	1090–1080	34·6	30	
20	52	28	—	—	—	898–879	50·4	25	
20	—	55	10	15	—	1105–1060			
20	80	—	—	—	—	1175–1070			
20	75	—	5	—	—	1120–1000	28·3	11	cobalt alloys, as Nimonic
21	—	—	31	48	—	1120–1120	83·5	7	high-alloy steels and Ni-Cr super-alloys
24	—	—	—	39	33Cr, 4Si				
25	54	21	—	—	—	950–901	52	15	stainless steels
25	—	—	—	—	75 Au	1400–1364			
27	—	—	—	22	41Au, 10Cr	1206–1180			
30	70	—	—	—	—	1234–1160			
33	64	—	3	—	—	1200–1180	50·4	25	
34	30	—	—	36	—	1169–1134			
35	—	50	—	15	—				
40	—	—	—	—	60 Au	1444–1415			
54	—	—	—	36	10 Cr	1260–1230			
59·7	—	—	—	40	0·2Li, 0·05B	1210–1210			
60	—	—	—	40	—	1237–1237	78·7	30	Hg-resistant steel

TABLE 3.8

Vapour pressure (mm Hg) of some vacuum solder metals

	at 20°C	at 500°C	at 750°C	at 1000°C	at 2000°C
Silver	$<10^{-8}$	10^{-7}	5×10^{-6}	10^{-1}	1000
Tin	$<10^{-8}$	$<10^{-8}$	7×10^{-7}	4×10^{-4}	80
Gold	$<10^{-8}$	10^{-7}		10^{-5}	40
Copper	$<10^{-8}$	10^{-7}		10^{-4}	15
Nickel	$<10^{-8}$	$<10^{-8}$		10^{-7}	3
Platinum	$<10^{-8}$	$<10^{-8}$		10^{-8}	$<10^{-2}$
Indium			10^{-4}		

TABLE 3.9

Some vacuum solders

Composition by weight (%)	Melting traject (°C)	Application	Notes
100In	156	metallized glass	
100Sn	232	Fe, Cu, Pe	
61·5Ag-24Cu-14·5In	705–625	Cu, Ni, Cu-Ni, radio valves	brittle; fluxless in H_2 or inert gass
60Ag-30Cu-10In	720–600	Cu, Ni, Cu-Ni, radio valves	
64Ag-26Cu-10In	723–698	Cu, radio valves	
63Ag-27Cu-10In	736–655	Cu, Ni, Cu-Ni	
68Ag-24Cu-8Sn	746–672	Cu alloys, stainless steel	
64Ag-26Cu-In, Mn, Ni	750 à 777–725	stainless steel, Mo, W	
72Ag-28Cu	779	Cu, also to Be (R.F.); Ag to Cu, Ag, Ni; CrNi, FeNi, brass: kovar; radio valves; Au to Au	eutectic, very popular
72Ag-20Cu-8Ti	780	metal to ceramics	
71·5Ag-28Cu-0·5Ni	795–780	radio valves	
5Pd-68Ag-27Cu	810–807	silver coated ceramics	Pallabraze 810
60Ag-40Cu	825–779	Cu	
95Au-5In	830–647	all metals	
65Ag-28Cu-21Ni-5Mn	850–750	kovar and invar to Cu, radio valves	
50Ag-50Cu	870–779	Cu, Ni	
15Pd-Ag-Cu	880–856		Pallabraze 880
85Cu-15Mn	880	Cr, Mo, Cu, fernico, Fe-Ni-Cr	in O_2-free atmosphere!
90Ag-10Sn	885–770	Cu, radio valves	
90Ag-10In	887–850	Cu, radio valves	

TABLE 3.9—*continued*

Composition by weight (%)	Melting traject (°C)	Application	Notes
80Au-20Cu	889–889	Fe, Cu, fernico, kovar, Ag, W	brittle
75Au-5Ag-20Cu	902–890	ferrous and non-ferrous metals	no Mo, no W
40Ag-60Cu	910	Cu	
58Au-2Ag-40Cu	935–910	ferrous and non-ferrous metals	no Mo, no W
70Au-14Ni-16Cu	935	Fe, Ni, Fe-Ni, Cu-Ni, Mo, fernico	
62·5Au-37·5Cu	940–930		
82Au-18Ni	950–950	C-steel, alloyed and stainless steel, Ni, Fe-Ni, Cu-Ni, fernico, kovar, Mo, W	highly corrosion-resistant, heat-resistant
42Au-3Ag-55Cu	960–940	universal, especially Cu-kovar	no Mo, no W
5Pt-95Ag	980–960	Mo, W	
7Ag-85Cu-8Sn	985–946	Cu, kovar	
75Au-25Ni	990–950		
37·5Au-62·5Cu	998–980	Cu, Fe, Ni, fernico, kovar	
43Ni-57Mn	about 1000	Ti to other metals	
5Ag-95Cu	1025–960	Cu to Cu	
97Cu-3Si	1025–970	Cu to Cu	
35Au-62Cu-3Ni	1030–1000	Cu, Fe, Ni, Mo, kovar	
100Au	1065	Mo; Pd to Pt	
28Ni-32Cu-23Mn-17Fe	1074	Fe, Ni, Fe-Ni-Cr, Mo, fernico	in O_2-free atmosphere!
100Cu	1083	Fe, Ni, Fe-Ni, fernico, monel to W	
62W-35Cu-3Ni	1084	Fe, Ni, kovar	
21Pd-48Ni-31Mn	1120		in O_2-free atmosphere!
25Pd-50Au-25Ni	1121–1102	stainless steel, Ni, Mo, W	heat-resistant
65Pd-35Co	1235–1230	stainless steel, Ni, Mo, W	
60Pd-40Ni	1237–1237		
25Ni-75Cu	1240–1180	Fe, Fe-Ni, Mo, W	
92Au-8Ni	1240–1200	stainless steel, Ni, Mo, W	
53Ni-47Mo	1320–1320	Fe, Mo	
100Pd	1550	Mo, W	
100Pt	1770	Mo, W	loadable up to 1500°C
100Rh	1970	Mo, W	very quickly, with little solder, in H_2 or A
00Nb	2500	Mo, W, W to Ta, Ta	braze in (high) vacuum
100Ta (foil)	2996	W	

because it alloys with the solder. In and Sn lower the melting point but also reduce the toughness of the solder. For vacuum solders, see [32], [46] and [49].

In general, vacuum-tightness increases with the strength of the joint if it is sufficiently tough to withstand the thermal stresses due to heating and cooling, when the component parts have unequal coefficients of linear expansion.

REFERENCES

[1] E. LÜDER *Handbuch der Löttechnik*, Berlin 1952.
[2] A. KEIL *Legierungen mit extrem niedrigen Schmelzpunkten*, Metall, No. 13/14, p. 515, 1954.
[3] R. C. JEWELL *Brazing and soldering*, Metal Industry, **89**, No. 5, p. 83, 1956.
[4] W. J. SMELLIE *Soldering and brazing*, Metal Industry, **79**, No. 4, p. 76, 1951.
[5] W. R. LEWIS *Notes on soldering*, London 1948.
[6] *Metal Industry*, Handbook and Directory, London, 1957.
[7] E. HERMANN *Das Hartlöten von Aluminium*, Aluminium, No. 4, p. 139, 1953.
[8] *Brazing Manual*, American Welding Society, New York, 1955.
[9] H. SPENGLER *Niedrig schmelzende Metalle und Legierungen*, Metall, p. 682, 1955.
[10] A. KEIL *Weichlote für Sonderzwecke*, Metallwissenschaft und Technik, p. 689, 1955.
[11] *Metals Handbook*, The American Society of Metals, Cleveland, Ohio.
[12] R. J. NEKERVIS *Tin and its alloys*, Industrial and Engineering Chemistry, p. 2253, 1953.
[13] *Reports of the conference on reliability of electrical connections*, April 15/16, 1954, RETMA Engineering Office, 1954.
[14] A. B. KAUFMAN *Selecting solders for low temperature service*, Materials in Design Engineering, p. 114, November 1958.
[15] *Reports of Symposium on Solder*, A.S.T.M.-S.T.P. No. 189, 1956, p. 129–158.
[16] H. H. MANKO *How to choose the right soft solder alloy*, Product Engineering, March 6, 1961.
[17] F. GORDON FOSTER *How to avoid embrittlement of gold plated solder joints*, Product Engineering, August 19, 1963.
[18] *Zeitgemässe Lötverbindungen*, Bulletin des Schweizerischen Elektrotechnischen Vereins, **54**, August 10, 1963.
[19] K. M. WEIGERT *Zur Entwicklungsgeschichte der amerikanischen Silberlote*, Metall, Heft 15/16, p. 721, 1956.
[20] K. M. WEIGERT *Ag-Cu-Zi brazing alloys*, Journal of Metals, No. 2, 1954.
[21] G. H. SISTARE, J. J. HALBIG, L. H. GREHELL *Silver-brazing alloys for corrosion-resistant joints in stainless steels*, Welding Journal, **33**, No. 2, p. 137, 1954.
[22] K. M. WEIGERT *Physical properties of commercial silver-copper-phosphorus brazing alloys*, Welding Journal, **53**, No. 7, p. 672, 1956.
[23] W. ESPE *Lote undLöten in der Hochvakuumtechnik*, Feinwerktechnik, **58**, H10, 1953.

[24] E. LÜDER — *Studien über die Anwendbarkeit der Kupfer-Phosphorlote*, Schweisstechnik, Heft 5, 1957.

[25] N. BREDZS, D. CANONICO — *Lithium additions to brazing alloys*, Welding Journal, **34**, No. 11, p. 535, 1955.

[26] N. F. LASHKO, S. V. LASHKO-AVAKYAN — *Brazing and soldering of metals*, London 1961.

[27] J. SAGOSCHEN — *Palladium in der Hartlöttechnik*, Metallwissenschaft und Technik, Jrg. **15**, Heft 9, September 1961, p. 870.

[28] D. C. HERRSCHAFT — *The evolution of ductile high-temperature brazing alloys*, Metal Progress, September 1961, p. 97.

[29] M. J. STERN — *Brazing of components for small gas turbine engines*, Metal Progress, September 1961, p. 101.

[30] F. M. MILLER — *Importance of purity in manufacturing brazing filler metals for high temperature service applications*, Welding Journal, August 1961, p. 821.

[31] *Soldering Manual*, American Welding Society.

[32] W. H. KOHL — *Soldering and Brazing*, Vacuum, Vol. **14**, p. 175–198.

[33] ANON. — Technische Rundschau, March 4, 1966.

[34] R. L. PEASLEE — *Selecting high-temperature brazing alloys*, Machine Design, **33**, September 14, 1961.

[35] A. S. CROSS, J. B. ADAMEC — *New era brazing turns to filler metals with palladium*, Welding Journal, August 1963.

[36] ANON. — Technische Rundschau, April 1, 1966.

[37] *Dip brazing aluminium assemblies*, Machine Design, August 18, 1966, p. 158.

[38] H. H. MANKO — *Solders and soldering*, McGraw-Hill Book Company.

[39] *Materials Selector Issue*, Mid-October 1966, p. 558–559, Materials in Design Engineering.

[40] *Brazing and brazing alloys*, Technology Utilization National Aeronautics and Space Administration, NASA SP-5026, Washington, 1966.

[41] R. M. MACINTOSH — *Tin in cold service*, Tin and Its Uses, No. 72, p. 7, 1966.

[42] V. L. GRISHIN, S. V. LASHKO — *Specific features in soldering copper with gallium-base solders*, Clearinghouse for Federal Scientific and Technical Information, Department of Commerce, USA, AD 625 145.

[43] A. PRINCE — *The Au-Pb-Sn ternary system*, Journal of the Less-common Metals, **12**, p. 107–116, 1967.

[44] M. H. SLOBODA — *The selection of brazing alloys*, Welding and Metal Fabrication, p. 386, October 1966.

[45] D. C. HERRSCHAFT — *Ternary systems for ductile brazing alloys*, Metal Progress Data Sheet, p. 96B, September 1961.

[46] K. L. GUSTAFSON — *Development and evaluation of braze alloys for vacuum furnace brazing*, NASA-Report CR-514, July 1966.

[47] *Löten*, Vorträge der Sondertagung Augsburg 1964 des Deutschen Verbandes für Schweisstechnik e.V., Arbeitsgruppe 26 "Löten".

[48] A. VAN'T HOEN — *Einfluß des Antimons auf die Eigenschaften des Lötzinnes und der Lötnähte von Weissblechdosen*, Mitt. Forschungsgesellschaft Blechverarbeitung, 1963, Nr. 4, p. 55.

[49] J. T. KLOMP *Solderen in de hoogvacuümtechniek*, Mededelingenblad van de Nederlandse Vacuümvereniging, Jaarg, **4**, No. 4/5, October 1966, p. 48.
[50] R. J. KLEIN-WASSINK *Wetting of solid-metal surfaces by molten metals*, Journal of the Institute of Metals, Vol. **95**, p. 38, February 1967.
[51] B. KEYSSELITZ *Der Einfluß von Verunreinigungen in Weichloten*, Metall, Jahrg. **21**, Heft 6, p. 593, June 1967.
[52] H. LITTNANSKI *Hartlöten mit Silberloten*, Mitt. der Forschungsgesellschaft Blechverarbeitung e.V., No. 4, p. 57, February 20, 1965.
[53] J. SPERGEL *Tin transformation of tinned copper wire*, ASTM Special Technical Publication, No. 319, Papers on Soldering, 1962.
[54] D. T. HAWKINS, R. HULTGREN *Vapor pressure of lead and activity measurements on liquid lead-tin alloys by the torsion effusion method*, Trans. Metallurg. Soc. of AIME, vol. 239, p. 1046, July 1967.
[55] A. J. GUBIN, E. N. DOBKINA *Self-fluxing fillers for the brazing of stainless steels and heat resisting alloys*, Welding Production, **13**, No. 8, p. 48, 1966.
[56] W. WUICH *Einwirkung schädigender Gase auf den menschlichen Organismus beim Schweiszen und Löten*, Draht-Welt, **53**, No. 4, p. 273, 1967.
[57] K. LÖHBERG, P. PRESCHE *Beitrag zur β-α-Umwandlung des Zinns*, Zeitschrift für Metallkunde, Bd. 59, Heft 1, 1968, p. 74.
[58] S. W. ZEHR, W. A. BACKOFEN *Superplasticity in lead-tin alloys*, ASM Transactions Quarterly, vol. 61, No. 2, June 1968, p. 300–313.
[59] J. COLBUS *Festigkeitsänderungen von Kupfer-Zink-haltigen Loten infolge des Lötvorganges*, Metall, Jahrg. 22, Heft 12, November 1968, p. 1090.
[60] H. R. THRESH, A. F. CRAWLEY *The viscosities of lead, tin and Pb-Sn alloys*, Metallurgica Transactions, vol. 1, June 1970, p. 1531.
[61] F. R. N. NABARRO, P. J. JACKSON *Growth and perfection of crystals*, Proceedings, 1958, p. 14, New York.
[62] W. C. ELLES *et al.* *Growth and perfection of crystals*, Proceedings, 1958, p. 102, New York.
[63] R. V. COLEMAN *The growth and properties of whiskers*, Met. Reviews, **9**, No. 35, 1964.
[64] S. M. ARNOLD *The growth of metal whiskers on electrical components*, Proc. Electronic Components Conference 1959, Bell Monograph 3304, 1957.
[65] S. C. BRITTON, M. CLARKE Publication No. 341 (1964), Tin Research Institute.
[66] S. M. ARNOLD *Repressing the growth of tin whiskers*, Plating, January 1966.
[67] M. ROSEN *Practical whisker growth control methods*, Plating, November 1968.
[68] W. P. MCQUILLAN *Guide to soldering*, Welding Engineer, April 1965, p. 64.

3.5 Base metals

3.5.1 Introduction

Most metals are used because of their technological, physical and chemical properties, and their price. The degree to which metals can be soldered usually plays a subordinate part in their choice, and generally the soldering process has to be adapted to the base metal. If there is a certain freedom, the base metal with the best solderability will be chosen. The solderability of a product is considered good when, within a permissible time and at a permissible temperature, with an acceptable flux and a prescribed solder, a joint is obtained that reasonably satisfies the requirements (as detailed in Section 3.6.1). Solderability, therefore, is not a term excluding all ambiguity. In fact, each material can be soldered with many fluxes and solders, and the sequential order of solderability varies for each flux and for each solder. Such sequential order should contain every combination between metal, flux and solder, and each material should then have to be investigated under equal conditions as regards:

(1) *the preparation*: degreased, etched, aged or with residual stress due to deformation (to which base metals seem to be more sensitive than precious metals).

(2) *the roughness*: very rough and very smooth surfaces are less suitable for soldering than surfaces with a roughness of about 60 Ru, which provides the correct capillarity.

(3) *the time* elapsed after preparation: temperature, humidity, oxygen, corrosive vapours, liquids or solid matter (for instance, sulphur in paper, water in wood) reduce solderability. Of course, base metals are more sensitive to these environmental influences than precious metals (aluminium loses its solderability in a fraction of a second, gold only after years).

(4) *the way* in which soldering takes place. Rapid cooling, for instance, prevents crystal growth, diffusion and brittle fracture, but is conductive to irregular cooling, which results in stress peaks. The way the soldering has been applied plays a role too.

(5) *the certainty* that only the intended material is tested and not at the same time the base material below a layer. Thin finishes should therefore be tested separately, taking into account the way they have been applied (electrolysis, dipping, chemically, etc.) and also the layer thickness. The purity, porosity and adherence (which is an indication for the activity of the base metal) play a role, too. Not to forget the composition, which should be such that a stable oxide does not form on the top layer.

(6) *the way of testing* solderability: the equipment, the time, the temperature, the properties to be determined (for instance the maximum flow or the flow rate), etc.

Setting up a sequential order of metals according to their solderability is thus practically impossible. As regards this solderability, only a few pointers concerning the removability of contamination, such as oxides, sulphides and carbonates, will be given in the following.

For a few metals, the sequential order, as regards the increasing *removability of their oxides*, is:

Be, Mg, Al, Zr, B, Ti, Si, V, Ta, Nb, Cr, Zn, Mn, Sn, W, Mo, Fe, Cd, Ni, Co, Cu, Pb, Pd, Ag and Au.

This sequence is roughly determined by the formation heat of the oxides; as some metals form different kinds of oxides with unequal formation heat, the sequential order is not a hard-and-fast rule. In practice, the physical character of the metal and the kind of medium affecting it also have a certain influence on the order.

This order cannot readily be maintained for alloys. Bronze is somewhat easier, brass somewhat more difficult to solder, than copper.

A more accurate sequence follows from the variation in free-energy during the formation of oxides.

Sometimes, it is possible to adapt the temperature in such a way that the influence of oxygen, sulphur, moisture, temperature, etc., is reduced.

More important than the sequential order of solderability is to improve and maintain it, by adequate pretreatment.

(a) *Preliminary treatment*

A good joint can only be made when the surfaces to be joined are mechanically pure, so all grease, dirt, paint, oxide, etc. is removed in advance.

Grease, dirt and paint are removed by rubbing and rinsing in ethyl alcohol, tri (degreasing in condensing vapour is better than dipping in liquid tri, because the component parts come into contact with non-contaminated tri), tetra, per, aircraft petrol, alkaline (salt) solutions, sometimes borax solutions and/or ultrasonic vibration in a bath. Ultrasonic cleaning is used especially to remove the last remnants and gives better results than chemical solutions. Oxides are removed mechanically and chemically (see also Volume 2, Chapter 10 and Volume 4, Chapter 5).

Mechanical treatment: chipping, scraping, flame, brushing, filing, scouring (for instance with steel wool), grinding, sand-papering, and shotblasting. When sand-papering and shotblasting with silicate sand, non-metallic particles are forced into the surface layer, from which they should be thoroughly removed. When possible these two techniques are not applied; when inevitable sand-papering must be performed wet, with light pressure, and the abrasive particles are to be removed by chemical etching[80], polishing, etc. As lead dust is poisonous, lead should be treated by techniques that produce chips, e.g. scraping. In general, the aim is to machine the material as much as possible, instead of forcing it aside.

The temperature must be kept so low that no oxidation occurs.

The roughness should not be less than 10–15 μm, for press fits 80–150 μm.

Very delicate component parts cannot be treated mechanically. Often too much material is removed, there is always the risk of embedding loose and even foreign particles, and the result is not always uniform. Mechanical methods are especially used for heavy parts and metals; for the lighter ones, use is made of:

Chemical treatment: the best treatment is *annealing* in a reducing atmosphere, if the products fit in the furnace and are not likely to distort or lose their hardness due to the heat.

A second method is to *pickle* in strong inorganic acids and salts. The material is then often so active that it must be immediately dried and processed or neutralized. Due to the constantly changing composition of the products, the composition of the bath changes too, which sometimes has an unfavourable effect on the solderability. The method may affect tolerances, depending on solution composition and time of immersion. Design should be adjusted for such material removal. Over-pickling will give rise to hydrogen absorption and pitting corrosion (inhibitors are used to prevent this). The method is not suitable for products with crevices, etc., which retain chemicals: this leads to corrosion.

Pickling is more suitable for mass production than mechanical methods and the surface has a better solderability; handling aggressive chemicals has little attraction.

Pickling agents are H_2SO_4 (for mild steel, copper alloys, nickel, aluminium and magnesium), HNO_3 (for stainless steel, nickel and magnesium), HCl or NaCl (for mild steel, lead, tin and zinc), HF (for magnesium, titanium and stainless steel), NaH, NaOH and Na_2CO_3, (for aluminium, titanium and tungsten) and Cr $(OH)_3$, H_2CrO_4, H_3PO_4 and CH_3COOH, all for magnesium.

The seldom-applied electrochemical pickling (product to anode) produces a very pure surface, which must be coated with tin immediately afterwards, preferably by dipping in liquid tin.

Pickling must be followed by thorough rinsing and drying.

It is also possible to pickle with gas: $HCl(+CO_2+N_2)$ at about 650 °C. Degreasing is not necessary; there is no hydrogen embrittlement and the working rate is high.

Electrochemical deoxidizing with simultaneous tinning of steel objects in an H_2SO_4 bath with tin anodes [2] is seldom applied. Metallizing is far more important.

(b) *Coating techniques*

Sometimes the coating is so thin that, locally, the solder is dissolved and the base metal comes to the surface. To prevent dewetting when the solder comes into contact with the base metal, the metal should be thoroughly cleaned before coating.

(i) *Dipping in tin-lead 60/40.* After tinning, this is almost certainly the method for ensuring the best solderability, even when the coat is less resistant against vapours from organic acids. The coat is a little cheaper and has a lower melting point than tin. The layer thickness varies (especially when wire is coated) so greatly that it is sometimes extremely difficult to form a thickness of at least 7 μm, and at the same time meet reasonable fitting tolerances. Brass products are sometimes first galvanically copper plated (3 μm) or nickel plated (3 μm) to prevent zinc diffusing towards the surface, which happens rapidly at dipping temperature. In that case, a solder thickness of 3 μm is sufficient. On solid nickel and copper, 3 μm solder 60/40 or pure tin is adequate.

(ii) *Thermal drum solder coating.* Solder powder, flux and sand (as a separating medium) act on the products (solder tags, etc.), in the first phase of the process below the melting point of the solder, and later on, when the product has become flocky with solder, 20 secs above the m.p. For every batch of products, fresh solder is added, to keep its purity at a constant level. Layer thickness and homogeneity are reasonable, holes in the products remain open, no sand inclusions occur, it is claimed, but great skill is essential.

(iii) *Centrifugally solder coating* (*solder slinging*): See Fig. 3.5. Cycle time is about 60 secs, a well-solderable solder layer of about 5 μm thickness is applied.

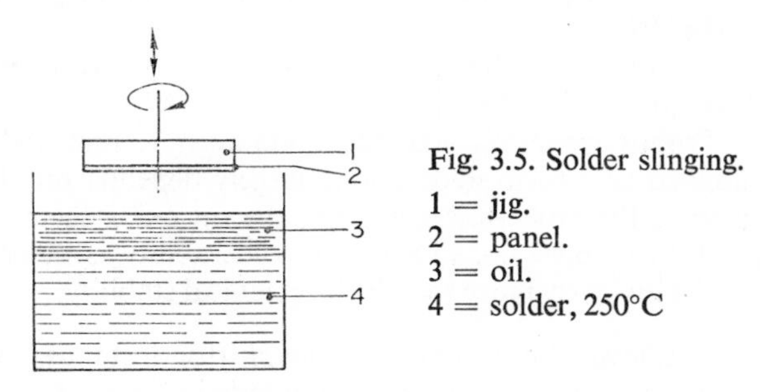

Fig. 3.5. Solder slinging.
1 = jig.
2 = panel.
3 = oil.
4 = solder, 250°C

(iv) *Spraying-on a solder coating* Only mentioned for the sake of completeness.

(v) *Roll solder coating* See Section 3.7.2.

(vi) *Electroplating* (the product must be well accessible to the electrolyte, but must not hold it in capillaries). The layer thickness is more uniform and the process is more difficult to control. The generation of gas bubbles often renders the surface somewhat porous, therefore bath salts are absorbed, so that corrosion results. Remelting the layer, the so-called flow-brightening, is a good remedy and prevents whisker formation. Adhesion is sometimes unsatisfactory, and sharp edges are insufficiently covered. Tin (8 μm) and tin-zinc are used on steel [44], 60/40 tin-lead (8 μm) is used for copper* and iron alloys, cadmium (at least 8 μm) on steel (cadmium is poisonous [70]). Silver (8 μm) migrates through printed wiring panels and even causes short-circuits; on a substrate of solid nickel or on a 1 μm thick nickel or on a 2 μm thick copper intermediate layer on brass and nickel silver, a coating of 0·5–1 μm silver will suffice. 1 μm Ag directly applied to copper and brass also provides good solderability. Ag and Au are used only when Sn, Pb, Cd, etc. do not meet the requirements. Galvanically applied pure gold (0·1–0·2 μm) is still porous, so the substrate material is important. To brass, nickel silver and nickel-iron, an intermediate layer of 1 μm thick Ni or 2 μm thick Cu is applied, which is—electrochemically—plated with 0·2 and 0·5 μm gold, respectively. Pure copper and nickel only receive 0·5 and 0·2 μm gold respectively. Gold layers thicker than 1 μm result in the formation of brittle gold-tin compounds (see Section 3.5.2(d)). Such layers of gold should be removed or nickel plated.

Gold alloys are only well solderable when they are freshly-applied; one favourable exception is Autronex CI.

* Copper diffuses towards the surface, where it oxidizes and reduces solderability. Since copper will not alloy with lead, a coat of solder is to be preferred over a coat of tin. An intermediate layer of 1 μm thick nickel plating or 2 μm thick copper plating on brass is excellent for maintaining the solderability of the 4–5 μm thick solder plating for a long time.

Tin-nickel 67/33 (5 μm) has excellent flow characteristics, is rather hard (700 Vickers), has a high melting-point, does not melt or dissolve during soldering, prevents the formation of whiskers and diffusion from the base metal, and is highly resistant to chemical atack, since it is protected by an oxide layer. This layer, which is unfavourable for soldering, is prevented by applying an 0·1 μm thick layer of tin or gold, or it is removed with activated resin [51].

For the solderability of electrochemically and otherwise applied protective coats, see [56] and [57].

From copper coatings, the oxide must be removed chemically, i.e. by the flux, so that their solderability largely depends on the type of flux. In the case of the much-less-noble tin, the tin oxide seems to float away over the underlying pure tin, so here the solderability is determined by the thickness of the tin layer, more than by its chemical stability.

(vii) *Chemically coating with metal* (for instance, nickel, gold and copper) by the reduction of metal compounds in water. For the application of a tin layer see [50]. For the remainder, the solderability of electroless nickel is no better than that of galvanically-applied nickel.

(viii) *Coating* (with nickel, gold, silver) of silicate-containing materials, by heating metal salts (burning-in).

(ix) *Rolling* metals of good solderability, such as copper, brass, silver, alloyed steel or nickel, on to practically any other metal.

(x) *Ultrasonic and mechanical tinning of aluminium* (see Section 3.7).

(xi) *Applying an etch coat*, functioning like a pickling inhibitor and, in a later stage, covering the product with a layer of resin.

(xii) *Coating with resin* deposited from alcohol or water displacer.

If the coating is not followed immediately by soldering, the application of a "water repellent", usually a resin solution, can be useful to conserve the product for longer periods, say several months. The water repellent has no impeding effect on soldering (Volume 6, Chapter 5).

(xiii) Packaging the products in plastic bags. These should not release gases or chemicals affecting the solderability. The ambient temperature should not fall below the dew-point of the atmosphere in the bag, so it is advisable to add some hygroscopic material.

3.5.2 Various base metals

(a) *Iron and its alloys*

The oxides can be removed by means of chloride fluxes (soft soldering), boric fluxes (brazing) or by mechanical treatment.

In principle, it is possible that the vapour pressure of the cadmium used to plate steel exceeds the permissible value of 0·1 mg/m^3. In practice, this risk only exists when larger areas are heated to a high temperature for a longer time (furnace soldering, for instance). In that case, a fan or exhaust system is required.

- Carbon steels and low alloy steels are soldered with tin-lead, either with the soldering bit or by dipping; hardened steel then becomes very brittle.
  Brazing is done with copper, copper-zinc, copper-zinc-silicon, copper-zinc-nickel (46/40/10) or silver alloys, e.g. 40Ag/19Cu/21Zn/20Cd but not with Cu-P, as this results in brittle Fe_2P, with borax, boric anhydride or fluorides. The following methods are used: flame, vacuum or reducing furnace, high frequency heating [16].
  Dissolving iron in copper solder facilitates the copper brazing of iron to tungsten, for tungsten will not alloy with copper.
- Galvanized steel is flame soldered with 60Sn/40Pb solder and a suitable flux. HCl in the flux dissolves the zinc layer. Brass solder gives a stronger joint than silver-brazing alloy.
- Corrosion-resistant chromium steels are susceptible to intercrystalline corrosion due to the soldering temperature, especially when they have been hardened or cold deformed [2]. A uniform and very short heating can prevent a great deal of damage, as can low temperatures; it is best, therefore, to use low melting solder, not high melting solder. A better solution is to temper the components completely before soldering, or choose a filler alloy with a melting point so high that the base metal is already wholly tempered during heating, so that the filler applied later on does not meet non-tempered susceptible material.

The relatively poor heat conduction of stainless steel easily causes local overheating.

Soft soldering with $HCl+ZnCl_2$, with a soldering bit or with H_3PO_4 by dipping; hard soldering with fluxes containing boron and metal chlorides, fluorine compounds, dry hydrogen (dew point −60°C) or vacuum. Chloride-containing flux residues can cause stress corrosion. H_3PO_4 is less suitable for flame heating, has a short active life, and strongly attacks many soft solders and copper bits.

As working temperatures increase, a change is made from silver base filler metal to copper, to palladium base, gold base and nickel base alloys. Silver in the solder causes intercrystalline attack and brittleness in stainless Cr-Ni steel under tensile load. Cd and Zn, Si and B from the solder have the same effect.

Both soft soldering and brazing are done by bit, by dipping, by induction and resistance heating and in a furnace.

Soft solder alloys use tin with up to 50% Pb, if need be after a nickel flash followed by tin plating. A good brazing alloy is 85Cu/10Ni/5Fe, but the filler metal should not contain S or P (see Table 3.5).

Soldering is facilitated by first applying a copper layer 15 to 20 μm thick or using a solder containing nickel, for instance 27Ag/40Cu/6Ni/10Mn/17Zn [3], or 10 to 15% Ag, about 42% Cu, about 43% Zn, 2 to 3% Ni; the melting point of these two solders lies above the dangerous range between 500 and 900 °C.

Ferritic and austenitic chromium steels must be brazed with a solder containing nickel (50Ag/15Cu/16Cd/16Zn/3Ni, for instance) if they are to be used in humid environments [4]. Another way to join ferritic chromium steel is by nickel plating followed by brazing with 65Ag/28·5Cu/6Sn/2·5Ni or 60/28/10/2, for instance.

- Heat-resistant steels are brazed with solder containing nickel and/or palladium and/or manganese (for instance, 75Ag/20Pd/5Mn, melting point 1 120°C). A reducing atmosphere is desirable, to remove the oxides. Previous nickel plating is recommended, if need be followed by tempering to prevent hydrogen brittleness.
- Kovar is joined to itself or to Cr-Ni steel with pure copper in a hydrogen-filled furnace at 1 100–1 150°C for 5–15 min. Kovar is brazed to iron with 55Cu/45Ni (1 300°C).
- High strength steels can be brazed with solders containing Ag, Cu, Pd, Mn, Ni, Cr, B, Si and Li [45].
- Tool steel is brazed with for example 75% iron manganese and 25% borax (with Cu-Ni-Fe-Mn-Si). Stress between tip and base metal is reduced by: low brazing temperature, tough solder, slow cooling and heating, intermediate layers, for instance permalloy, if necessary as a gauze (foil or sandwich brazing), and a thicker layer of solder.
- Hard metal tool tips, etc., are joined with a filler metal from Table 3.10. The flux practically always contains borax. Here again thin metal ply is sometimes used, i.e. a strip of Cu, Ni or Fe with a rolled-on coat of solder on both sides to limit a shrinkage stress caused by different coefficients of linear expansion (resulting in minute cracks or even loosening of the tool tip). The contact area between tip and shank should be kept to a minimum and is best restricted to one plane. Here, too, slow cooling is advised, when possible in electro-carbon powder.
  Soldering methods are: salt bath, flame, electric resistance, induction and furnace with a reducing gas, the latter especially when the brazing temperature equals the one where the carbide starts being oxidized (about 800°C). See [71] and [72].
- Sometimes cast iron is difficult due to its graphite content and Si-containing surface layer. These components are removed by sand blasting, an oxygen flame or borax. The solder is Sn-rich Sn-Pb, Cu-Zn or Ag-Cu-Zn, if needs be with an addition of Ni.
- Fernico is brazed with 85Ag/14Cu/1P, 42Au/55Ag or 7Ag/85Cu/8Sn.
- For vacuum applications, use is made of solder on a copper, silver or gold base, [3] and [77].

(b) *Nickel and its alloys*

Are soldered with a flux on an alkali or fluoride basis, if they contain chromium or aluminium, otherwise with HCl, NH_4Cl or $ZnCl_2$. The solder used is copper (no Cu/P for high-nickel alloys; as copper reacts very rapidly with nickel, brazing must be done quickly), silver base, copper-zinc (for instance 61Ag/28·5Cu/10·5Zn, melting traject 735–690 °C and 60Ag/15Cu/25Zn, melting traject 667–674 °C), nickel and tin-lead (63Sn/37Pb on a galvanically- or thermally-applied 4 μm thick Sn/Pb layer, the latter on a 5 μm thick nickel plating).

TABLE 3.10

Some brazing alloys for hard metal

Composition by weight (%)						Brazing temp. (°C)	Working temp. (°C)	Notes; application
Cu	Zn	Ag	Ni	Mn	Other elements			
77–88	—	—	10–20	2–8	—	1200–1150	900	σ_B at 600°C abt. 7·5 kg/mm²; although tough, not for brittle, hard metal
96·86	0·04	—	2·5	—	0·6Si	1100	900	σ_B at 600°C abt. 7·5 kg/mm²; as above
99·9	—	—	—	—	—	1090	450	σ_B about 21 kg/mm²; not for brittle, hard metal
86	—	—	—	10	4Co	1020	900	σ_B at 600°C abt. 8 kg/mm²; not for brittle, hard metal
80	—	—	—	—	20Sn	1000–950		
38	20	27	5·5	9·5	—	840	350	σ_B 15–30 kg/mm²; not for brittle, hard metal
16	23	49	4·5	7·5	—	690	—	σ_B 25–30 kg/mm²; for any hard metal
62	38	—	—	—	—	—	150	
88	—	—	12	—	—	—	—	
28	—	72	—	—	—	—	—	σ_B 15–30 kg/mm²; steel and hard metal previously copper or silver plated
27·5	20·5	49	0·5	2·5	—	—	—	σ_B 15–30 kg/mm²
58	40	—	—	—	2Pb	900	—	
—	—	—	—	—	50Sn/50Pb	250	—	reamers etc.

Inconel and nimonic to be used at high temperature are commonly furnace-brazed in H_2 or cracked NH_3, with Pd/Ag/Mn, Pd/Ag/Al, Cr/Ni/B/Si, Cr/Ni/Mn, etc. Nichrome is sometimes copper-plated before soldering.

Nichrome and monel must be annealed before soldering, to avoid stress corrosion when using Ag, Cd or Zn. In brazing, even the jigs must not introduce stress into the component parts, lest they form cracks close by the joint. Highly oxidation-sensitive metals are first nickel-plated (5 μm) and then brazed in a high vacuum (10^{-5} mm Hg) or in very dry hydrogen. Manganese reduces the melting point of Ni solder; Ni-Mn solder is, moreover, tougher and shrinks less when solidifying than Ni-Cr solder.

Chemically-applied Ni-P layers [9] ensure a firm joint when heated up to about 880 °C.

Methods used are: flame, resistance, induction, salt bath and dip soldering.

(c) *Copper and its alloys*

Are soldered in all ways [12, 13] and [14], but the zinc should be prevented from evaporating out of the brass (pre-copper plating). See also [43].

After it has been stored for some time, tinned brass is difficult to solder, due to the corrosion of zinc diffused towards the surface. A layer of galvanically applied copper or nickel about 2 μm thick, between the brass and tin, solves this problem. Free-cutting brass contains lead, which greatly reduces adherence in brazing. Limited brazing time and temperature reduce the dissolving of the lead in the solder.

In bronze, the elements tin and copper are already fully reacted and redissolve very slowly in molten tin. As copper coated with a *thin* layer of tin only has a bronze coating, the solderability is equally bad. The remedy here is to provide a thick layer of tin or tin-lead, as lead retards the reaction between tin and copper.

Copper oxides are removed mechanically or by means of $ZnCl_2$ or NH_4Cl; the latter is preferable, especially for low melting solders.

In a hydrogen-containing atmosphere (flame soldering, for instance), copper containing oxygen may give rise to the formation of water vapour and other gases, rendering the surface layer brittle (hydrogen brittleness). A reducing atmosphere consequently requires copper free from oxygen ($<0{\cdot}02\%$ oxygen).

For electrical parts, the use of non-acid fluxes is recommended. Borax and boric anhydride are used for the brazing of brass.

The solder used for copper is, for instance, copper-zinc, silver base (e.g 50Ag/50Cu or 72Ag/28Cu), copper-phosphorus (e.g. 91·6/8·4 or 80Cu/15Ag/5P), tin-lead, 97Cu/3Si, 70Cu/30Au, 42Au/55Cu/3Ag, 75Au/20Cu/5Ag, 81·5Au/18·5Cu, zinc or low melting solder with Sn, Pb, Cd, Bi, In, Se. The four types of solder first mentioned are also used for brass.

Tin-lead solder gives a mechanically-weak joint, which becomes weaker still as the temperature increases, and also shows creep at room temperature.

As tin dissolves copper, low-tin solders are used in soldering very thin copper component parts.

Silver-cadmium solder is much stronger, at higher temperatures too; it is heated by means of electric resistance, induction and dip soldering. Cold deformed brass and nickel-copper are susceptible to stress corrosion when using solders [15], so annealing is carried out, if need be. Besides, $(\alpha+\beta)$ brass, as 60 Cu/40Zn, is susceptible to intercrystalline attack when brazed with P-containing filler metal.

Copper tubing for sanitary hot water equipment is soldered with 95Sn/5Ag; in the case of cold water, 95Sn/5Sb or 50Sn/50Pb is applied.

Lead oxidizes rapidly, thus reducing the flow.

Components containing aluminium and silicon are fluxed with HF, HCl, and HNO_3 before soldering, and HCl plus $ZnCl_2$, or H_3PO_4, during soldering. Borax, etc. and fluorine compounds are also used. Aluminium diffuses into silver solder and is oxidized, resulting in a weak joint: previous copper coating is a satisfactory remedy. Soldering time should be kept to a minimum.

Aluminium bronze is susceptible to stress corrosion in contact with low melting solder. Brazing is done with Easy-flow, for instance; for joining to steel, type no. 3 is used.

Cu-Ni-Zn alloys (nickel silver) is soldered as brass. Its colour matches that of 42Cu/58Zn and 30Ag/28Cu/21Cd/21Zn and 40Ag/19Cu/20Cd/21Zn.

Cu-Ni alloys are processed as pure copper.

In copper manganese compounds, the manganese is inclined to evaporate.

Due to the soldering temperature, beryllium bronze, etc. can lose its hardness, so a solder with a safe, low soldering temperature should be selected*. Beryllium oxide must be removed with $ZnCl_2$ or with fluorine containing fluxes [16] or with activated resin [13]. In cases where these fluxes cannot be removed, the component parts are previously pre-soldered galvanically, by hand (both with chloride flux) or thermally, and afterwards soft soldered by bit or flame, with resin and 50Sn/50Pb, 95Sn/5Sb or 97·5Pb/1·5Ag/1Sn. Brazing can be done without previous tinning: with Easy-flow 45, m.p. 620 °C, or with 72Ag/28Cu, m.p. 778 °C, in a furnace, to give homogeneous heating and hence structure of the base metals, even if they differ in thickness.

When soldering solderable insulated copper wire (Posijn wire) isocyanates are produced, which have an irritating effect on the lungs. Soldering should therefore take place under an exhaust system.

Copper is brazed to Fe-Ni with 81.5Au/18.50Cu, 50Ag/50Cu or 72Ag/28 Cu.

Copper is brazed to kovar with 80Cu/20Au or, when kovar is pre-copper plated, with 72Ag/28Cu or 80 Cu/15Ag/5P; soft soldering is accomplished with Sn/Pb eutectic.

(d) *Precious metals and their alloys*

Silver and its alloys are soldered with, for instance, Ag-Cu-Zn or with Sn-Pb solder [3], the flux being borax compounds and $ZnCl_2$ respectively.

* Soldering at the solution treatment temperature, followed by normal ageing, is not attractive, as filler and base metal will alloy too much during the time needed for a solution treatment (> 30 min.). When soldering at a low temperature it is advisable, in view of the remaining strength of the base metal, to braze in the solution treated condition, followed by standard ageing.

Gold and its alloys are brazed with Au-Cu-Ag-Cd(-Zn) solder, but it should be borne in mind that, in some cases, there must be no colour difference between soldered seam and the base metal. The flux is usually borax.

Soft soldering gold plated or gold components (with 60Sn/40Pb for instance) very rapidly results in a hard, brittle, lead-free gold-tin alloy with a higher melting point. This alloy seems to envelop the gold in layers: Au, AuSn, $AuSn_2$, $AuSn_4$. This skin is not wetted and leads to porosity, pits, roughness and embrittlement, which increase with the gold content. When all the gold has been converted into gold-tin compounds, which are then removed, the now bare metal will be excellently wetted by the tin, provided it contains no gold; [47] and [48]. Thin films of gold, short soldering times, dipping baths of large content, low temperatures, wide capillaries and a large amount of lead in the solder, slow down the undesirable reaction. Cadmium in the solder is claimed to suppress enbrittlement; 35Sn/29Pb/17In/0·5Zn appears to be a satisfying filler metal. However, see [64], and especially [66].

A thin gold layer is porous and suppresses oxidation insufficiently; a thick and dense layer leads to brittleness, so gold is always an expensive means of obtaining small results. In general, 2·5% Au in the joint is still permissible.

Platinum is brazed with gold or gold chloride yielding gold. The flux, if any, is borax.

(e) *Molybdenum and tungsten*

Molybdenum can be soft soldered without difficulty, with tin or tin-lead when it is copper or nickel plated, or provided with a layer of Woods metal or 40Sn/60Pb applied with a grindstone (see Section 3.7.2(q)).

Molybdenum oxidizes strongly above 500 °C, resulting in brittleness; hence, in brazing, the heating rate is high (up to 1 150 °C) and the flux is a neutral atmosphere or hydrogen [18,19]. When the metal is well-cleaned, a flux is often superfluous. Techniques used are: furnace, high frequency and resistance soldering, if needs be, flame. A new technique is described in [74].

Filler metal is copper, the wetting power of which is improved by Ni, P (1% substantially increases wetting and strength), B or Fe [3,16,20]. Also used are Cr, Co, Ge, Ta, Ti, Mo, Ru, Ag, Pd, Pt, Si and Zr [21,77]; filler alloys are, for instance, Cu/4–5Ge/2–3Si for brazing to Fe, Ni and Cu; 50Ag/15·5Cu/15·5Zn/16Cd/3Ni, 80Au/20Cu, 82Au/18Ni, Cu/10–30Ni. For high working temperature are supplied: 60Pd/40Ag (m.p. 1 332 °C), 60Pd/40Ni (m.p. 1 237 °C), 54Pd/36Ni/10Cr (m.p. 1 260 °C) and 60Pd/40Cu (m.p. 1 196 °C). 20Pd/75Ag/5Mn is quite satisfactory in hydrogen, see [37], giving many alloys. See also [46]. For very high working temperatures, Mo (and W) is brazed with pure Pt (m.p. 1 769 °C).

Brazing to graphite is done with 72Au/22Ni/6Cr in inert gas, hydrogen or vacuum.

Tungsten is better wetted by silver- or nickel-base solder when pre-nickel coated; nickel is the base of many filler metals for molybdenum and tungsten. Tungsten also can be joined with Pt/2B [52] or 90W/6Ni/4Fe. Other alloys for joining W to W are: Zr (>1 863 °C), 55Cu/45Ni (1 300 °C), 75Cu/25Ni

(1 205 °C), 73Ag/27Pt (>1 185 °C) and 82Au/18Ni (950 °C). For brazing W to itself, to M and their alloys, [53] cites 40–87Mo/5–40Ru/8–55Re and 42–95Mo/5–44Rh/0–45Re. It also informs on filler metals silver, gold and platinum.

Tungsten is brazed to steel with copper (capable of dissolving some iron) with or without some iron.

Tungsten is joined to copper by nickel or by copper plating the former and brazing with Ag, or brazing, without pre-coating, with 80Au/20Cu, 72Ag/28Cu or 78·5Cd/16·5Zn/5Ag.

Oxides are removed by metal fluorides, NH_4Cl, $ZnCl_2$ and MoCl, for instance. The atmosphere is reducing or inert.

Apart from flame brazing, resistance brazing is carried out with Ta foil, for example.

Soft soldering with tin or tin-lead is done after copper [45] or nickel plating, as Mo and W practically do not dissolve in liquid tin.

(f) *Titanium, zirconium and tantalum*

They oxidize quite easily and the oxides are very difficult to remove. Moreover, the formation of nitrides and hydrogen absorption occurs when exposed to the open air, so that brazing and cooling must take place in vacuum or in an argon or helium atmosphere. Heating should take place rapidly; furthermore, pre-tinning or special fluxes must be used, such as chlorides of lithium, sodium, potassium, magnesium and silver and fluorides of lithium and potassium [22]. Inert gas is better than any flux [3]. For brazing alloys, see [46] and [67].

Silver solder (pure, with 15% manganese or with 7% Cu+0·25% Li) is the most suitable [23].

The tendency of titanium to form brittle compounds with many materials, calls for very short soldering times, and hence induction or resistance heating. Eutectic titanium-nickel solders give a strong but porous joint. 95Ag/5Al is claimed to cause the least corrosion to titanium.

Titanium can be soft soldered (for instance with 90Sn/10Zn or 95Sn/5Ag) after pre-tinning (for instance, 15 min dipping) or after having been coated with Ag, Ni/Co, Cu, In, Sn, Pb, Ni, Ni-P, Ag, Al, Mo, Au or Zr to prevent brittle compounds with the solder.

Zirconium can be brazed fluxless by an arc torch with pure silver in an argon atmosphere. For the soldering of zirconium see [46].

Tantalum is coated with Ni, Cu or Pt. See also [45] and [77].

(g) *Aluminium and its alloys*

Are difficult to solder owing to their stable oxides, their loss of strength when they have been hardened, their heat conduction, requiring an enormous heat supply, their sometimes low corrosion resistance, and the low melting points of many eutectic alloys. For the remainder, it is possible to solder aluminium, even aluminium castings (except die castings, that blister when subjected to brazing heat), provided the melting temperature lies above that of the solder. Therefore, there are very active fluxes on a chloride and fluoride basis of Li, Na, K, Ca, Zn, and Sn (working temperature 250–350 °C). An organic flux (8·6% ammonium-fluoborate, 5% cadmium-fluoborate and 86·4% triethanolamine) deposits cadmium and disintegrates above 275 °C, with a carbon deposit; hydrazine compounds when heated sufficiently disintegrate into N_2 and NH_3. All these fluxes are so corrosive that, after use, they must be removed with HNO_3, HF and Na_2CrO_3, which in turn have unfavourable effects on health and the base metal.

Preparation involves a few minutes immersion in an NaOH or silicate solution, a hot water bath, immersion in HNO_3 to neutralize any alkali remnants and a rinse in hot water. Another method: degreasing in tri and mechanical cleaning with steel wool or a steel wire brush or wheel, followed by swabbing with (or dipping in) a 5% by weight solution of Na_3PO_4 in water, till the surface becomes bright; washing; drying in warm air.

Magnesium in aluminium reduces solderability due to its increased oxide formation compared with aluminium. The material suffers from inter-granular penetration by the solder, particularly when the component part is prestressed by cold working. Stress relieving or soldering at 370 °C or more reduces the penetration danger. Zinc and copper in aluminium also cause the solderability to decrease. (When possible the amount of Zn is $<6\%$, Mn <3, Mg <2, Si <2, Cu $<0{\cdot}5$.) The melting points of aluminium alloys with a large content of copper, magnesium or zinc lie below the m.p. of the brazing solders commercially available.

Aluminium with more than 5% Si is so difficult to solder with flux that abrasion or ultrasonic soldering (see Section 3.7.2(q) and (r)) are preferred.

Solder consisting mainly of Sn, Pb and Cd, which dissolve badly in aluminium, gives low adherence. Ag-Cd and Ag-Cd-Zn solders are better. Adherence increases with diffusion, so with soldering temperature, hence zinc and silver improve tin solders. Bismuth is unsuitable.

Zinc-aluminium-copper solder gives a reasonable, maximum corrosion-resistant joint, especially if a small quantity of Mg or Si has been added[3]. Tin-aluminium solder is also used, in which zinc (10%) and copper (1%) improve the flow and corrosion resistance respectively. Solders for aluminium are, for instance, Al-Zn 60/40 (0·1% Be improves solderability), Al-Zn-Cu-Si (0·3% Mn improves corrosion resistance), Al-Cu-Si (for instance, the eutectic composition of 50/40/10) and Al-Si, preferably 88Al/12Si (eutectic, very low viscosity, good mechanical properties, scarcely any shrink cracks, reasonably corrosion-resistant, colour nearly matches that of the base metal. Suitable for pure Al, Al-Mn, Al-Cu-Mg, Al-Mg-Si, cast Al-Cu and cast Al-Si. The filler metal reacts very quickly with the base metal, so flows fairly badly, and thus requires large gaps. Al-Si has about the same solution potential as the parent alloys; Al-Si-Cu filler metal is even galvanically protected by the base metal).

This does not overcome the fact that there is not a single soft soldered joint in aluminium that is corrosion-resistant in humid environments and under load. So, the environment must be kept dry or the joint insulated by wrapping or paint or protected by plating (see also Section 3.6.3, first footnote). The chemical stability of the joint is about proportional to the melting point of the filler metal.

A good joint requires the gap (0·15–0·25 mm) to be kept heated for a few minutes. To obtain the correct temperature, it is often necessary for thick components to be preheated to about 400 °C, due to the great heat conduction of aluminium. For the sake of good solderability, aluminium is sometimes coated with silumin (7·5 % Si), with Zn-Al, with copper, or it is ultrasonically tinned, an operation that also improves the corrosion resistance. For the same reason, Al-Mg is sometimes coated with Al, particularly as Al-Mg is sensitive to stress corrosion.

A plating of tin, copper, zinc, silver or electrochemically- or chemically-applied nickel, or: activating the base metal, plus plating with cobalt and pre-tinning by dipping, makes it possible to solder aluminium to monel, copper, nickel, silver, and steel. In the manufacture of capacitors, soft solder is sometimes sprayed on; both adhesion and corrosion resistance are moderate. Soldering aluminium to another metal requires attention, because of the unequal coefficients of expansion and possible corrosion in a humid environment. The choice of flux is also more important. Steel and titanium can also be joined to aluminium by dipping them previously in aluminium (for ultrasonic soldering see Section 3.7.2 (r)), by silver plating the steel or by applying a zinc layer to it.

Fluxless joining of aluminium is generally accomplished by treating it with a steel wire-brush or with an ultrasonically vibrating soldering bit. Such tools break up the oxide skin of the base metal, which is covered with a layer of liquid solder. In friction welding, the component parts are heated and afterwards pre-soldered by rubbing them in firmly with a stick of solder with a large melting traject. For the fluxless joining of aluminium to corrosion-resistant steel, see [41]. Both components are pre-tinned, cleaned with tri, (ultrasonically or by friction), pre-tinned with 50Sn/50Pb or 1Pb/1Ag/1Fe/0·1Si/0·01Al/0·01Cu/rest Sn, hot compressed and slowly cooled.

As Al-$CuAl_2$ is very brittle, silver is best applied to the base metal copper in advance; the solder can be 75Ag/25Zn or 65Ag/20Cu/15Zn [16]. Another way is to make a transition joint, where an intermediate piece of a third metal is applied. When joining Al to Cu, this piece is from steel covered with Al at the Al side; the Cu side is silver brazed.

Solders with a low melting point and good corrosion resistance are Al-Si-Ge alloys [26], which unfortunately are expensive. Solders on a basis of aluminium are seldom used: zinc and solders on a basis of zinc and tin are usually preferred, e.g. 80Sn/20Zn (265–200 °C) to 60Sn/40Zn (340–200 °C). Sn/Pb solder causes serious corrosion. For solder alloys, see Table 3.4.

The overlap must be small (less than 4 mm for sheet, less than 12 mm for pipe), as otherwise the seam will contain flux. In big overlaps, flux inclusion is prevented by little holes in the lap. A short overlapping seam is already stronger than annealed aluminium base metal. Pipes are sometimes turned conically for better access of the solder; in principle, the now thicker seam is weaker. With flame soldering, the pipes fit accurately into each other; furnace soldering allows a certain amount of play. The third widely used technique in aluminium soldering is that of soldering in molten salt (chloride and fluorides) or solder. Also, induction heating and special soldering irons are applied.

Flux removal after brazing aluminium parts is carried out as follows:

Immerse the warm parts in boiling water or apply a pressure spray washer. Remove as much flux as possible by this method.
Immerse for 15 min in a nitric acid solution. The fumes so produced are dangerous both to health and to the products.
Or immerse for 15 min in a HNO_3-HF solution.
Or immerse in a HF-solution, no longer than 10 min, as Al is dissolved.
Or immerse for 5–10 min in a HNO_3-$Na_2Cr_2O_7$ solution at 65 °C. For thin-walled parts and maximum corrosion resistance.
Rinse in hot water.
Dry in warm air.

(h) *Magnesium and its alloys*

Are difficult to solder due to their great solidifying traject, their low solidification point and their stable oxide, which can only be removed by rubbing and fluxes. These fluxes are chlorides and fluorides of lithium, sodium, calcium and potassium, which must be thoroughly removed after soldering. Sodium phosphate, dichromate and nitric acid are also used. Sometimes, the specific mass of the flux exceeds that of the solder, which results in flux enclosure. Protective SO_2 gas (in a furnace) reduces the flow of the solder.

The solidifying point of many magnesium alloys lies below that of otherwise useful solders. Low melting solders often give brittle and weak joints and are therefore used only to correct surface imperfections in components subjected to light loads. (The electromotive force between magnesium and soft solder often leads to corrosion.) For example: 60Cd/30Zn/10Sn (lowest melting point 157 °C); 90Cd/10Zn (lowest melting point 264 °C); 40Zn/60Sn (lowest melting point 198 °C) [14]; 9Zn/91Sn (lowest melting point 198 °C); 72Sn/28Cd (lowest melting point 175 °C). Zinc-copper plated magnesium forms an excellent base for soft solder. Tin, cadmium and silver also provide a good base.

Brazing alloys contain magnesium, e.g. Mg/9Al/2Zn/0·1Mn (lowest melting point 383 °C); 83Mg/12Al/5Zn (melting traject 550–410 °C); 43Mg/55Zn/2Al (the latter for dip soldering only).

Heavy components must be preheated to 300–350 °C for flame soldering.

Dip soldering always requires preheating, to about 460 °C, also for the removal of liquid and solvent. Dip soldering is the most commonly employed technique, followed by flame and furnace soldering. An alloy like Mg/12Al/5Zn, for instance, is preheated to 480 °C and then immersed for some minutes in a salt bath of 600 °C under argon atmosphere.

(j) *Beryllium*

Is tinned or thermally gold-plated plus copper-plated in the presence of NH_4Cl and other halogenide salts of Na, K, Li, Be, Mg and Ba. Afterwards, it is brazed (in a vacuum, hydrogen or inert gas) with Al, Ag, Al-Si, Cu-Si, Cu-Ag, for instance Ag/7·5Cu+Ni or 72Ag/28Cu or 50Ag/50Cu [32], Cu-Ag-Li, 72Ag/28Al, 94Ti/6Be, to aluminium, monel, nickel and stainless steel.

Another method is described in [61]: zinc—copper—nickel plating the beryllium, nickel plating the corrosion-resistant steel and brazing with 45Ag/15Cu/16Zn/24Cd, for instance. [73] describes an almost similar method.

When joining Be to Cu, an intermediate piece of steel is used (Be and Fe have about the same coefficient of linear expansion). Firstly, Be is high-frequency brazed to Fe with pure Cu, after which steel and copper are joined with 72Ag/28Cu filler metal in a hydrogen-filled furnace.

Heating is carried out by means of induction, in a furnace, or by means of resistance; the atmosphere is neutral, inert or vacuum. For literature investigation, see [45].

Beryllium oxide is poisonous.

(k) *Zinc*

Is soldered with Cd-solder, or Sn ($>33\%$)-Pb solder. If the zinc contains aluminium, as is the case with castings, it is soldered with 82Cd/18Zn, or nickel or copper plated and afterwards soldered in the normal way, or ultrasonically pre-soldered, as Zn-Cd and Pb-Sn diffuse into zinc castings. For repair work, 60Sn/40Pb is used. Zinc oxide is removed with HCl and $ZnCl_2$.

3.5.3 Soldering metal to glass and ceramics

The poor heat conductivity of glass and ceramics results in long heating times and heat stresses; the latter are increased by the difference in the coefficient of linear expansion with the metal parts to be joined. A third drawback is that glass and ceramics are not wetted by the solder; this contracts into droplets.

Here, the bonding mechanism differs from that between metals: no alloying takes place, so adhesion and consequently the strength is low, as the latter is based on anchorage and the formation of an intermediate layer of oxides. It is therefore not necessary to remove oxide with a flux.

The intermediate layer can consist of glass or enamel sintered to the ceramic. This method employs borax plus all kinds of metal oxides, and can hardly be looked upon as soldering.

Far more important is the application of a metal film on the non-metallic product, so that conventional soldering may follow.

Glass is provided with a film of platinum, silver or chromium by means of heating an emulsion, reduction of oxides, vapour depositing, cathode sputtering [42], or chemical (electroless) depositing, if necessary followed by copper plating.

This is followed by ordinary brazing, with brazes containing chromium, nickel, cobalt and iron for joining glass with a great coefficient of linear expansion, or with low melting solder not dissolving the substrate. (The coefficients of linear expansion of glass and metal should preferably be matched. Where this is impossible, filler alloys of low strength are used.)

Ceramics

A metal film can be applied to ceramics by adopting one of the following techniques:

- The active metal process: a zircon, titanium or niobiumhydride slurry is applied to the ceramic (the metals used are described as active). By heating in vacuum at about 1 000 °C, a layer of pure metal is deposited on the substrate. This operation is followed by hard soldering in a reducing atmosphere.

  Another possibility is to deposit and attach a combination of slurry and solder in one heating cycle. A second variation is to place a foil of active material between the ceramic and metal component; when heated, the foil melts, reacts with the metal part, adheres to the ceramic part and forms a joint after solidification. See [3, 49] and [69]. In [60], it describes how a mixture of 75% copper dust and 25% titaniumhydride is sintered to the substrate, after which the resulting film is copper plated before hard soldering in hydrogen.
- Fused active spray transfer process [69]: in this technique, also applicable to glass and graphite, a slurry containing active metals is sprayed (usually through a mask) on the substrate, after which it is dried, fired at about 800 °C in a non-oxidizing atmosphere, cleaned and pickled in HF. Sometimes, the film will then be nickel plated as a preparation for soldering and brazing with, say, Sn/Pb, Ni/Au, pure silver, pure copper or 72Ag/28Cu in a non-oxidizing atmosphere.
- Fired-silver process [69]: a mixture of silver powder and powder of low-melting glass, applied between the components, is heated until the glass melts; after solidification the components are joined together. The presence of silver is not strictly necessary.
- Sintered metal powder process: a suspension of very fine platinum, silver, tungsten, molybdenum or their respective oxides, plus other ingredients such as Al_2O_3, titaniumhydride, silicon, manganese, etc., is applied to the ceramic and sintered at about 1 500 °C in a reducing atmosphere. This is followed by dip soldering preparatory to soft soldering; if silver was sintered on, there will be less risk of silver dissolving, if the solder used afterwards contains silver. If the product must be brazed, the sintering process is first followed by nickel coating, which can be done either by plating or by sintering. Now the product is ready for brazing. Brazes for aluminium oxide are, for instance, pure copper, 35Au/65Cu, 50Au/50Cu, 49Ti/49Cu/2Be (carbide forming) and 72Ag/28Cu [58]. By vapour depositing a 4 μm thick film of titanium or zirconium by means of an electron beam, aluminium oxide can be brazed with Al/5Si. For brazing directly to metal, eutectic Ag/Cu with a 5% core of titanium at 900 °C is used. This combination also wets graphite and diamond and is, furthermore, used to attach sapphires in pick-up heads. A commonly used solder is 2–9Ti/30–75Pd/rest Ni (U.S. patent 3 277 150).

  In a seemingly allied technique, a mixture of 75% Cu and 25% glass is sprayed on with the plasma torch; this layer is claimed to be suitable for soft and hard soldering to metals.

REFERENCES

[1a] H. RICHAUD — *Surface treatment of titanium*, Metal Industry, **89**, No. 4, p. 496, 1956.

[1b] G. W. SEVLEN — *Induktionslöten von Fahrradrahmen*, Zeitschrift des Vereins deutscher Ingenieure, **92**, p. 337, 1950.

[2] T. H. BOHN — *Silver brazing lap joints in stainless steel tubing*, Welding Journal, **95**, No. 9, p. 884, 1956.

[3] N. F. LASHKO, S. V. LASHKO-AVAKYAN — *Brazing and soldering of metals*, London 1961.

[4] G. H. SISTARE, J. J. HALBIG, L. N. GRENNEL — *Silver-brazing alloys for corrosion-resistant joints in stainless steels*, Welding Journal, **33**, No. 2, p. 137, 1954.

[5] E. LÜDER — *Handbuch der Löttechnik*, Berlin 1952.

[6] G. W. HINKLE — *How to solder stainless steel*, Metal Products Manufacturing, April 1957, p. 64.

[7] *Proceedings of the second RETMA conference on reliability of electrical connections*, September 11/12, 1956.

[8] *Working instructions for hot-tinning cast-iron*, Tin Research Institute, November 1958.

[9] A. W. GOLDENSTEIN, W. ROSTOKER, F. SCHOSSBERGER, C. GUTZEIT — *Structure of chemically deposited nickel*, Transactions of Electrochemical Society, **104**, No. 2, p. 104, 1957.

[10] H. A. SALLER, J. T. STACY, H. L. KLEBANOW — *Brazing nichrome V with GE-81 alloy*, U.S. Atomic Energy Commission BMI-947, August 27, 1954; BMI-933, August 2, 1954.

[11] *Joining nimonic*, Aircraft Production, **12**, 143, p. 265, 1950.

[12] I. T. HOOK — *The welding of copper and its alloys*, Welding Journal, **34**, No. 7, p. 321–337, 1955.

[13] R. M. MACINTOSH — *Technical aspects of soldering practice*, Welding Journal, **31**, No. 10, p. 881–897, 1952.

[14] H. R. CLAUSER — *How to select brazing and soldering materials*, Materials and Methods, **35**, No. 3, p. 105, 1952.

[15] *Welding Handbook*, 3rd edition, AWS, New York 1950.

[16] C. H. CHATFIELD — *Silver brazing of refractory metals*, Welding Journal, **33**, No. 9, p. 864–867, 1954.

[17] W. J. SMELLIE — *Soldering and brazing*, Metal Industry, **79**, No. 4, 1951.

[18] T. PERRY, H. S. SPACIL, J. WULFF — *Effect of oxygen on welding and brazing molybdenum*, Welding Journal, **33**, No. 9, p. 4425–4485, 1954.

[19] J. H. JOUSTON, H. UDIN, J. WULFF — *Joining of molybdenum*, Welding Journal, **33**, No. 9, p. 449–458, 1954.

[20] A. KEIL — *Ueber die Benetzungsfähigkeit von Löten*, Zeitschrift für Metallkunde, Heft 7, p. 493, 1956.

[21] H. R. CLAUSER — *How to select brazing and soldering materials*, Materials and Methods, **35**, No. 3, p. 105, 1952.

[22] W. R. LEWIS, P. S. RIEPPEL, C. P. VOLDRICH — *A preliminary report on the brazing of titanium, mild and stainless steels*, Sheet Metal Industries, **32**, No. 343, p. 833, 1955.

[23] N. A. TINER *Metallurgical aspects of silver brazing titanium*, Welding Journal, **34**, No. 9, p. 846, 1955.

[24] J. D. DOWD *Soldering aluminium*, Welding Journal, **33**, No. 3, p. 113, 1954.

[25] J. H. DUNN, E. P. WHITE *Aluminium advances in brazing castings*, SAE Journal, No. 9, 1952.

[26] P. T. STROUP (Alcoa) USA 2, 659, 138. Chemical Abstracts, **48**, No. 7, p. 3886, 1954.

[27] M. A. MILLER *Select the right method*, Materials and Methods, **38**, No. 3, p. 96, 1953.

[28] *Soldering aluminium*, The Aluminium Development Association, November 1957.

[29] *British tool solders aluminium without flux*, Modern Metals, September 1955, p. 106–107.

[30] W. J. SMELLIE *Soldering joints in aluminium: mechanism of corrosion*, Light Metals, July 1956, p. 210–214.

[31] J. C. BAILEY, J. A. HIRSCHFELD *Soldering aluminium*, Research, p. 320–326, 1954.

[32] D. W. WHITE, J. E. BURKE *The metal beryllium*, ASM, Cleveland, 1955.

[33] H. C. WHELPS *Soldering flaws in zinc die-castings*, Welding Engineer, **41**, No. 3, p. 18, 1956.

[34] W. R. LEWIS *Notes on soldering*, Tin Research Institute, 1948.

[35] G. W. ELDRIDGE *Aluminium soldering*, British Welding Journal, October 1965, p. 488–495.

[36] *How to solder aluminium*, Modern Metals, December 1962, January, February, March 1963.

[37] F. G. COX *Joining molybdenum*, Welding and Metal Fabrication, September 1961, p. 371.

[38] M. A. MILLER *Joining aluminium to other metals*, The Welding Journal, August 1953.

[39] H. BENDER *High-temperature metal-ceramic seals*, Ceramic Age, **63**, No. 4, p. 15–17, 20–31, 46–50, 1954.

[40] H. J. NOLTE, R. F. SPURK *Metal ceramic sealing with manganese*, Television Engineering, November 15–19, 1950.

[41] Beilage Mitteilungen der Deutschen Forschungsgesellschaft für Blechverarbeitung und Oberflächenbehandlung e.V., Band 17, No. 16/17, August 25, 1966.

[42] J. M. SEEMAN *Ion sputtered coatings provide multi-functional finishes*, Materials in Design Engineering, p. 102, November 1965.

[43] S. C. BRITTON, M. CLARKE *Effect of diffusion from brass substrates into electrodeposited tin coatings on corrosion resistance and whisker growth*, Tin Research Institute, Publication No. 341.

[44] *The trend to tin-zinc coatings*, Tin and Its Uses, No. 59, p. 12, 1963.

[45] *Brazing and brazing alloys*, Technology Utilization National Aeronautics and Space Administration, NASA SP-5026, Washington, 1966.

[46] R. E. CURRAN, L. I. MENDELSOHN *Brazing molybdenum and zircon sensor components*, Metal Progress, p. 94, December 1966.

[47] F. G. FOSTER *Embrittlement of solder by gold from plated surfaces*, ASTM Special Technical Publication, No. 319; Papers on soldering p. 13.

[48] J. D. Keller — *Printed wiring surface preparation methods—elimination of gold plating as a surface preparation for printed circuits and development of a contamination-free surface.* ASTM Special Technical Publication, No. 319; Papers on soldering, p. 3.

[49] J. Kramàř — *Joints metal-ceramique soudés avec le titane dans le vide poussé*, Le vide, No. 125, p. 402, September–October 1966.

[50] W. R. Lewis — *The action of fluxes that assist tinning and soldering*, Tin and Its Uses, No. 72, p. 3, 1966.

[51] J. W. Price — *Tin en tinlegeringslagen in de electronische industrie*, Tijdschrift voor Oppervlaktetechnieken van Metalen, November 1966.

[52] A. G. Metcalfe — *Method of brazing tungsten*, To U.S. secretary of the Navy, USA 3.276.113, October 4, 1966.

[53] — *A. E. C. Brazing alloys for tungsten and molybdenum*, Belgian Patent 665.860 (June 24, 1965) and Belgian Patent 669.448 (September 10, 1965).

[54] F. Gordon Foster — *How to avoid embrittlement of gold-plated solder joints*, Product Engineering, p. 59, August 19, 1963.

[55] U. Harmsen, C. L. Meyer — *Über Weichlötungen an Gold*, Zeitschrift für Metallkunde, Bd. **56**, Heft 4, p. 234, 1965.

[56] J. M. Thompson, L. K. Bjelland — *Evaluation of solderability of electroplated coatings*, Proceedings of the American Electroplaters' Society, 1961.

[57] C. J. Thwaites — *Solderability of coatings for printed circuits*, Transactions of the Institute of Metal Finishing, vol. **43**, p. 143, 1965.

[58] D. K. Davis, L. de Give — *High-reliability Ceramic-to-Metal Joints*, Machine Design, vol. **39**, p. 133, January 5, 1967.

[59] G. V. Browning, M. H. Baster — *Experimental evaluation of reliable soldering processes*, Proceedings 9th National Symposium on reliability and quality control, p. 211, January 22–24, 1963.

[60] — Copper, p. 32, May 1967.

[61] W. Kudenov — *Joining beryllium to stainless*, American Machinist, p. 118, September 12, 1966.

[62] — *Löten*, Vorträge der Sondertagung Augsburg 1964 des Deutschen Verbandes für Schweißtechnik e.V., Arbeitsgruppe 26 "Löten".

[63] M. F. Jordan, D. R. Milner — *The removal of oxide from aluminium by brazing fluxes*, Journal of the Institute of Metals, vol. **85**, p. 33.

[64] E. V. Walker, F. A. Waldie — *Soft-Soldering to Gold-Plated Surfaces*, Post Office Electrical Engineering Journal, vol. 58, No. 4, p. 268, 1965.

[65] — *Brazing alloys for tungsten, molybdenum and their alloys*, U.S. patent 3.292.255 and British patent 1,063,274.

[66] A. Prince — *Solderability of Gold Plating*, The General Electric Company Ltd., Hirst Research Centre, Wembley, Middlesex, January 27, 1963.

[67] K. Rüdinger, A. Ismer — *Löten von Titan und Titanlegierungen*, Schweißen und Schneiden, Jahrg. **19**, Heft 2, p. 71, February 1967.

[68] W. B. Harding, H. B. Pressley — *Soldering to gold plating*, Techn. Proceedings American Electroplaters' Soc., 50, p. 90, 1963.

[69] R. F. Karlak — *Metallizing Ceramics*, Machine Design, May 11, 1967, p. 160.

[70] W. WUICH — *Einwirkung schädigender Gase auf den menschlichen Organismus beim Schweissen und Löten*, Draht-Welt, vol. 53, No. 4, p. 273, 1967.

[71] G. WEIRICH — *Das Löten von Hartmetallwerkzeugen*, Löten, 41, No. 6, p. 18, 1964.

[72] J. VAN ROOY — *Construeren met hardmetaal*, Philips Technische Bibliotheek, 1968.

[73] L. MISSEL, R. K. TITUS — *Plating of beryllium for brazing*, Metal Finishing, vol. 65, 65, No. 10, p. 59, October 1967.

[74] D. E. SALOMON — *Joining dissimilar metals by gas tungsten-arc braze-welding*, Welding Journal, p. 181–191, March 1968.

[75] — Tin and Its Uses, No. 81, p. 14, 1969.

[76] M. M. KARNOVSKY, A. ROSENZWEIG — *The gold-tin-lead alloys—The gold-tin-lead system*, Trans. Metallurgical Society of AIME, Vol. 242, p. 2257, November 1968.

[77] ANON. — *Vacuum brazing fills big vacuum*, Iron Age, January 9, 1969, p. 64.

[78] — *Brazing Alcoa Aluminum*, Alcoa, Pittsburgh, 1967.

[79] L. E. HELWIG, P. R. CARTER — *Solder flow on galvanized surfaces*, Metal Finishing, February 1969, p. 63–68.

[80] C. J. THWAITES — *Some effects of abrasive cleaning on the solderability of printed circuits*, Metal Finishing Journal, September 1968, p. 291.

[81] W. E. HOARE — *Hot tinning*, Tin Research Institute, 1948.

[82] C. J. THWAITES — *Some experiments on the hot-tinning of small parts*, Metallurgia, September 1961, p. 117.

3.6 Soldered constructions (see also Section 3.7.2(g))

3.6.1 Introduction

The quality of a soldered joint depends on the requirements made as regards current transmission, electrical resistance [3]*, heat to be dissipated, porosity, strength, toughness, corrosion resistance, appearance (especially with jewellery) and the possibility of visual inspection. It also depends on the degree to which these requirements are met by the loading (if possible, the position of the joint should be outside the stressed area, to avoid stress concentrations in the joint), the flux (good wetting), the solder and its impurities, and the component parts to be connected, as regards composition and therefore the possibility of alloying with solder; and as regards surface roughness and purity and residual stress. It depends too on the correct atmosphere, soldering time and temperature (decisive for among other things the viscosity of the capillary-active solder), the correct soldering method and positioning of the solder, and the correct construction of the soldered joint.

* Although the specific resistance of tin-lead solder is considerably higher than that of copper, the resistance in wire joints is equal, because the soldered joint has a larger cross-sectional area than the wire itself.

When considering the construction of the joint, we think of shape and dimensions of the seam, accessibility, produceability, the heat dissipated per time unit, the dimensions of the capillaries at soldering temperature*. Also, the possibility of filling them properly and of letting flux, air and other gases escape, via a separate channel if need be.†

At the position of the soldered joint, the components parts must match each other's profile, so that the seam is equally thick along its whole length. Besides, the flux layer should be so thick that the tops of the roughness profile of the mating surfaces do not come into contact, so that the channel in between is wide enough for the flux to be quickly removed by the solder. Even joints with an excellent appearance often have many flux inclusions. To avoid this situation, the joint is often designed so that the distance to be covered by the solder is as small as possible, as for instance in Fig. 3.27(m).

A narrow gap requires a solder with low viscosity, a low melting point and/or a short melting traject, good wetting capacity and little tendency to alloy with the base metal. Solder satisfying these demands is not suitable for filling a wide gap.

A wide gap is required when the atmosphere, or the base metals, give rise to the formation of oxides. In such cases, a larger quantity of flux is wanted, which together with the oxides must find room in the gap.

Other points to be borne in mind are: the stresses caused by unequal thermal expansion and conduction, non-homogeneous heating and unequally heated volumes. Sometimes, even stress corrosion due to stress must be taken into account.

The strength of soft solder used for electrical joints is roughly 1/10 that of the base metals, at least for loads of short duration: 60Sn/40Pb solder has a tensile strength of 4×10^7 N/m^2 for loads of short duration, after 10 h: 1×10^7 N/m^2, after 1 000 h: $0{\cdot}4 \times 10^7$ N/m^2, after 100 000 h: $0{\cdot}25 \times 10^7$ N/m^2 for lap joints. At 70 °C the strength has been reduced to 50 %, at 125 °C to 25 %, to reach zero at 175°C. See also [20].

In telescopic pipe joints three sources of failure must be avoided: eccentric positioning of one pipe, out-of-roundness, and non-parallelism of the centre-lines. In this sequence, they seem to affect the strength of the joint in an increasing order. When the mating surfaces of the component parts are heated homogeneously, the molten solder exerts a pressure in all directions, thereby centring the components and rendering centring jigs superfluous, in principle. To obtain homogeneous heating just for this type of joint, shrink fits are often used (with special means for the flux to escape, as, for instance, via a separate channel).

A general mechanical advantage of soldered joints is that they are not weakened by holes needed for fasteners, and that the stress is uniformly distributed over the whole area of the joint.

* Note: The coefficient of linear expansion is not equal at all temperatures, and certainly not if a change of structure takes place.

† Flux entrapment is mainly caused by incorrect clearance, blind joints, too-long joints, filler metal entering from both ends of a seam, a too-short heating period, insufficient filler metal and incorrect flux or filler metal.

3.6.2 Construction forms

The following forms can be made with
plate and pipe:

- The components are placed parallel to one another: butt joint, scarf joint, lap joint (Fig. 3.6).
- The components intersect: angle and T-joints (Fig. 3.7).
- The components are in contact with each other (Fig. 3.8): all kinds of construction used in radiator design, honeycombs and filling pieces.

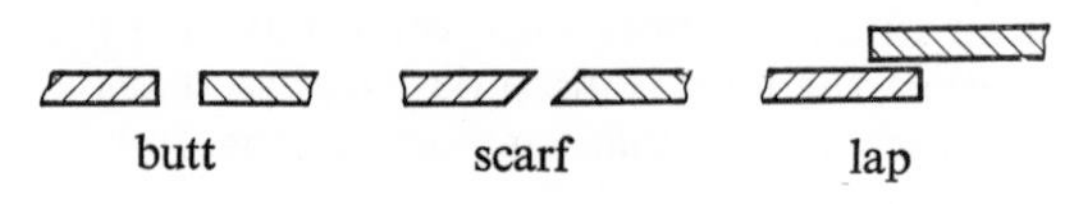

Fig. 3.6. Seams of components placed parallel to one another.

Fig. 3.7. Seams with intersecting components: T-joints and angle.

Fig. 3.8. Seams with contacting components.

Wire joints form a separate group.

In principle, the force a soldered joint can take is proportional to the soldered area, and consequently to the solder consumption. Point-contact types take relatively much solder.

At the fringe areas of the component parts, stress concentration occurs. In general, this is caused by a sudden change in cross-sectional area, a change in the direction in which stress is transmitted (hub and lap joints) and a change in the stiffness of the construction at the position of the joint. Many of these faults can be avoided by a suitable choice of material and by proper design: Fig. 3.9.

So it is necessary to avoid roughly-shaped joints, cracks, porosity, flux inclusions; to keep brackets, fittings, baffles or openings away from regions of concentrated stress; to provide stiffness to reduce secondary bending as in Fig. 3.10, to increase the area of the joint (larger lap, for instance) or to add relief devices. This is mainly in soft soldering; in brazing, the main load, especially if it is a bending load, should be taken, too, by the component parts, not by the solder. So, *not* according to Fig. 3.11, but according to Fig. 3.12(a) or to Fig. 3.12(b) if need be. Wire is sometimes looped to prevent it from becoming too taut, thereby stressing the wire and thus the joint.

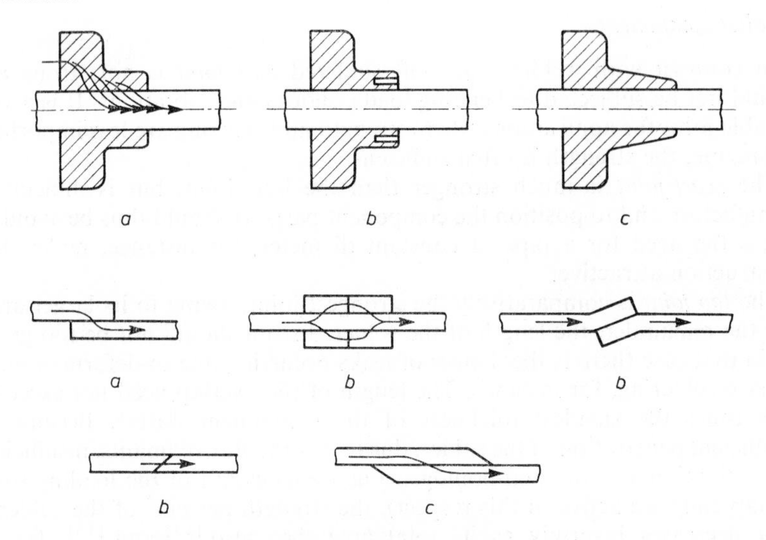

Fig. 3.9. Means of avoiding stress concentrations.

a to be replaced by *b* or *c*.

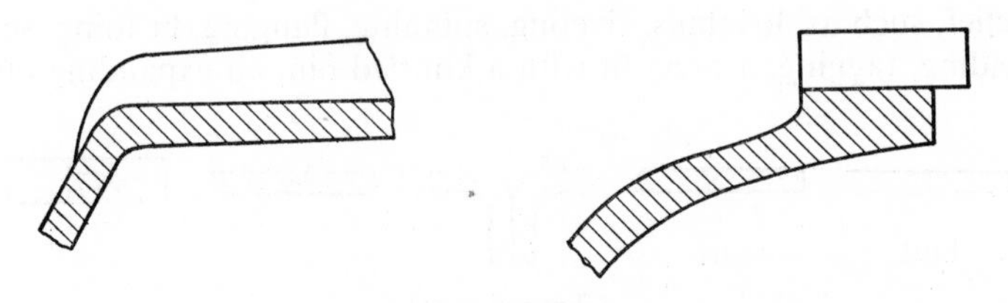

Fig. 3.10. Increasing stiffness of the joint.

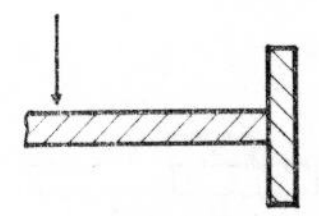

Fig. 3.11. This joint is not stress relieved and should be avoided.

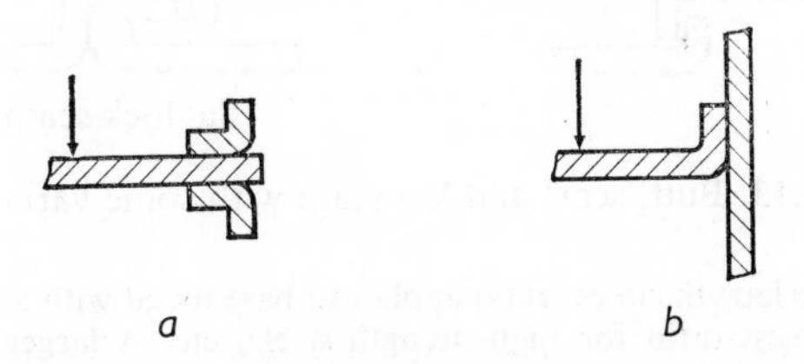

Fig. 3.12. Stress relieved and reinforced joints.

Parallel components

In contrast with welding, the soft soldered *butt joint* is not strong and should not be subjected to bending loads, shocks and vibrations. It is a very suitable joint for applications where strength and centring are less important. In brazing, the strength is often sufficient.

The *scarf-joint* is much stronger than the butt joint, but is difficult to manufacture and to position the component parts. It should thus be avoided, unless the need for a pipe of constant diameter, for instance, makes this construction attractive.

The *lap joint* is comparatively the strongest joint, owing to its larger area. For the remainder, the length of the soldered seam should not be too great, for in that case there is the danger of leaks occurring due to deformation, in furnace soldering, for instance. The length of the overlap need not exceed 4 to 8 times the smallest thickness of the component parts*. Because of insufficient penetration of the solder, due to gas and flux remnants, insufficient flux activation and the non-homogeneous transmission of the loading (only the lap ends are active in this respect), the strength per cm^2 of the soldered joint decreases inversely as its total area. See also [13] and [15]. Hence, combinations of butt and lap joint are often found without great length (Fig. 3.13); by hard soldering, these joints can easily be made as strong as the base material.

Soft soldered constructions subjected to stress require special means of stress relief, such as dovetails, riveting, spinning, flanging, beading, seaming, spot welding, tagging, a press fit with a knurled pin, an expanding pipe in a

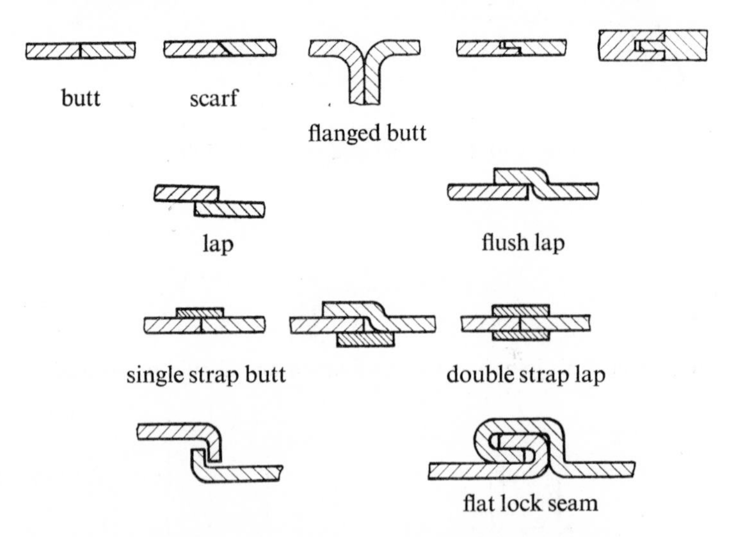

Fig. 3.13. Butt, scarf and lap seam with some variations.

* The smallest overlap: thickness ratio applies to base metal with a low tensile strength, e.g. copper; the largest ratio for high-strength steels, etc. A larger overlap just wastes filler and base metal.

cylindrical or countersunk hole, upsetting a pin, blows of a centre-punch in the pin or in the metal surrounding it, etc. These techniques are applied independent of whether the components are parallel, intersecting, etc. (see Fig. 3.14).

Some pipe joints are shown in Fig. 3.15.

Intersecting components

T-joints and L-joints are usually not strong and are therefore avoided, or reinforced when the direction of load is unfavourable: The construction of

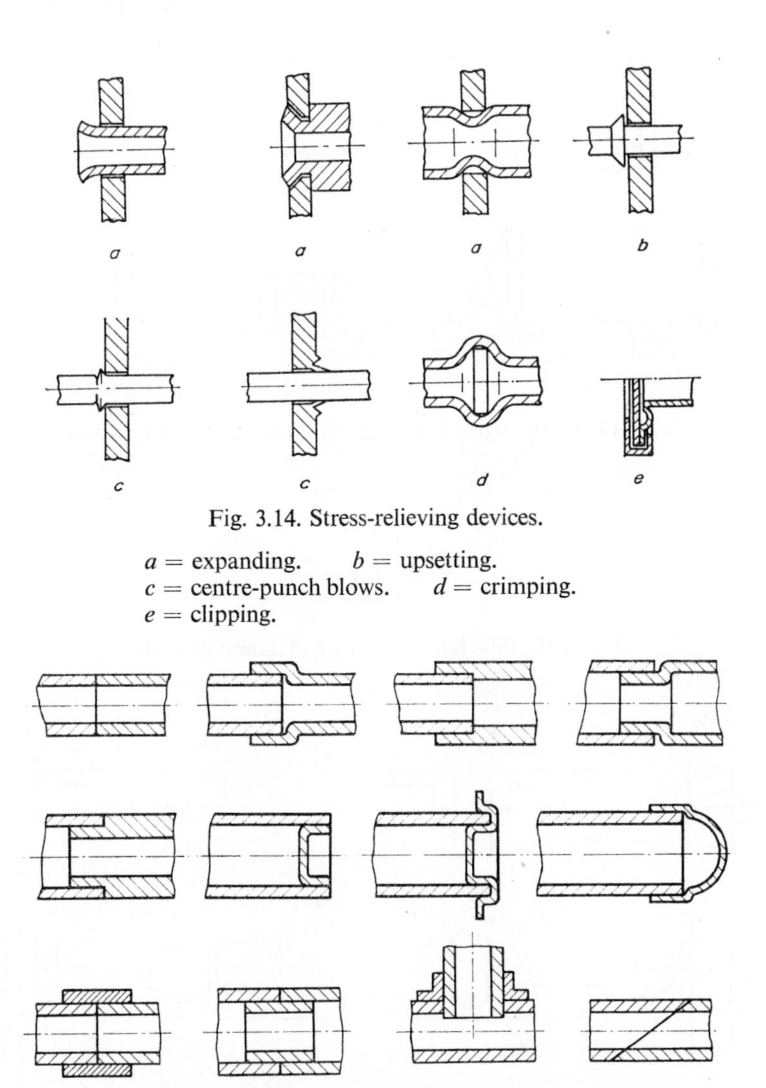

Fig. 3.14. Stress-relieving devices.

a = expanding. b = upsetting.
c = centre-punch blows. d = crimping.
e = clipping.

Fig. 3.15. Construction of pipe joints, including end sealing.

Fig. 3.16 becomes that of Fig. 3.17. The same principle is applied to the joint between plate and pipe or rod; the construction of Fig. 3.18 changes to that of Fig. 3.19.

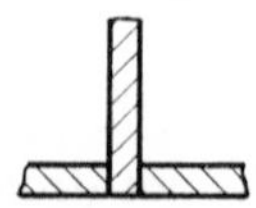

Fig. 3.16. Joint with a small adhesion area.

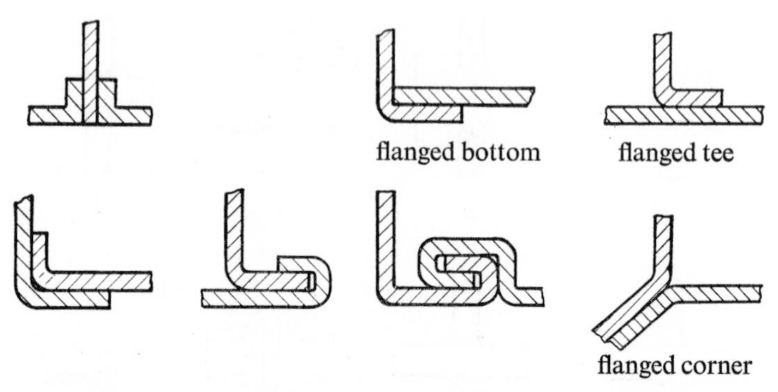

Fig. 3.17. Joints with a large adhesion area and better shape.

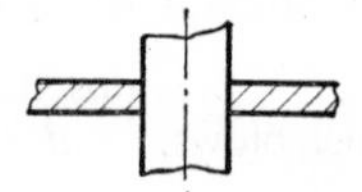

Fig. 3.18. Pipe joint with a small adhesion area.

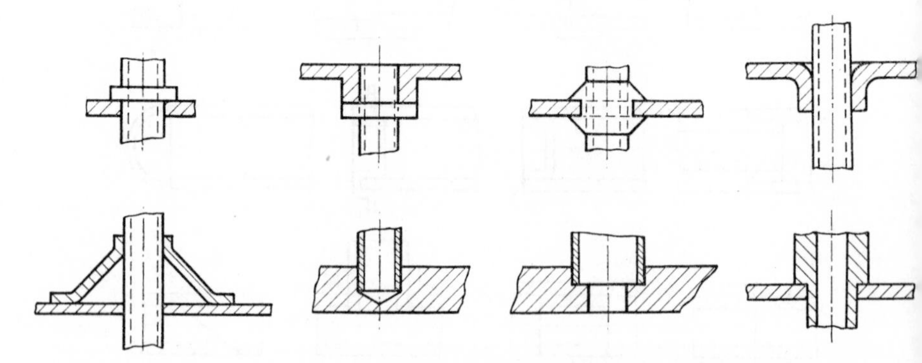

Fig. 3.19. Pipe joints with a large adhesion area and better shape.

Wire joints (at least one component is wire)

Means of increasing the soldered area are indicated in Fig. 3.20. If only one wire is led through the hole, the hole is usually round. The wire flattened on the strip provides sufficient capillary effect for bit soldering, so the ratio of wire diameter: hole diameter is not very critical. The point then deserving most attention is: easy insertion of the wire. With dip soldering (see Section 3.7(e)) this ratio is far more important; in that case an 0·8–1·1 mm wire is used for a 1·3 mm hole or an 0·5–0·8 mm wire is used for a 1 mm hole. The clearance around the wire should be approximately 0·1 mm.

If two wires are inserted in one hole, the situation changes to that of Fig. 3.21.

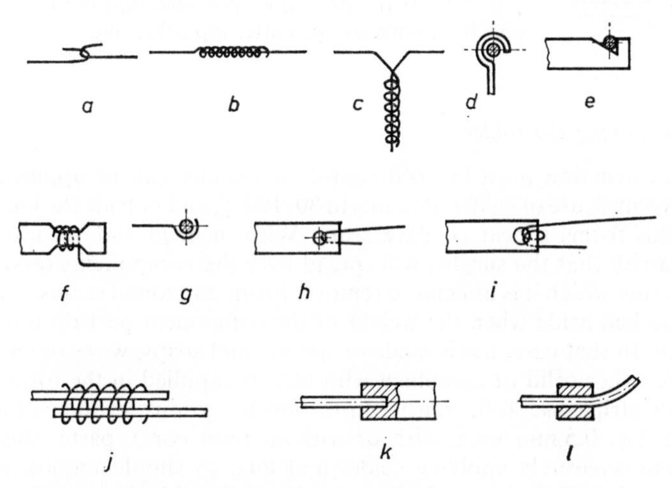

Fig. 3.20. Means of increasing the adhesion area of wire joints.
a, b, c, d: two wires.
e, f, h, i: wire with strip.
g: wire with strip or wire.
j: coil is slipped over wire ends prior to soldering.

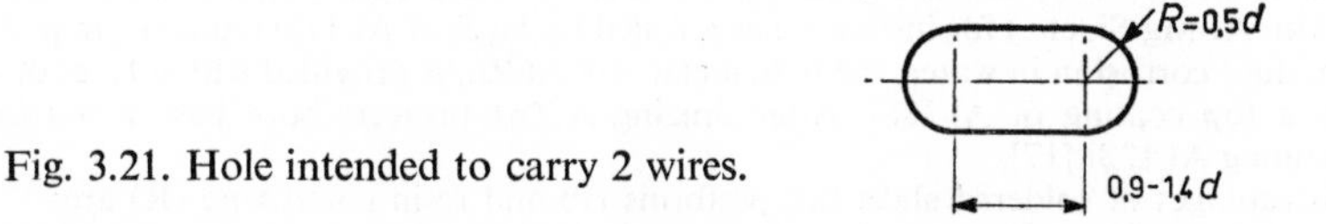

Fig. 3.21. Hole intended to carry 2 wires.

The PVC insulation must be removed over such a length that its new end on the metal wire has a distance to the joint of 5–6 times the wire diameter. This is done to prevent decomposition of the PVC during soldering, which might set free chlorine and give rise to corrosion, also outside the joint.

For printed wiring, the circuit is sometimes intentionally broken, to test a certain circuit after dip soldering: then the bridge is soldered with the soldering bit. The gap should not be wider than 0·35 mm (Fig. 3.22). If, when dip soldering printed wire panels, a hole must remain open, the print is so designed that it does not completely surround the hole (Fig. 3.23).

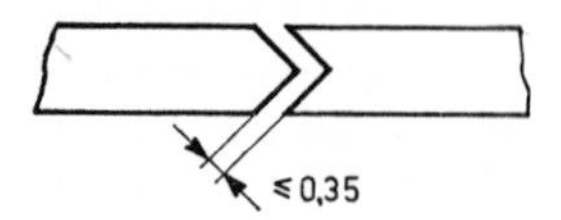

Fig. 3.22. Bridge in printed circuit.

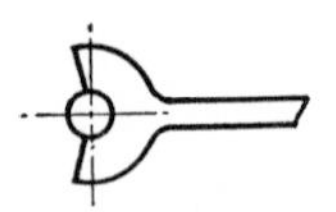

Fig. 3.23. If the print does not surround the hole completely, it remains open after dip soldering.

3.6.3 Positioning the solder

The construction must be so designed that solder can be applied.

The overmeasure of solder amounts to 30–100 %, and outside the joint areas, the surplus forms a neat concave seam. With inclined and vertical seams, there is a risk that the surplus will spread over the components (assisted by gravity) from which it is difficult to remove. From horizontal seams, the solder can be pushed aside when the weight of the component part on top of it is too much. In that case, use is made of spacers and strips, wires or shoulders.

Solder, either solid or combined with flux, is supplied in the form of rod, wire, ring, strip, disc, foil, transfer tape, powder, grains, cylinders (upwards from 0.5 dia; 0.5 mm long, with or without resin core), paste (this works faster than separately applying solder and flux, so should reduce the total joint cost), syrup and as soldered sheet*. Fig. 3.24 shows what can be

* Soldered plate is advantageously applied where no practical means of positioning the filler metal is available.

Sheet covered with soft or hard solder is coated completely or in strips.

Soft solder (thickness 50–125 μm) such as Sn, Sn-Pb, Sn-Pb-Sb, Sn-Ag, Sn-Zn [12] and Sn-In is available on a backing of steel, stainless steel, copper and nickel and their alloys and silver.

Hard solders too, are available on all sorts of base metals [8]. Aluminium, (pure, Al-Mn, Al-Mg-Si, etc.) for instance, has a rolled-on layer of Al/12Si (eutectic, m.p. 577°C). To reduce corrosion in water, the base metal, say Al-Zn, is provided with a layer of AlZn1 with a top coating of Al/12Si. After brazing AlZn1 protects both base metal and the remaining Al/12Si [17].

Advantages of soldered sheet (S), preforms (P) and resin cored wire (R) are:

- dependence on capillary action is not required (S,P).
- no need for clearance between the component parts, constant filler metal thickness (S,P).
- no need for solder preforms (S), resin cored wire (S), separate flux (P,R) and their preplacement (S) (time consuming, risk of faults).
- skilled help is unnecessary (S,P) (e.g. the solder is already in the correct position).
- constant ratio of flux and solder (P,R).
- measured amount of flux (S) and solder (P,R) is applied after experimental determination or calculation: too much solder costs money, too little gives a weak joint. Soldered

Continued at foot of page 189

achieved by heating such sheet. Fig. 3.25 depicts an application of foil: in the manufacture of the product (right) it was found attractive to do away with the need for jigs. So the two component parts were joined in a flat position, after which the resulting product was bent into its definite shape.

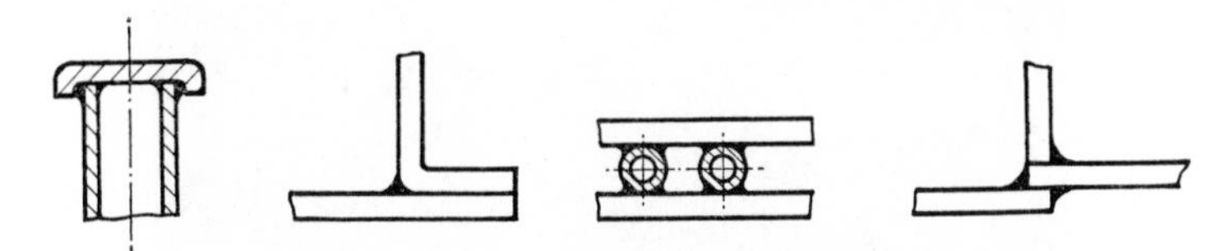

Fig. 3.24. Pre-soldered plate is versatile and yields good joints.

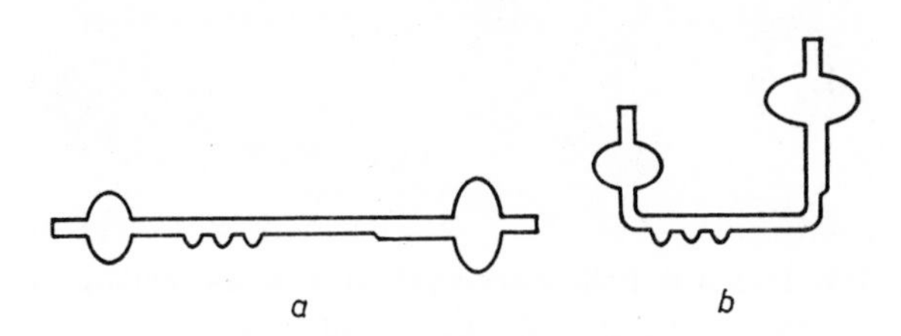

Fig. 3.25. Application of foil.
a. solder shape. b. definite shape.

Solder can also be applied galvanically, chemically or by spraying.

The form in which solder is supplied depends, amongst other things, on the soldering method. It is therefore important to know whether the solder is already applied as a semi-product or that it must be supplied simultaneously with the heat. A typical solution for a specific problem is pins with hollow heads filled with a low-melting solder. They make it possible to remove worn-out components from printed-circuit boards and to replace them by new

---

sheet has an adequate amount of solder; for preforms and resin cored wire, the quantity of solder equals the volume of the seam to be filled, on the understanding that tolerances are dealt with in such a way that the volume is a maximum. To that quantity of solder is added 30% for every concave seam between the components to be joined.

- blind, large, intricate and poorly accessible joints are easily accomplished as flux (P) and solder (S,P) are already distributed and come into play at the right time.
- no more material is heated than consumed (contrary to dip soldering).
- solder bridges (as in dip soldering) scarcely occur (S,P).
- no contamination of the solder by the base metal (as in dip soldering).
- joints are made at a lower temperature (S,P), thus avoiding warpage.
- by eliminating guesswork and variables (S,P,R) quality and appearance are improved, less checking, greater uniformity, less rejects, less cleaning.
- total joint cost is often reduced (S,P,R).
- favourable structural and technological solutions possible (S,P,R):
  (1) component parts can be stamped and drawn (S,P);
  (2) as all flux is heated to soldering temperature (P,R) and thus all corrosive volatiles are removed, a somewhat higher degree of acidity of the fluxes can be tolerated;
  (3) mechanisation is easily accomplished (S,P,R).

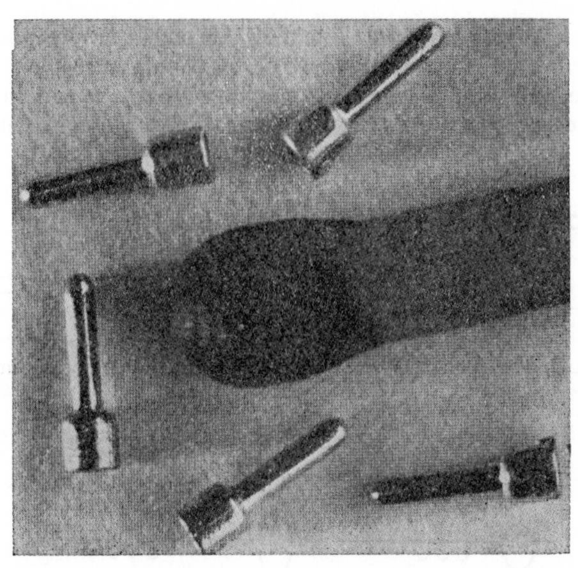

Fig. 3.26. Pins with hollow heads, filled with low-melting solder.

ones, without risk to the soldered connections nearby. Fig. 3.26 (from Tin and Its Uses, No. 80, 1968) gives their shape and dimensions (smaller than a match head).

When soldering pipes, flanges, rings, etc., especially in large batches, such as in furnace, induction and flame soldering, preforms,* brought into position before soldering, will be preferred. The joint should be designed so that the preform can be mounted in the right position and cannot shove away.

* Preforms are pieces of solder adapted to the shape of the product to be soldered, and are distinguished by the following types:

- rings of round resin core solder or massive wire, the ends touching, slightly overlapping or slightly apart; helices consisting of several turns (to administer sufficient solder in a narrow gap). The rings are supplied separately or on mandrels, with a wire gauge upwards, from 0·25 mm, stress relieved, if required, so that they will not expand too much when heated, thereby wetting the wrong places when melting.
- the same material is used to bend wire figures of any shape and dimension.
- (stand up and turn-down) rings are also made from strip; such rings have a greater contact area with the product, take less heating time, melt sooner and increase the soldering rate.
- punched solid flat plates, with a thickness upwards from 40 μm, any shape, separately or on mandrels.
- punched flat plates, with an intermediate layer of resin, any shape, and a thickness upwards from 0·125 mm.
- punched and bent solid plates, used, for instance, for the soldering of pipes positioned at right-angles.
- rod with or without resin core, straight, bent to any angle or arc.
- pellets, of any diameter.

These forms are available in any alloy. Most instructive is [16].

Since 1969, Wall Colmonoy have supplied pressure-sensitive tapes of 0·002, 0·003 and 0·05 in. thick Nicrobraz alloy, covered with a polyethylene film which is peeled off after application.

Fig. 3.27 gives a few examples:

In Fig. 3.27(a), the solder is applied from two sides (if the area to be soldered is too large), thereby entrapping flux and air.

Apart from the outside ring of solder, it is principally wrong (at least in soldering by hand, where the quality of the joint depends on the skill of the operator) to place the solder in a position where it is invisible from the

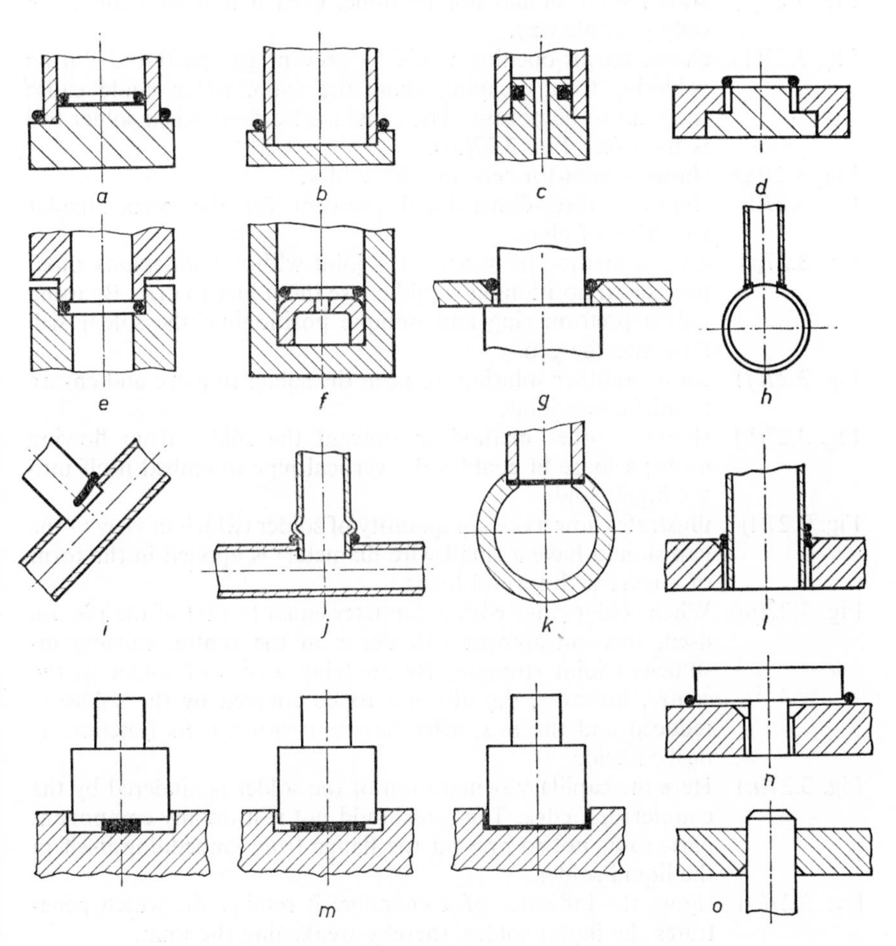

Fig. 3.27. Examples of how to place the solder.

outside. Hence, the design of *a* (inner ring) should be replaced by *a* (outer ring) or by *b* (outer ring). Contrary to this philosophy is the opinion of those who do not stop the soldering process before solder appears from the narrowest capillary, farthest from the source. They want to see the solder appearing on the outside, and therefore place it at the inside, confirming *a* (inner ring).

Fig. 3.27(c) shows a separate chamber for the solder when the components fit accurately together, and the resistance the solder meets is too much for it to penetrate along a great distance: one long way is divided into two distances half as large.

Fig. 3.27(d) shows that both gaps are filled when the solder is positioned at the widest gap, from where it is drawn into the narrowest capillary.

Fig. 3.27(e) shows what should not be done, even if it is sometimes the only possible way.

Fig. 3.27(f) shows an air-opening made to prevent gas produced during soldering from escaping along the seam, which could cause the joint to be porous. The same is also done when solder foil is used (see Fig. 3.27(a)).

Fig. 3.27(g) shows a pilot for centring the solder.

Fig. 3.27(h) shows a three-dimensional preform for the perpendicular soldering of pipes.

Fig. 3.27(i) gives a method of making a T-joint with a countersunk pipe; horizontal positioning would cause the solder to drip. By using half a preform ring and oblique positioning, the solder will flow into the gap.

Fig. 3.27(j) shows another solution, to keep the solder in place and ensure a satisfactory joint.

Fig. 3.27(k) shows another method to prevent the solder from flowing away; a loose fit enables the vertical pipe to embed itself into the liquid solder.

Fig. 3.27(l) illustrates how a certain quantity of solder (which in view of the gap should have a small wire diameter) is applied in the form of a helix with several turns.

Fig. 3.27(m) When solder rings with a diameter equal to that of the pin are used, flux entrapment will occur in the centre, causing insufficient joint strength. By applying a disc of solder in the centre, however, the distance to be covered by the solder is reduced and the flux, after having performed its function, is pushed aside.

Fig. 3.27(n) Here the capillary penetration of the solder is hindered by the countersunk edge. The pin should not rest on any component part, so as not to prevent the flange from embedding itself in the liquid solder.

Fig. 3.27(o) shows the influence of a chamfer: it retains air, which penetrates the liquid solder, thereby weakening the joint.

3.6.4 *Gap width and gap strength* (see also Sections 3.2, 3,6,1 and 3.6.2)[20]

In principle, the aim is to have the narrowest possible gap between the component parts to be joined. A wide gap consumes more solder, requires more time to fill, runs more risk of emptying itself when the temperature is not controlled well, increases the risk of shrink holes forming, promotes an as-cast structure (thereby reducing the strength), has a lower capillary fill pressure (see Fig. 3.28) so there is more need to position the workpiece such that the gravitation promotes, or at least does not counteract, the filling of the gap, and gives a joint which contains not only the alloy of solder and base metal, but also the less-strong pure filler metal.

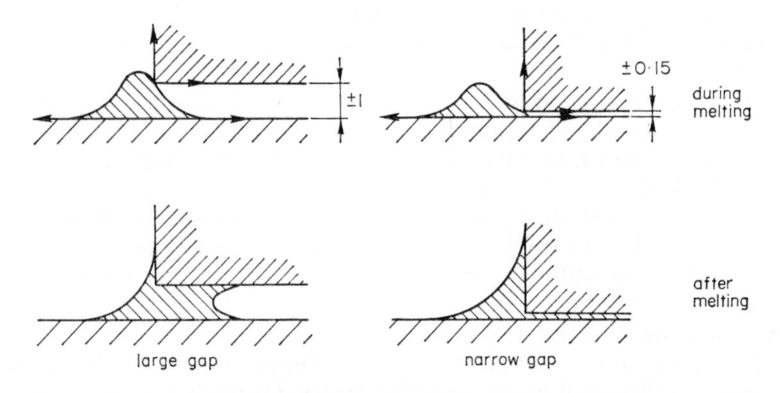

Fig. 3.28. Influence of gap width on joint quality.

On the other hand, a too narrow gap impedes the escape of gas and flux and the penetration of the solder. A normally good joint contains 20% by volume inclusions of flux, oxide, water vapour, gas bubbles and the remains of pickling agents.

The choice of gap width depends on the following influences:

- the flux or the atmosphere.

  The type and quantity of flux used affect both the thickness of the oxide layer on base and filler metal at a certain moment, and the spreading of that solder. When the flux is gaseous the gap width can be as small as 1 μm.
- the filler metal.

  In principle, the strength of a good solder joint equals that of the cast solder, when no alloying occurs, and is as strong as the cast alloy when alloying takes place. Intense alloying requires a wider capillary than less-strong alloying.
- the material of the component parts to be joined, including their strength.

  For a certain filler metal and a certain gap width, the strength of the joint increases about proportionally with the tensile strength of the parts joined, a phenomenon that is attributed to the supporting action of these parts.
- the cleanliness of the component parts to be joined.

  A thicker oxide layer requires more flux to remove it, hence a wider gap to house the flux and to remove the products of decomposition.
- the roughness of the component parts to be joined.

  A too-rough surface results in insufficient local capillary action and in stress concentration; a too-smooth surface has insufficient contact area for mechanical and metallurgical adherence. The capillary action is promoted when the direction of the machining grooves coincides with the flow direction of the solder.

- shape and dimensions of the component parts to be joined, and their design.

The distance to be covered by the liquid solder should be limited, to prevent premature solidifying, for instance; there is a relationship between this gap length and the soldered area on the one hand, and the gap width on the other.

To maintain a good fit and, notwithstanding alloying and segregation, still retain enough gap width for the solder to penetrate, round components are lightly knurled. Flat parts can be treated in a like manner.

- the technique.

Soldering by hand offers the possibility of making and checking the joint simultaneously, so any joint can be improved immediately and individually. Hence, the gap width can be larger than in mechanized soldering, but should not exceed 0·5 mm.

- the soldering temperature and time.

When one component part is heated earlier or more intensely, or when it has a coefficient of linear expansion differing from that of the other part (important for parts fitting together), the gap width at soldering temperature can vary from that at room temperature. In the case of different coefficients of linear expansion, a larger gap width is generally chosen, thereby reducing the stress in the seam.

The soldering temperature also determines the viscosity of the solder, and hence its penetration rate; temperature and time influence the degree of alloying between base and filler metal, and thereby the gap width.

Moreover, they have an effect on the quantity of the oxides formed during soldering, and on the spreading and activity of the flux; consequently its quantity, and the space it occupies.

In actual practice, where all these factors play their role, experience will, in general, dictate the best gap width; its value is between some μm's and some tenths of a mm. For many applications, a gap of 0·075–0·125 mm is the most suitable one. Brass filler alloy requires a capillary of 0·08–0·25 mm, as do nickel silver and aluminium solder too. Tin-lead solder requires 0·025–0·125 mm; silver base solders, 0·025–0·075 mm (maximum 0·5 mm); and pure copper, 0·001–0·01 mm (gas flux when 0·001).

REFERENCES

[1] E. LÜDER *Handbuch der Löttechnik*, Berlin 1952.

[2] W. J. VAN NATTEN *Design data for brazing*, Welding Journal **31**, No. 11, p. 1023, 1952.

[3] H. H. MANKO *How to design the soldered electrical connection*, Product Engineering, June 12, 1961.

[4] J. COLBUS *Verbindungen durch Löten*, Lastechniek, July/September 1962.

[5] G. L. J. BAILEY, G. P. MCKNIGHT *The influence of joint design on solderability*, Sheet Metal Industries, January 1955, p. 47–57.

[6] H. Bühler, J. Colbus — *Die Festigkeitssteigerung von Spaltlotverbindungen bei Abnehmen der Spaltbreite, als Erscheinung der Festigkeit bei behinderter Verfirmung*, Zeitschrift für Metallkunde, **48**, Heft 2, p. 66, 1957.

[7] N. Bredz — *Investigation of factors determining the tensile strength of brazed joints*, Welding Journal, **33**, No. 11, p. 545 s, 1954.

[8] Materials in Design Engineering, p. 104, July 1966.

[9] *Dipbrazing aluminum assemblies*, Machine Design, p. 158, August 18, 1966.

[10] H. Zürn, Th. Nesse — *Die metallurgischen Vorgänge beim Weichlöten von Kupfer und Kupferlegierungen und das Festigkeitsverhalten der Lötverbindungen*, Metall, 20. Jahrg., Heft 11, November 1966.

[11] H. H. Manko — *Solders and soldering*, McGraw-Hill Book Company.

[12] *Tin-zinc plating*, Tin Research Institute, Fact Sheet E 6.

[13] K. F. Zimmermann — *Schweissen und Schneiden*, 18. Jahrg., No. **9**, p. 467, September 1966.

[14] H. Zürn, Th. Nesse — *Beitrag zum Zeitstandverhalten von Lötverbindungen aus Zinn-Weichloten bei Raumtemperatur*, Schweissen und Schneiden, Jahrg. 18 (1966), Heft 1, p. 2.

[15] J. Colbus — *Die Scherfestigkeit von Silber-, Messing- und Neusilberloten auf St* 37.11 *in Abhängigkeit von Spaltbreite und Lötfläche*, Mitteilungen der Forschungsgesellschaft Blechverarbeitung e.V., No. 13, p. 141, July 1, 1957.

[16] Designing for Preforms. Lucas-Milhaupt Engineering Co., Cudahy, Wisconsin, USA.

[17] F. E. Faller — *Neue Aluminium-Plattierlötwerkstoffe mit verbessertem Korrosionsverhalten*, Zeitschr. f. Metallkunde, **58**, Heft 10, p. 676, 1967.

[18] E.-A. Cornelius, J. Marlinghaus — *Gestaltung von Hartlötkonstruktionen hoher Tragfähigkeit*, Konstruktion, **19**, Heft 8, p. 321, August 1967.

[19] K. F. Zimmermann — *Hartlöten-Regeln für Konstruktion und Fertigung*, Deutscher Verlag für Schweisstechnik G.m.b.H, Düsseldorf, 1968.

[20] W. P. McQuillan — *Guide to soldering*, Welding Engineer, April 1965, p. 112.

[21] H. Schwarzbart — *Metal fibre reinforced soldering tape*, British Welding Journal, November 1968, p. 538–542.

3.7 Soldering techniques

(see also Section 3.5.1(a), Preliminary treatment)

3.7.1 Introduction

On no account should the components that are being joined be displaced before the solder has solidified; otherwise, the joint will be ruined.

The soldering technique to be used is chosen in the first place for its economy, and hence by the number of workpieces to be soldered, their dimensions, and the construction of the joint. Another factor is the active life of the flux, which determines the heating rate. Mechanization is useful only for large batches of equal products that are easy to handle and position, and have a simple soldering seam (see Section 3.6.1). Another factor is the required joint quality.

3.7.2 Methods

The most important methods of heating the soldering place are given below. (Items (r) and (s) are not heating methods, but are included for the sake of completeness):

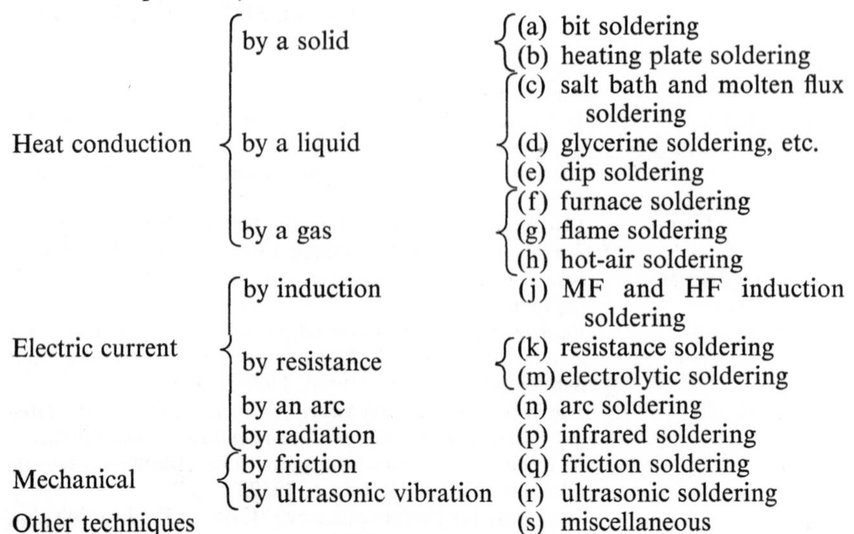

Supplying heat to and removing heat from the workpiece take time and consequently cost money, so the heat should be supplied economically and used efficiently. Consequently, low-melting solders are generally preferred and jigs, clamps, etc., are made from materials having small heat conduction and capacity, small (heat-removing) cross-section areas at the joint and small contact areas with the workpiece.

These jigs and clamps (see also Section 3.7.2(f) and (g)) must:

- be mounted and removed quickly and simply (no welding to workpiece);
- position the component parts strongly and accurately;
- resist repeated heating and cooling;
- not affect heating and cooling of the workpiece or of the soldering bath;
- not obstruct the operator;
- not take out a large quantity of flux when being removed from the salt bath.

(a) *Bit soldering*

The tip of the bit transfers its heat to the base metal and to the liquid solder with which it is in contact. A larger bit can store and transfer more heat per unit time and per joint, has a more constant temperature variation (350–400 °C) and a longer life, works faster and more efficiently, but becomes heavy for manual use and is therefore not suitable for continuous work. Another function of the bit is to pick-up, store and deliver solder. The bit is used for soft soldering small, light components. The (usually) copper soldering tip has a very good heat conduction and a good heat capacity, is well wetted by the solder (thus has good heat transfer), but copper oxidizes rapidly at higher temperatures and dissolves in active flux and in solder containing tin or zinc. Remedies are: the addition of 1% chromium to the copper, or copper to the solder [3], a copper core with an iron and/or nickel coating, solid monel or nickel. See [4] and [43]. All these methods reduce the thermal conductivity, however.

Deep-drawn iron caps on the tips give better results than a plated-on or thermally-applied iron coating, but it is difficult to fit them in a satisfactory long-lasting way. Lightly (<0.3%C) carburizing the plated-on iron film, followed by diffusion heating, gives a good result at a rather low price.

The shape of the tip is adapted to the shape of the product to be soldered (Fig. 3.29).

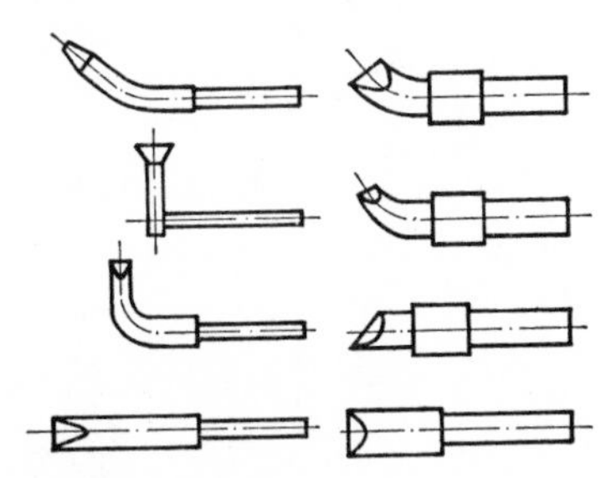

Fig. 3.29. A few soldering-iron tips.

It appears that solder spattering occurs when the bit is removed, so this removal should be done very slowly. Using less solder is another, principal, remedy. Oxide-free solder has a higher surface tension and hence is more inclined to form drops. Consequently, spattering is fought by oxidizing the solder, by longer soldering times and the application of less flux, for instance.

The bit is heated in a flame, in a furnace or by resistance (round or in the tip of the bit). Continuous heating is preferred over intermittent heating; electric heating is used most; simple and light, but vulnerable. The temperature, which is roughly proportional to the voltage applied, is adjusted by controlling the supply voltage with a variable transformer. But overheating still occurs, so a thermal protection device, based, for instance, on the Curie effect of a magnetic switch, is sometimes built in [55]. Another solution is the application of a heating element with a positive temperature coefficient (resistance increase with rise in temperature).

There are soldering bits fitted with a manually-operated solder wire supply mechanism [5]. Other bits, fed from a rechargeable nickel-cadmium battery, are suitable to work at remote locations. See Fig. 3.30, for instance.

Unsoldering tools or tin cleaners, loose or mounted on the bit, remove surplus liquid solder from the soldering locus via a vacuum, supplied by a hand, foot or electrically operated pump (Fig. 3.31), by the action of compressed air passing along an orifice, or by a releasing spring.

There are also bits fitted with a lamp.

To increase the working rate when soldering bits are used, a low melting solder is usually preferred, i.e. a practically-eutectic solder, for instance 60Sn/40Pb or 50Sn/50Pb. The bit temperature is chosen about 50–100 °C higher than the melting point of the solder involved. Depending on the

Fig. 3.30. A 40W battery-fed soldering bit, supplied by Picard et Frère, London. ("The Engineer", December 21, 1968).

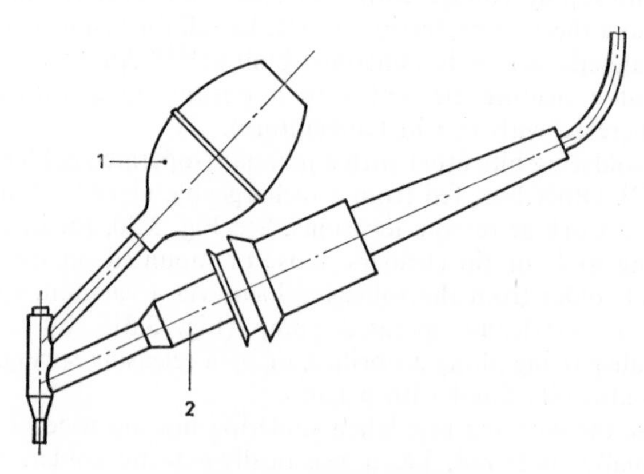

Fig. 3.31. Tin sucker.
1 = bag. 2 = soldering bit.

application, the soldering bit is designed for a power of 6 to 1 000W. The bit power depends on the heat to be supplied per unit time, and hence on the soldering rate, amongst other things.

The warm tip is cleaned by wiping with a wet synthetic sponge (no foamed plastic) or with a sponge soaked in colophony and alcohol; immediately afterwards, it is retinned to prevent corrosion. Corrosion reduces heat transfer, prevents the tip from being wetted by the solder and can release corrosion products in the liquid solder. When necessary, pre-tinned tips are sandpapered with, say, aluminium grain 320, but they are never filed.

There are also soldering needles, weighing only a few grammes and suitable for soldering microcircuits, etc.

So far the equipment needed for this technique is inexpensive.

The soldering bit after the heat-pipe principle has a pipe-shaped, closed-end, evacuated soldering element, in which a heat transport medium absorbs heat through the wall, hence evaporates, moves to the colder, solder spot with lower vapour pressure, precipitates and throws out heat. Then by capillary action it runs back through a porous mass in the soldering element to the starting point, where the condensate absorbs heat, etc. (Fig. 3.32).

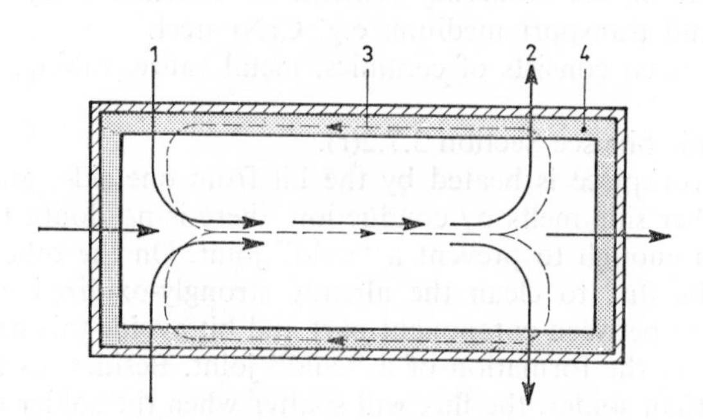

Fig. 3.32. Principle of the heat pipe.
—— = heat flow. ---- = medium flow.
4 = porous mass.

Because of the relatively large heat of evaporation of liquids, and the heat transport capacity of the medium, the weight and dimensions of the soldering element can be small, and there is practically no temperature-drop between the heat absorbing and the heat releasing wall of the element. A heat source of small dimensions can heat a large soldering plane, or several solder spots in one element (Fig. 3.33) or in several elements, with similar or different positions, shape, dimensions and temperature.

Other advantages claimed are: a more homogeneous tip temperature and a greater choice of shape and dimensions of the soldering element.

The choice of the heat transport medium is determined mainly by the working temperature: water (50–100 °C), diphenyloxide (150–375 °C), cadmium, caesium (425–700 °C), potassium (450–1 000 °C), sodium (500–1 000 °C), lithium (for higher working temperature), zinc chloride, zinc and aluminium bromide, cadmium and calcium iodide.

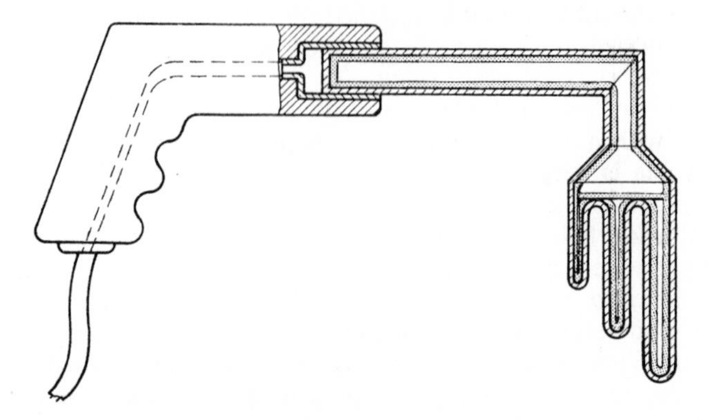

Fig. 3.33. Soldering bit, after the heat-pipe principle.

The material of the soldering element is determined by the working temperature and transport medium, e.g. CrNi steel.

The porous mass consists of ceramics, metal gauze, tubing, grooves and so on.

For ultrasonic bits see Section 3.7.2(r).

When the workpiece is heated by the bit from one side, and the solder wire at the other side melts by conduction, there is no doubt that the base metal is warm enough to prevent a "cold" joint. On the other hand, it is difficult for the flux to clean the already strongly-oxidized contact area. Placing the wire between component part and bit avoids this oxidation, but does not prevent the formation of a "cold" joint. Besides, as flux expands more rapidly than solder, the flux will spatter when the solder melts.

In some cases, heat-absorbing clamps (heat sinks) are placed between a heat-susceptible component and its end to be soldered (Fig. 3.34). The place to be protected is also cooled down by a sponge dipped in alcohol, or by a jet of gas (too-fast cooling may introduce brittleness).

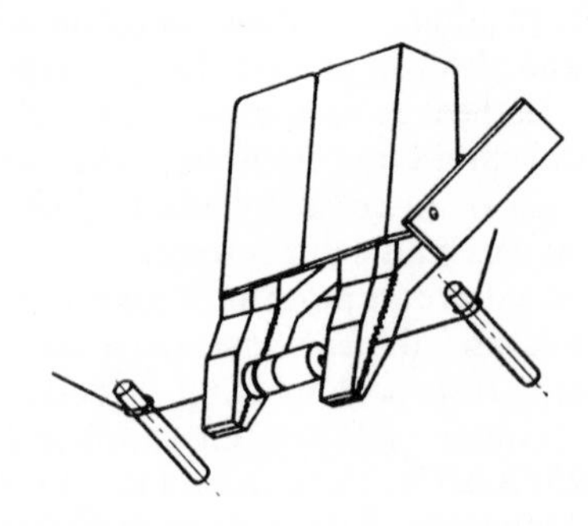

Fig. 3.34. Heat sinks protect heat-sensitive component parts.

(b) *Heating plate soldering*

The use of a heating plate, receiving heat electrically or via a flame, is mentioned only for the sake of completeness: the method is very seldom used.

(c) *Salt bath and molten flux soldering*

(i) *In liquid salt baths*

Dip heating usually shuts out the atmosphere, works very rapidly (there is no crystal growth, decarburizing, distortion, etc.) and gives a rather uniform heating, which can be accurately controlled and can sometimes be combined with hardening. Furthermore, it allows many joints to be made simultaneously, while the process can be mechanized (blind holes can retain salts and should therefore be avoided).

The baths consist of alkaline chlorides, barium salts, borax, etc. The composition depends on the objects to be treated: for copper brazing the composition is, for instance, 20–30% NaCl and 80–70% $BaCl_2$; carbon steel need not be fluxed in advance. Steel is carbo-nitrided besides, if the bath contains sodium cyanide; in this case, brass brazing alloy is used. The bath temperature lies above the melting point of the brazing alloy, that is, between 485 and 1 300°C. The baths are heated externally or internally by a.c. (current through the bath). If internally, the bath contains the electrodes; their position is decisive for the flow of the liquid and therefore for possible overheating of the products. Volume and temperature of the bath are such that its temperature does not fall intolerably when a new batch of products is immersed.

After degreasing and rust removal, when necessary, the component parts are preheated to remove all moisture; sometimes the surfaces to be joined are preheated to just below the melting point of the solder, so that the solder melted in the bath will find sufficiently heated metal, the flux will not solidify and the dipping time will be short.

After pre-placing the solder (powder, wire, shims, paste), the components are fixed by spot welding, wire constructions and clamping. These clamping parts should be made from inconel, stainless steel, etc., to prevent welding, and such jigging should be avoided as much as possible, for it can cool the bath and result in a heavy salt loss. Sometimes, however, jigging is necessary to eliminate the buoyant effect of the bath on the parts, causing dislodging.

Decarburization of steel by oxide and oxygen in the bath is limited by adding cyanide or SiC. Sulphur in the bath causes cracking of nickel component parts and must be removed. The bath must be purified each day of copper and zinc, nickel and iron (from the electrodes and the clamps for example), which otherwise will adhere to the products and hamper the reduction of the oxides and the access of the solder; water and SO_4, originating from the flux, should be removed, too. The products are rinsed after soldering to remove the salt film. For the remainder, this film protects the product when taken out of the bath against renewed oxidation.

(ii) *In molten flux* (*flux dip brazing*)

There is no principal difference between this method and the other one: flux dip brazing just employs somewhat more aggressive agents. It is mentioned separately, as a concession to the commonly-accepted classification.

Because the open air is shut out, the method is particularly suitable for soldering aluminium. Fluxes used are: NaCl, KCl, LiCl, KF and AlF_3 (a proper exhaust system is required); the baths are of ceramics, stainless steel or nickel (alloys) [6].

Pre-assembled aluminium components must be preheated to about 400 °C (with the additional advantage of drying them); after soldering, the products are thoroughly rinsed, pickled in, say, sulphuric acid and chromic anhydride at 50–60 °C for 5–7 minutes, and thoroughly rinsed again.

The baths are purified by adding strips of aluminium, which react with the water in the bath.

(d) *Soldering in more neutral media like glycerine or oil*

The (commonly brass) component parts, their mating surfaces plated with a 10 μm-thick tin layer and if need be provided with flux, are clamped together in a jig, immersed in a bath of glycerine, 250 °C, to which some flux ($ZnCl_2$) is added. The results are claimed to be excellent [56]. In another variation, an assembly of board and component parts, both quite heavily pre-tinned or provided with close-fitting solder washers, is placed on the surface of a bath heated to about 240 °C.

(e) *Soldering in molten solder* (*dip soldering*)

The pre-assembled components are fluxed* and dipped in liquid solder or liquid flux plus solder. (For brazing aluminium, the preheated assembly is dipped in molten flux and then in molten Al-Si alloy.) Usually, soft soldering is carried out, preferably with tin-lead solder 65/35, since the low melting point protects the components against excessive temperatures. As the immersed components cool down the bath, the temperature of the bath is usually about 100 °C above the melting point of the solder.

The components often come from the electrical industries and because of the quality requirements, they are often thermally pre-coated in tin-lead or pure tin. In doing so, the bath is contaminated and the products are marred by the copper, zinc and iron of the components alloying with the solder. This process is accelerated by the fluxes, $ZnCl_2$ and NH_4Cl. Appropriate mixing, replenishment of the components and purification of the bath are called for. Copper is removed by employing the difference in specific gravity of the solder; zinc is bound by palm oil or tallow and skimmed off; iron is skimmed off after blowing through with air or mixing with potatoes or carbon powder which gives carbides. See also [7] and [8]. The surplus of liquid

* To avoid solder splatter, and the formation of gas bubbles that prevent the solder from wetting the area to be covered, the flux should be nearly dry before the assembly is dipped into the bath. As the drying of water-based fluxes takes too long, an alcohol is used as a solvent.

solder clinging to the workpiece is removed by centrifuging or an electrolytic bath.

When necessary, a "stopping off" material (a stable oxide, e.g. titanium oxide, in the form of a stick, for instance), is applied to prevent the solder from wetting certain areas of the product.

Soldering printed wiring requires heating times short enough to protect the print and its components against damage. Compared with bit soldering, the soldering time is less and the temperature lower in dip soldering. As the flux cannot be removed, its aggressiveness should be limited, so the solderability of components to be dip soldered must be better. (For solderability see Section 3.5.1.) Hence, the solder alloy must meet stringent requirements (see pp. 131, 132). Besides, remnants of flux or oxide can also effect the electrical diagram. Other soldering techniques are therefore used in this case. The panels with mechanically fixed components are provided with a flux film 7 to 10 μm thick and are soldered by:

(i) *Manual dip soldering*

In this process, the panel distorts and the enclosures of flux and vapour, as well as the formation of solder bridges between the print tracks, largely depends on the skill of the worker (Fig. 3.35).

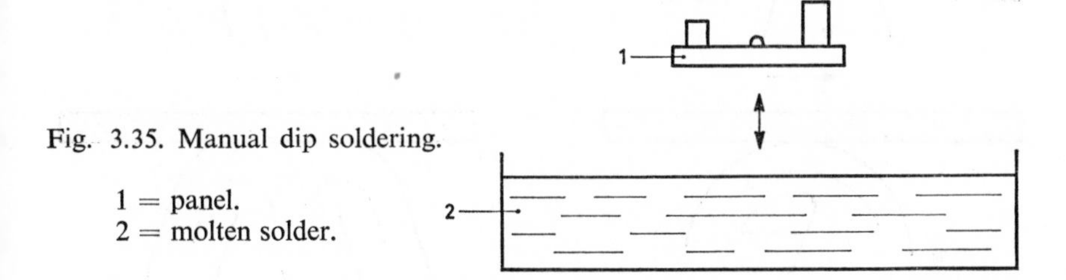

Fig. 3.35. Manual dip soldering.

1 = panel.
2 = molten solder.

(ii) *Machine dip soldering*

The same as (i) but the machine vibrates the panel to ensure better contact and gas removal. The panel remains in a horizontal position and is clamped to prevent warping. The speed at which the workpiece is withdrawn is decisive for the formation of solder bridges.

(iii) *Machine dip soldering*

The same as (ii), the panel moves vertically but is held at an angle of 3–5°, which improves the removal of flux, vapour and surplus solder. The panel is clamped.

(iv) *Train dip soldering*

A large bath is required; the flux forced aside impedes the flowing down of surplus solder. Vibration is required. Mechanization is possible (Fig. 3.36)

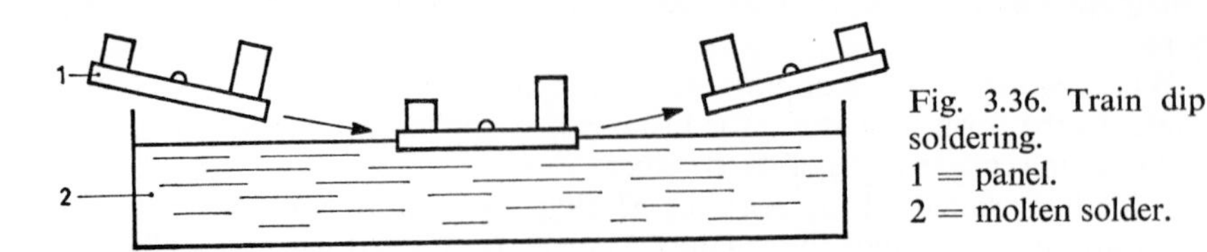

Fig. 3.36. Train dip soldering.
1 = panel.
2 = molten solder.

(v) *Wave dip soldering*

The panel moves along a wave of solder which is pumped up. Mechanization is possible, there is no limit set for the panel length and the result is a good product. The bath oxidizes rapidly (fought by covering the bath with Shell Peablum A*), there is a strong inclination towards icicle and bridge formation. Icicles can be avoided by lacquering the places not to be soldered, or by creating an extra, asymmetric wave to absorb the icicles (Fig. 3.37).

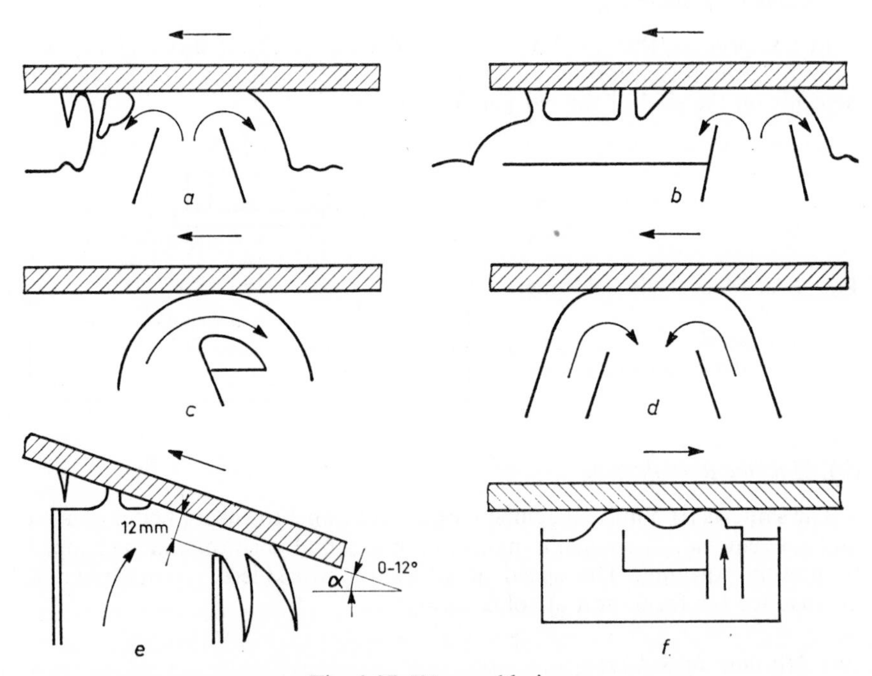

Fig. 3.37. Wave soldering.

a. large contact area, heavy icicle forming.
b. large contact area, icicles drip down better.
c. small contact area, but no icicles.
d. large contact area, no icicles.
e. icicle formation is fought by ensuring that the dripping speed of the solder is larger than the vertical component of the panel withdrawing speed. This requirement can effect α.

* Hence the need to degrease the panels afterwards.

(vi) *Cascade soldering*

This method is less suitable than wave dip soldering (Fig. 3.38).

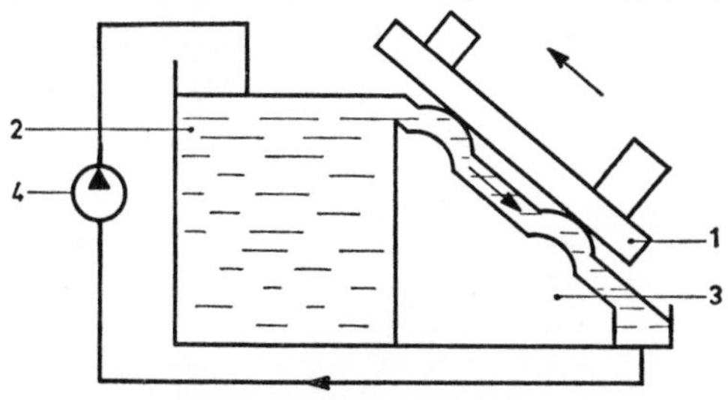

Fig. 3.38. Cascade soldering.
1 = panel. 2 = solder. 3 = block. 4 = pump.

Note:

With dip soldering of printed circuit panels, hundreds of joints are made in a few seconds by a mechanized technique. Testing and fault finding (the circuit can be broken down into convenient, pluggable units) is made more easy, too; and because of increased accuracy and reliability, there are less faults compared with other techiques. This method saves labour and space and operates with less-skilled labour. As Cu-Sn and Ag-Sn compounds float to the bath surface, and Cd oxidises rapidly, only Sn/Pb solders are used. To avoid segregation, giving variable tin content at the bath surface, the tin content is not less than 60%, in general. Common compositions are: 60–65 Sn/40–35 Pb, or 48·5Sn/48·5Pb/3Bi. Al and Zn make the solder stringy, the content of Zn, Mg and Al should be less than 0.001%, of copper less than 0·2%.

When the board is dipped into static solder, the latter must be scraped free of dross immediately before dipping.

To avoid the formation of solder icicles or solder bridges between neighbouring conductors, the board is removed slowly from the solder, and/or under a small angle and/or under vibration.

Another problem is presented by the "solder slivers": when the soldered prints are etched (solder resist process), copper is removed from under the solder. On further processing the print panel, the resulting solder "awnings" (which can be up to 10 mm long!) might bend or even break, and shorten the breakdown path between the tracks far too much. By immersing the horizontally clamped pre-fluxed panel for about 15 sec in oil with a temperaure of 235 °C, the protruding solder will contract [47] (hot oil fusing, flow melting, oil flowing).

Some people advocate preheating (e.g. with hot air) of the panels prior to dip soldering: flux solvents are removed (hence no formation of bubbles at soldering temperature which cause the solder to spatter and prevent it from wetting the board), by the increase in temperature the flux has become active and has cleaned the print metal before it is dipped, consequently the net soldering time can be reduced, and preheating lessens the thermal shock to the panel.

(vii) *Ladle soldering*

This is another type of dip soldering where a measured quantity of flux, followed by a measured quantity of solder, is applied to the soldering locus. The angle at which the materials are applied plays a certain part (Fig. 3.39).

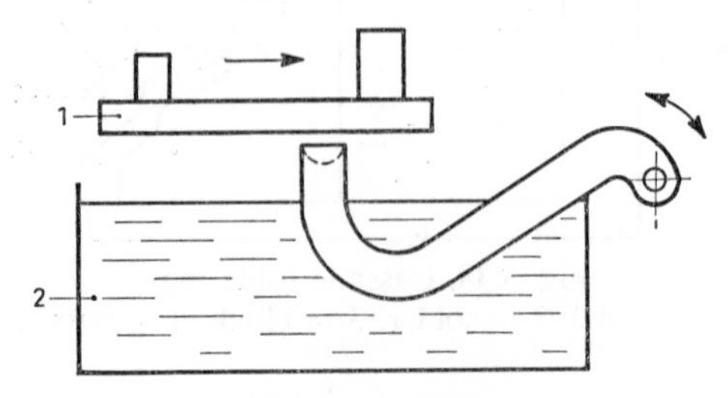

Fig. 3.39. Ladle soldering.
1 = panel. 2 = flux, followed by molten solder.

(viii) *Dimple strip soldering*

A mechanized method of ladle soldering, useful for multilayer panels without extensions, or other protruding parts on the soldering side. This technique is preferred when wave dip soldering takes too long or involves great difficulty.

A polyolefin sheet is impressed to contain a large number of dimples, each containing a solder pellet. The pattern formed by the dimples corresponds with the soldering-hole pattern of the panel, so that a dimple is located over each hole. When the sheet is heated, it straightens itself, thus forcing the molten solder into the panel holes. To prevent the dimples from shifting with respect to the holes, the dimpled sheet is clamped in a jig (Raychem Corp., Menlo Park, California).

(ix) *Roll soldering*

Print panels are fed horizontally to the machine, and led between two pairs of rollers. The bottom roller of the first pair rotates partially through the flux, which is thus applied to the panel. The bottom, heavily-tinned roller of the second pair transfers solder from a thermostatically controlled reservoir to the print tracks. Variation in panel thickness is overcome because the rollers are resiliently mounted (Fig. 3.40). Nevertheless, the non-uniform layer thickness is the great drawback of this technique. On the other side, the liquid solder alloys so well with the copper of the prints that an excellent, dense layer is formed.

This method is also applied to sub-assemblies that must be pre-tinned. The layer of solder deposited on the panel is 1–2·5 μm thick; where the roller gets out of contact with the tracks, and the pressure between roller and panel is zero, lumps appear about 20 μm high.

The panel holes often close with a membrane of solder, the formation of which is prevented by sealing-up the panel side, thereby creating air buffers

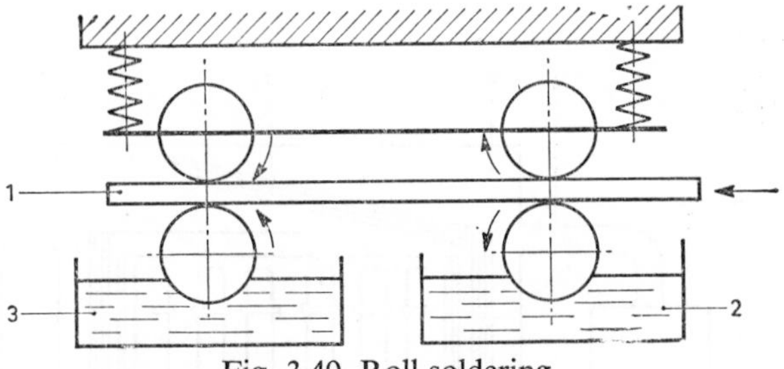

Fig. 3.40. Roll soldering.
1 = panel. 2 = flux. 3 = molten solder.

(f) *Furnace soldering*

Using a furnace, many joints, including inaccessible ones, can be soft or hard soldered at the same time. There are scarcely any restrictions on the number of products soldered per unit time, or their dimensions, shape and weight. The quality of the joint is practically independent of the operator's skill; capital investment is relatively high. Furnace soldering sometimes gives less distortion than does local heating, as in the case of welding, for instance. In gas-fired furnaces, the burnt gas should not be allowed to come in contact with the workpieces. In cases where it must be possible to change the atmospheric conditions, therefore, electric heating is to be preferred. Electric furnaces can be accurately controlled, which opens the way to mechanization of the process. The "atmosphere" used is: flux, reducing or inert gas and vacuum. Flux is preferred. The other atmospheres call for particular care as regards the sealing and corrosion resistance of the furnace (although the flux used in soldering aluminium attacks the steel of the furnace and penetrates the insulation); they are used when the application and removal of flux entails difficulties or if the flux is not effective enough. (Hydrogen is expensive and explosive. Cracked ammonia is explosive, too, but when partially burnt and dried is safe and cheaper; sometimes it is used again after reconditioning. See Section 3.3(f) (iv)(2) "Soldering in a reducing gas".)

A vacuum furnace is quite a capital investment. Its working temperature ranges from 800–1 350 °C, it is controllable to an accuracy of 5 °C in the heating zone, and its operation is more complex than that of a normal furnace, the commercially available vacuum range being from 10^{-3}–10^{-5} mm Hg.

The vacuum furnace leaves a clean joint and product, and no flux is necessary, so it need not be removed or fought. Degassing of air pockets is an extra advantage; the absence of oxygen and the possibility of high temperatures enable is to be used to braze refractory metals, ceramics, graphite, Ti, Be, Co, Al and steel. Operation is safer than with flux.

Cooling can be very slow. That is why an inert gas (argon, dew point −55 °C) is sometimes used, cooled in a separate system (Fig. 3.41). Cooling can be achieved in a few minutes, not only improving the efficiency of the total equipment but also the structure of hardened products, etc.

It is necessary to differentiate between the hot wall and the cold wall furnace. In the generally-preferred cold wall type, the heating element is

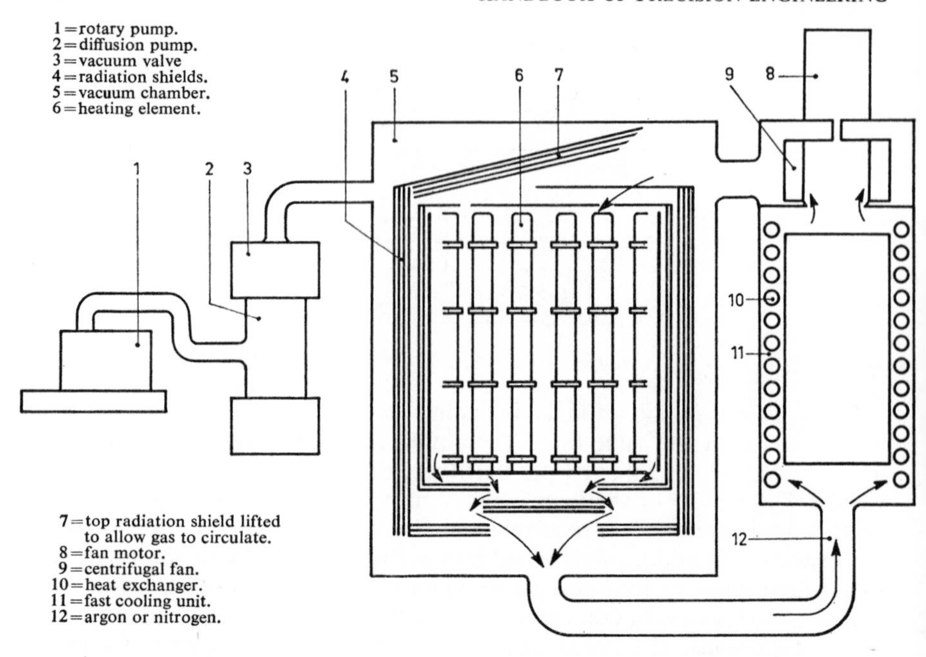

Fig. 3.41. Vacuum furnace ("Engineering", 9.1.1969).

mounted within the vacuum chamber, surrounded by insulation as metal radiation shields.

If a vacuum is used, heat is transferred by radiation, which heats the product rapidly, sometimes even too quickly, thereby causing irregular heating of the workpiece. (The distance between workpiece and furnace wall plays a role here, and consequently the shape and sort of the furnace: horizontal and vertical, bell type, oven and pit type, carry-through type.) Therefore, inert gases are best used.

Wet workpieces should never be placed in the furnace, as water may decompose, to form explosive hydrogen. (Gas formation is prevented by drying the fluxed products or by replacing the flux solvent water by alcohol.) If the atmosphere must meet stringent requirements, the workpieces are packed in a gas-tight container, which is then placed in the furnace.

High-carbon steels tend to decarburize when being furnace brazed with copper. Pre-plating with copper, and decreasing the amount of water and CO_2 in the atmosphere, reduce this tendency. Copper plating is also carried out for chromium steel and aluminium bronze, to minimize oxidation.

In view of the long heating time in furnace soldering, a solder alloy is chosen with a small melting traject.

After cleaning, the components are centred and fixed to each other, preferably by utilizing their shape, otherwise by means of clamps (cylindrical or conical), grooves, stitch and spot welding, knurling, slitting and earing, screwing (with ample play in the thread), riveting, dowelling, wires (made of Ni-Cr, or pre-oxidized, or coated with ceramic or graphite to avoid welding and soldering), belling, beading, even by means of over-pressure and vacuum equipment (see Fig. 3.41). (The elimination of all jigs and fasteners whatsoever saves on capital outlay and maintenance, in assembling and removing, in heating and cooling and in the removal of flux residue.) The expansion pattern to be expected is decisive for shape and dimensions of the joint. If necessary, certain components are supported, depending on their strength at soldering temperature; the supports (see also the next section 3.7.2(g) "Flame soldering") should form no obstruction and should not be welded to the product. They are therefore made of non-metals (no corrosion and low heat content) or of chromium and nickel alloys. To reduce heat loss by jigs, they are made as small as possible, attached as far as possible from the joint, or insulated from the work with asbestos board. The coefficient of linear expansion must be practically equal to that of the products, or the jigs should be resilient.

The products should have an air escape, so that expanded air and flux vapours can get out, otherwise the solder might be blown away or the product distort. This escape channel should be so situated that the difference in density between the furnace atmosphere and the air in the products assists in removing the air, to prevent the product being oxidized.

The solder and flux must be so positioned (usually in the form of a wire) that it cannot be displaced when the products to be soldered are transported, and the solder must be able to expand without leaving its place (cutting into pieces).

If flux is used, the furnace must be ventilated and the heating elements should not be attacked by the corrosive flux gases.

A uniform through-heating of the products is of prime importance. If necessary, the atmosphere in the furnace is stirred or the products are kept in motion inside the furnace.

As the complete products are heated, power consumption is high and the soldering process takes a long time. Besides, to avoid oxidation, the products should cool down inside the furnace, with inert gas, of course. On removal from the furnace, the products are ready for the finishing treatment, without any need for intermediate pickling.

As the soldering temperature can affect the strength of hard(ened) workpieces, the product must be designed on the basis of the mechanical properties of the as-soldered unit, unless an extra heat treatment is given after soldering.

(g) *Flame soldering* [14, 15, 16]

If bit soldering is technically possible, but provides too little heat (as is the case when soldering aluminium), a soldering flame is sometimes used, in which propane, butane, town gas, acetylene, paraffin oil, petrol, alcohol, etc., are burnt in air or oxygen. An overdose of oxygen renders the flame oxidizing, and a more active flux will then be needed. (In manual soldering it more often happens that too much flux is used than too little; this hampers the flow of solder to the soldering locus.)

The temperature the soldered joint can reach depends on the type of energy used, the mixing ratio, the quantity per unit time, and the distance to the core of the flame.

The air-gas flame is the coldest, and heats the workpiece but slowly; oxygen-gas is better, oxygen-hydrogen is better still but more difficult to control; oxygen-acetylene is eminently suitable for heavy and well-conducting workpieces: there is even a risk of overheating.

Because irregularly-shaped workpieces cannot be uniformly heated, they tend to distort. If the workpiece is heated up too quickly, it will get overheated; if heating is too slow, there will be heavy oxidation of workpiece, solder and flux. The workpiece must be brought to the correct temperature in 0·5 to 5 min, and the flame must be kept moving all the time. In practice, the soldering time is determined by experiment and if needs be adapted to other factors involved, such as the speed of a conveyor belt on which the products will be brazed, etc. To ensure uniform heating, the thickest part of the workpiece and the material with the highest heat conductivity should receive most of the heat supplied.

Heat loss is kept to a minimum by means of asbestos shields and firebricks; especially with metal components, such as clamps and jigs, the heat capacity and heat conduction should be kept as low as possible. The cross-sectional area of the material thus has to be limited.

All burners are cheap in initial expense, use and maintenance, suitable for small batches of products, easily adapted mechanically (do-it-yourself), require little space and energy, several products can be made simultaneously, and are suitable for intricate products. They are commonly used for joints where at least one of the component parts has a small volume. Joint quality depends greatly on the skill of the operator; fixed and well-adjusted burners give a more homogeneous quality. See also under (ii) "Soldering torch", below.

Commercially, burners are available with a nozzle of ∅ 0·06 mm (Henri Picard Frère Ltd., La-Chaux-de-Fonds, Switzerland).

(i) *The soldering lamp*

Is used for soft soldering small products; it requires a great deal of skill in handling and even then is dangerous, due to the risk of its becoming filthy, of overheating and explosion. A water-jacket round the fuel tank reduces the danger.

(ii) *The soldering torch*

Is used for brazing low-carbon and structural steel, bicycle frames, pipelines and nickel, copper and lead components. Interchangeable parts make it possible to changeover from one gas to another. The temperature can be as high as 3 500 °C.

The torch is universal as regards the products made with it, easy to handle and to maintain, cheap to buy, but expensive in the manufacture of large batches of products, since it is hand-operated. If the products are placed on a turntable or conveyor belt, or if a soldering jig can be used, the soldering rate can be increased by using a stationary mounted torch (as the products are often heated completely, they corrode). See also the introduction of (g) "Flame soldering" and below.

When the soldering lamp or torch is used, the products are first fluxed (by dipping or brushing), then heated: afterwards, heat, solder and flux are continuously supplied, the solder and the flux manually. The solder rod is often dipped into the flux and then applied to the seam.

There are also gaseous fluxes (for instance methyl borate dissolved in methyl alcohol and acetone), which are mixed with other gases in front of the nozzle. These fluxes are not very active.

Product design requirements

In mechanical soldering in general, and in flame soldering in particular, product design must meet the following requirements (see also Section 3.6.3):

- component parts to be joined should be self-centring (preferably round). When possible, a stop on one of the components assures the right length of travel of one part into the other.
- eliminate jigs, or make them as simple as possible.
- the soldering locus should be accessible by the flame.
- solder and flux are applied before soldering: provide a special chamber, if needs be.
- maintain homogeneous wall thickness to obtain temperature homogeneity.
- when several joints are to be made in one product, the seams should either be so close together that they are heated by the same flame, or so far apart that they can be soldered individually.

Jig design requirements

As to the jigs (see also Section 3.7.2 introduction and Section 3.7.2(f), Furnace soldering), these requirements should be met:

- simple design, easy opening, firmly clamping.
- not exposed to the flame (to reduce heat loss and strength).
- little contact surface with the product (for the same reason).
- low mass, heat conductivity and capacity, and small cross-sectional area.
- good heat resistance.
- workpieces can be assembled in the hot jig, and taken out of it.
- when possible, workpieces should drop automatically from the jig at the end of the conveyor belt.
- solder or flux should not contaminate the jigs, for this affects the positioning of the product.
- the pitch of the jigs should be kept at a minimum to improve the efficiency of the flame.
- the jigs should be universal as regards the type of product they house.

The jig can be fixed to the conveyor belt and be filled there (less capital outlay, but cycle time is lost in filling), or be placed loosely on the belt (more jigs needed, no cycle time lost in filling them). The required quantity of products also plays a role.

Degree of mechanization

Production equipment depends for its degree of mechanization among other things on the number of products to be supplied.

- The simplest equipment is a table with burners on one or two side(s), upon (and from) which the product is slid. This technique is applied to small batches and to intricate parts, where time-consuming assembly must be done on a stationary product.
- Turntables rotate continuously or intermittently. The latter type is preferred
  where the joint requires long heating time, and the component must only be heated locally.
  when the prepared product can only be mounted in a stationary jig.
  when other operations have to be carried on at separate solder locations.

By providing conveyor belts, etc., with rotating discs containing a product moving along a stationary burner, or row of burners, the product is homogeneously heated.

Efficiency of the equipment, quality of the joint and time cycle are improved by push buttons, relays, temperature control, and by devices to load the products, and supply flux and solder.

(iii) *The blow pipe*

Is often used by jewellers and dentists. The higher temperature is obtained by leading air through the pipe (with oil or alcohol vapour). The older system of blowing with the mouth has been replaced by foot-operated bellows, and these by a compressor.

(h) *Hot air soldering*

This seldom-used method is a variation of flame soldering and is only mentioned for the sake of completeness.

Instead of hot air, hot hydrogen is sometimes used: no separate flux is necessary, so the soldering locus remains more visible, the joint is reliable, the process can be mechanized and is cheap. Hydrogen is explosive. In practice, the reducing action of this gas at soldering temperature sometimes proves to be insufficient; the resulting long soldering time can give rise to changes in material structure and promote the evaporation of highly-volatile metals like zinc and cadmium.

(j) *Medium- and high-frequency induction soldering* [17,41]

In this method, a coil consisting of at least one turn of (water-cooled) copper pipe is placed round (or in) the soldering seam. The current in the pipe induces eddy currents in the product which is thus heated and consequently causes the solder to melt [4,18,19], [60] for the electrical equipment.

Advantages of this method are:

- due to the great energy concentration, heating takes place very rapidly, so there is little oxidation, decarburizing, discolouring, volatilization and grain growth, involving loss of mechanical properties. There is also little heat loss in the ambient metal and a far more efficient use of heat than in furnace soldering, for instance.

- by using a protective gas, flux can often be omitted, so the surface remains clean over a longer period of time.
- the working cycle can be smooth, which means an improved production rate.
- the method is suitable for hand or mechanized operation.
- heating is very local and causes hardly any warping.
- the process can be accurately controlled (time switch, maximum temperature switch, stop for the product in the coil, etc.) and is therefore suitable for soldering aluminium [20].
- most economic in its use of filler metal.
- the flux is heated indirectly, and consequently not burnt or blown away by the solder flame.
- good work conditions for the operators: clean, and without a warm atmosphere.
- labour can consist of unskilled workers.
- less floor area required than in furnace soldering, for instance.
- the equipment is immediately ready for use, contrary to the furnace.

Disadvantages are:

- high capital outlay.
- high investment costs demand production in batches.
- maintenance requires skilled workers.
- the coil shape cannot always be suitably adapted to the shape and dimensions of the product.
- less suitable for joining components made from material with good current-conducting properties, hence most used for ferrous metals.

At higher frequencies, three effects play a part: *the skin effect* (the current density at the surface of the conductor is greater than in its core), the *proximity effect* (current density occurs in the sides of components facing each other) and the *coil effect* (the current density on the inside is greater than on the outside of the coil; so the efficiency of externally-mounted coils is greater than for internally-mounted coils). These three phenomena are turned to maximum use; in soldering pipe the coil is placed around it (coil and proximity effect), the soldering of thin components becomes easier at higher frequencies (skin effect), etc. Care should be taken that the skin effect does not heat only a layer with high specific heat conduction and high resistance.

Other factors of importance are:

A narrower *gap* between coil and workpieces gives higher temperatures, which are very local.

A great temperature gradient in a radial direction is obtained by *momentary* heating at great *power*.

A great temperature gradient in an axial direction is obtained by *momentary* heating with one turn.

Other factors are: the height of the coil, its position with respect to the soldering seam (the solder must melt by conduction from the heated material), the flux, the solder, the shape and dimensions of the soldering seam and the adjacent components (the part with the greatest mass is heated first or most*). Other factors in so far as the material is concerned are: the specific heat, heat conduction and electrical conductivity. The electrical conductivity, for instance, decreases for steel as the temperature increases, which is a clear skin effect. Hence, non-ferrous metals, with their better conductivity, require more windings and have a lower energy efficiency. To solder various materials together, the coil is placed closest to the material with the lowest resistance.

The frequency varies from 1 to 2 000 kHz, the power from 1 to 1 000 kW, the energy concentration from 80 to 1 500 W/mm^2, depending on the above-mentioned factors.

The shape of the coil is determined by the shape of the workpiece (Fig. 3.43). Where possible the pipe is round, but sometimes square (for instance, to prevent a coil with more windings from heating the product in "slices").

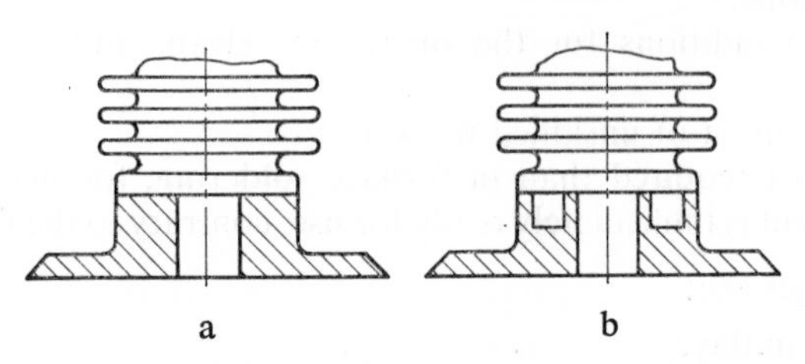

Fig. 3.42. Adapted shape prevents excessive heat flow to a small-volume component.*

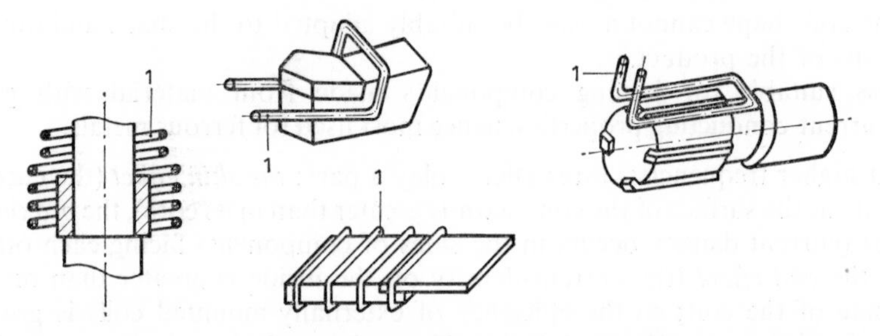

Fig. 3.43. The product determines the shape of the HF coil, denoted by 1.

When the pitch of the windings is constant, the highest temperature is found half-way down the coil. To obtain an even temperature distribution along the workpiece, the pitch of the middle coil windings is chosen somewhat more than the pitch at the end of the coil.

Sometimes, for workpieces of complex shape, the pipe *a* is replaced by a profiled copper plate *b* with cut-outs for the product (Fig. 3.44). Large flat surfaces are heated by means of a spiralled pipe. The coil is sometimes locally cooled or screened [7,20], in order not to damage a low-melting or hardened part. In that case, too, the gap between coil and workpiece is taken

* In joining thin, heat-sensitive components to thicker, less-vulnerable parts, e.g. bellows to a solid hub, the latter might heat the former too rapidly, inflicting thermal damage. To prevent excessive heat flow, the situation from Fig. 3.42a is then replaced by that of Fig. 3.42b.

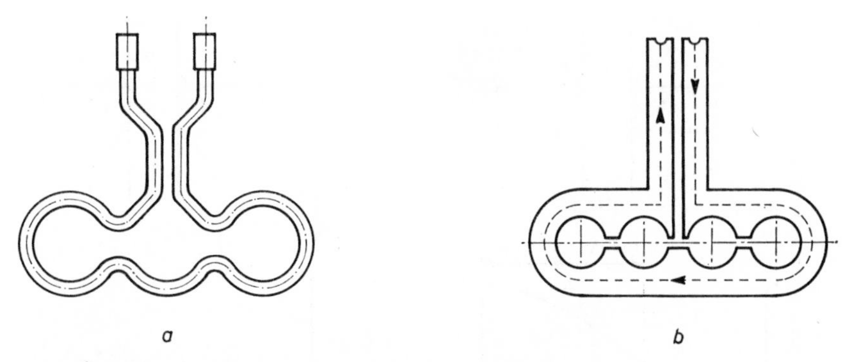

Fig. 3.44. A profiled plate follows the workpiece contours better than a coil.

somewhat larger, especially at protruding angles, where heat dissipation is less.

In practice, the clearance, which preferably should be zero as regards the electrical efficiency, lies between $1\frac{1}{2}$ and 5 mm (although 20 mm is no exception) to avoid short-circuits (large components, for instance, can expand so much as to come in contact with the coil). To avoid this, the coil is sometimes insulated, for example, with teflon, fibreglass, ceramics, enamel, asbestos cord with water glass or synthetic resin. A larger clearance allows for a certain difference in shape between product and coil, but leads to slower heating.

The position of the coil must give optimum heating; the plane of the coil is thus placed parallel to the plane of the soldering seam, and a spiral coil is avoided. If seams run close together, solders of different melting points are used.

For the same purpose, the coil is sometimes placed within telescopic pipes to be joined: due to the expansion of the inner pipe there is good heat contact with the outer one, so less solder is used.

The difference in the coefficient of linear expansion between copper and steel or steel and hard metal, for instance, should also be taken into account (in the case of, for example, chisels, the shank is slowly heated to about 750 °C: only then is the seam heated).

The products are supported by heat-resistant insulating material, such as asbestos-cement; the components are fixed together (see (f), Furnace soldering). For the remainder, the products must be locked-up to withstand electromagnetic forces.

The shape of the seam must be adapted to the soldering method: tightly-fitting components and a large contact area improve heat transmission. The axis must be parallel to that of the coil; a very complicated product can, if necessary, be rotated round its axis in the coil to ensure uniform heating. The solder should only be melted by the heat of the component parts, see Fig. 3.45. (This means that these parts are at working temperature when the solder melts, so that the solder does not contract into globules. Besides, it is then known that the moments of melting for solder and flux are mutually

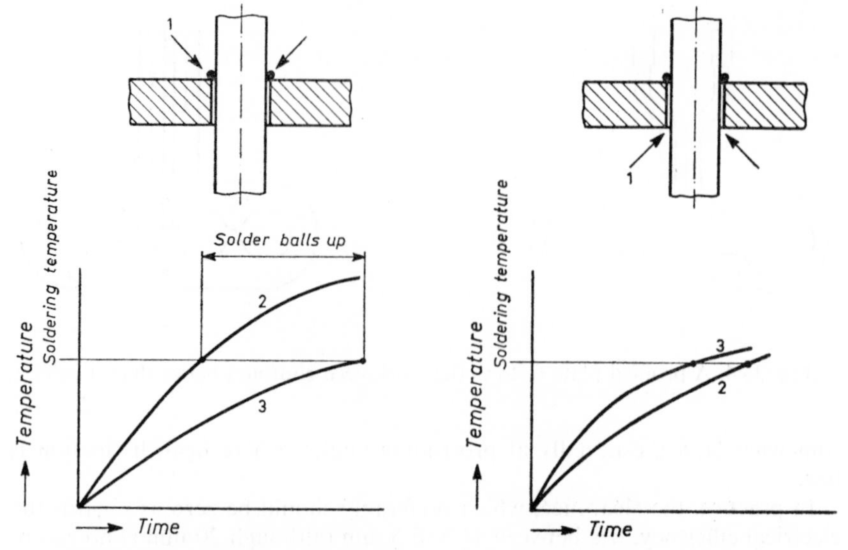

Fig. 3.45. The influence of heat flow on the ultimate solder temperature.
1 = direction of heat flow. 2 = solder. 3 = workpiece.

adapted, as in the case of resin cored wire. This applies to all soldering and brazing techniques.) The very popular soldering ring should therefore really be open, so as not to form a field and heat itself (apart from that, there are foil, wire, grains, powder, paste, etc.). Soldering grooves are often used, for example, so that the solder cannot be removed from it by electromagnetic forces.

In general, the use of (expensive) silver solder will prove advantageous because of its low melting point, rapid working and excellent results, while the flux is easier to remove than that used for high melting solder.

High-frequency soldering, with its high heat concentration, requires uniform and appropriate heating and cooling. Sometimes, the heating rate is so high that there is insufficient time for the flux to remove stable oxides (e.g. silicate grains embedded in the surface layer by spraying).

When using protective gas, the explosion risk must be taken into account, so the hydrogen coming from the furnace etc. is ignited and burnt, or a mixture of $10H_2/90N_2$, and sometimes an inert gas, is used instead. Vacuum is also employed [39].

There are high-frequency soldering installations with more coils for the simultaneous soldering of several seams (of the same product or of several products). Where possible mass soldering is preferred. It is better to insert and withdraw the product from a fixed coil than to insert a product and slide the coil over it: the flexible connections required for that purpose sometimes consume a lot of power and entail extra risks of short-circuit.

Preferably, rotational-symmetric products, which can be led through the fixed coils by a conveyor belt or a turntable, are chosen for this technique. Here, the products to be soldered are fed in on one side of the feeding device, while they are soldered simultaneously elsewhere.

One of the most recent developments is in soldering sleeves, that is, heat-resistant plastic sheaths filled with an annular charge of soft solder and flux. The sleeve is slid over parallel adjacent wire ends to be joined. The joint is made by heating the sleeve, which simultaneously shrinks tightly around the wires, as shown in Fig. 3.46, borrowed from [46].

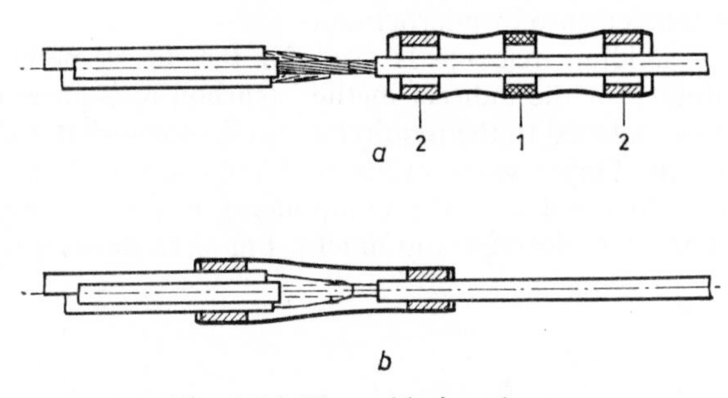

Fig. 3.46. The soldering sleeve.
a. before heating. 1 = solder ring.
b. after heating. 2 = sealing ring.

(k) *Resistance soldering*

In this method, the heat is developed by the resistance an electric current encounters in one graphite electrode (single point soldering) or two graphite electrodes (parallel-gap soldering), pressed against the two components (indirect resistance soldering or carbon resistance heating) or due to the transition resistance between solder and components (direct resistance soldering or interface resistance heating, with copper electrodes). Sometimes, a combination of the two methods is used; the graphite electrode then rests against the component part with the lowest resistance, and the copper electrode on the one with the greatest resistance.

By machining, the electrodes can be adapted to the shape of the components, so that clamps and jigs are often superfluous. On the other hand, the mechanical vulnerability of graphite requires the electrodes to be of simple shape, and thus the product, to some degree.

In resistance soldering, an exact dose of energy is only supplied when and where it is required, thereby reducing the risk of attacking nearby soldered joints or otherwise heat-sensitive components parts. The simple operations and low maintenance costs make this process suitable for rapid labour-independent soldering and brazing, with unskilled labour of large quantities of small products (cable tags, printed wiring boards, integrated circuits, contacts, springs, knives, saw blades, burner components, etc.), of very different material and dimensions, made of copper, brass, silver, titanium [21], molybdenum, tungsten [7,22], etc. Sometimes, several joints are made simultaneously (multiple lead soldering). If the component parts differ too much in thermal and physical characteristics, or in thickness, there is a danger of the soldering place shifting to one of the electrodes. Occasionally, this effect is aimed at; then one electrode is made from copper, brass, mild steel or tungsten, so that heat develops near the carbon electrode. When, however, a proper heat balance is required at the solder interface, an electrode with high conductivity or large contact surface area is placed on the component part with small conductivity or small thickness, so that the current density in the material of the component part decreases.

According to [44], thermocouples are even built into one of the electrodes to control the temperature in microcircuits.

Often a spot welding machine is used for the direct method (interface resistance heating); for the indirect method (carbon resistance heating), the machine must be matched to the requirements of a somewhat higher voltage* and lower current, longer work cycles and even lower electrode pressure (otherwise the solder between the components is pushed aside, and the interface resistance will decrease too much). Fig. 3.47 shows a few soldering

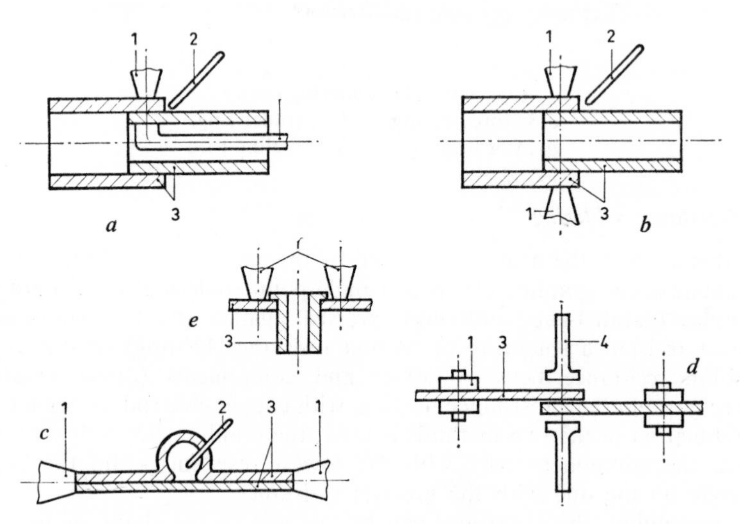

Fig. 3.47. Position of the electrodes with respect to the workpiece.
1 = electrode. 2 = solder stick. 3 = workpiece. 4 = insulated clamp.
In *b*, *c* and *d* not only the soldering spot is heated, but the part of the workpiece between two electrodes placed far apart. This is unfavourable for joints already made.

arrangements; in the lower-right diagram, the clamps (4) are placed directly on the soldering seam, and the contacts on either side of it. For very small products, use is made of hand-tongs, with automatically-fed resin core solder, for instance [42]. The solder is generally supplied in the form of a foil; borax is often the flux. Self-fluxing or pre-fluxed solder does away with the need to apply flux. From time-to-time, flux residue must be removed from the electrodes, etc., but this interruption of the work can be avoided by replacing the flux by reducing gas. Cadmium solder, copper-phosphorus 93/7, silver-copper-phosphorus (71/28/1), etc. are used in this method.

*For most applications voltage ranges from 2 to 12V.

When the component parts have been coated thermally or electrochemically with a layer of solder, and need at most some extra flux, the process is called *reflow soldering*. This method finds increased application in soldering printed wiring panels and integrated circuit chips, where all wire ends can be joined very reliably, simultaneously, and if needs be completely mechanized, including component positioning. Advantages are mentioned in the first note of Section 3.6.3. Often the flux can be omitted, in which case no absorbed gases and oxides are removed from the component parts.

Reflow soldering can be achieved by means of a gas jet, a furnace, a hot wire, induction, radiation or resistance heating. The last method is the one most used by far. Here, it is necessary to distinguish parallel-gap soldering (two parallel adjustable electrodes at the same side of the component parts, which provide the electrical resistance), single-point soldering (two parallel electrodes at the same side of the component parts are interconnected by a piece of molybdenum to provide the electrical resistance) and multiple lead soldering (a multiplication of the single point principle). (With these methods it is possible also to solder pre-soldered insulated wires without stripping them off.)

A new variation is to use a normal tungsten-arc welding machine which heats the base metal while adding extra argon, after which solder is applied and heated until it has wetted and spread. This technique is applied for the vacuum-tight joining of dissimilar base metals [54].

(m) *Electrolytic soldering* [7]

Here, the workpiece is connected to the cathode submerged in an electrolyte through which a d.c. current is led; the cathode is thus heated. The cathode/components are immersed in an electrolyte or the electrolyte is sprayed on them, or is supplied via a porous insulator making contact with the workpiece. Around the cathode, a coat of hydrogen builds up, with high resistance, which is repeatedly penetrated by powerful sparks. The correct hydrogen concentration, that is, the composition of the salt bath, is very important as regards its influence on the current density. The latter is higher at the sharp edges of the product, which must be locally screened. Moreover, the product dissolves and must be small and simple of shape. It is, however, possible to solder unequal materials rapidly, automatically and without oxide inclusions. The Russians are developing this process further.

(n) *Electric-arc soldering* [18,23]

In this process, an arc is drawn between a carbon electrode and the assembled components, which are provided with foil or wire solder. The process is sometimes adopted for electrical components, but there is a risk of a spark damaging the insulation.

A new application is described in [52]: pin soldering. This technique much resembles stud welding: here, too, a pistol and iron or copper base alloy

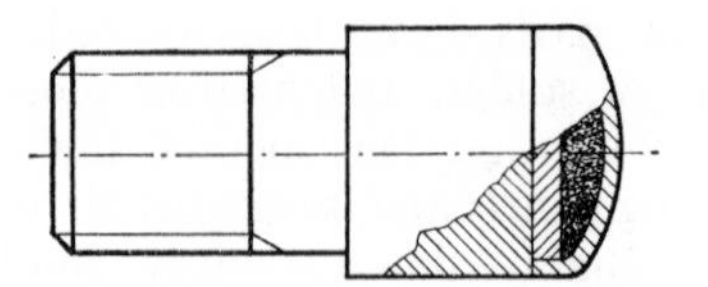

Fig. 3.48. Pin soldering pins.

pins, are connected to the anode and provided with flux and filler metal (see Fig. 3.48).

As the soldering temperature lies below the melting point of the base metal (steel and copper alloys), the method is suitable for soldering thin sheet, where holes would form in welding. The soldering temperature can be chosen below the hardening temperature of carbon steel.

During soldering, the pin-end is surrounded by a ceramic ring, which protects the arc, prevents spattering on the workpiece and excludes the air.

(p) *Soldering with infrared radiation or halogen lamp*

The infrared beams are focused by means of a gilt (silver oxidizes too rapidly) ellipsoid mirror behind the lamp. This is not a very quick method; besides, correct positioning of the workpiece takes time, and the distance between the lamp and each different product must be found experimentally, to arrive at the correct soldering temperature. There is no contamination by the heating medium, there is not even contact with the workpiece, and due to the absence of a soldering bit, no solder peaks can be drawn on solder which is too cold.

Other advantages are: a better reproduction of temperature (up to 1 600 °C) (by means of a transformer, the element has little thermal inertness), the possibility of soldering otherwise-inaccessible components and parts in closed rooms (with *consequent* freedom in atmosphere or vacuum; transmission of rays through thin-walled glass or quartz vessels does not lead to an appreciable loss in energy), the compactness and lightness of the heating source, the possibility of heating quite locally, and the ease of mechanization with a small capital outlay. The method is seldom used [40,45], but finds increasing interest.

There are point heaters with a capacity from 150 to 2 000W and a focal plane of 40 to 250 mm^2, and line heaters up to 5 000W with a band length from 18 to 260 mm. The energy density can amount to 200W/cm^2, which is comparable with that of a gas-oxygen flame.

Soldering by means of a laser is still in its infancy and is only mentioned here for the sake of completeness.

(q) *Friction soldering*

This is a technique where the solder, heated to above its solidus temperature, is "rubbed" into the surface of the components to be soldered, without using a corrosive flux. The result is a mechanical adhesion and sometimes (with metal components) the removal too of the oxides. The friction capacity is derived from a solid stick of solder or from solid particles of solder or solid particles of an abrasive in the liquid solder. The composition of the solder is, for instance, 30Sn/70Pb, with large melting traject, 40Sn/60/Zn, 60Sn/40Zn, 90Sn/10Zn, 85Zn/15Al or 56Zn/44Cd (DIN 1732).

The solder is rubbed-in with a piece of solder, a piece of pumice stone or something similar, or with a rotating grinding wheel (the friction produces heat) which has first been tinned itself by running it against a piece of solder [24,25]. The solder used here is 50Sn/50In, 50Sn/50Pb, 50Pb/50In or Woods metal. The presoldered component (usually made of a material difficult to solder, such as aluminium, stainless steel, titanium, tantalum, molybdenum, tungsten, ceramics or glass [26]) can then be soldered in the normal way (for instance, with a soldering bit). Friction soldering is also applied to fill up uneven parts in sheet and welded work [7].

Aluminium, for example, is soldered with a Belarc (England) apparatus, the steel wire brush of which vibrates at 100 Hz, fractures the oxide layer, and solders the underlying metal with the available liquid solder (80Sn/20Zn). For the soldering of cylindrical bodies, friction soldering is sometimes used in the way of friction welding; a piece of solder is placed between the rotating components, where it then melts.

(r) *Ultrasonic soldering*

This method is employed for the fluxless soldering of aluminium and its alloys, by means of an ultrasonic bit (Fig. 3.49). The point of the bit,

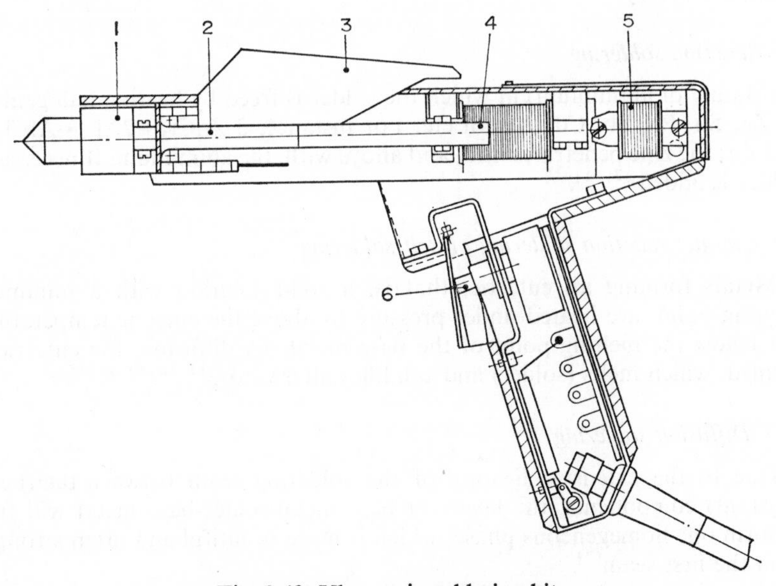

Fig. 3.49. Ultrasonic soldering bit.
1 = heating coil. 4 = magnetostrictive material.
2 = connecting rod. 5 = reaction coil.
3 = cooling fin. 6 = switch.

vibrating at about 20 kHz [27], is dipped into liquid solder lying on the aluminium, thus causing cavitation [28] and breaking the oxide film. Consequently, the bare aluminium is presoldered. A variation of the method is to connect a soldering bath to the bit, in which bath small products are then pre-soldered. This gives better results than the bit.

The aluminium is heated to just above the melting point of the solder. At increased temperature the aluminium itself is attacked. (Aluminium containing a large content of magnesium, but also magnesium, beryllium and titanium, raise difficulties in ultrasonic tinning; the same applies to the edges of aluminium products).

Operation and construction of the ultrasonic bit are described in [7]; the equipment is complicated, expensive and vulnerable, and not suitable for pre-soldering very thin foils and wire: these are damaged by cavitation.

Because of the absence of flux (residue), corrosion is prevented.

Since particles of aluminium oxide are embedded in the surface of the solidified solder, the solderability of the workpiece is not yet much improved.

The solder used is Zn-Sn(-Cd)(-Ag), for instance 42Zn/55Sn/3Ag; silver improves the corrosion resistance and the strength. Soldering takes place with an electric bit, the component parts being heated to a few dozen °C above the liquidus temperature of the solder: [29,30].

(s) *Other techniques*

For the sake of completeness, a few rarely-used processes are now discussed briefly.

(i) *Reaction soldering*

(Mainly of aluminium) in which the solder is freed by heating halogenides of Zn, Sn, Pb, Sb, Cd, Ag, Bi etc. For instance, $3\ ZnCl_2 + 2Al \rightarrow AlCl_3 \uparrow + 3\ Zn$; the zinc penetrates into, and alloys with, the aluminium. If necessary, solder is added [7,31].

(ii) *Contact reaction soldering or self soldering*

Metals forming an eutectic, that is, a solid solution with a minimum melting point, are heated under pressure to above the eutectic temperature, but below the melting point of the base metal, By diffusion, the eutectic is formed, which melts (solder) and solidifies afterwards. [7,32,33,34,35].

(iii) *Diffusion soldering*

Due to the extensive heating of the soldering seam between the (heat-resistant) components, the layers of base metal-solder-base metal will fuse to form one homegeneous phase, which is more beautiful and often stronger than the first seam[7].

The solder consists of an intermediate foil or of a metal that is vapour deposited on (or galvanically applied to) one of the components, to improve diffusion between the two base metals. Previous deformation of the component parts gives more dislocations, improving diffusion. Sometimes, pressure is applied to establish the atomic contact needed for diffusion. The factors of influence are then temperature, pressure and time, as well as the purity and smoothness of the mating surfaces, and the atmosphere.

Beryllium is joined to itself with a 2·5 μm thick aluminium foil heated to 725 °C for a period of 10 min or with 0·75 mm thick copper at 885 °C. Tungsten is brazed to tungsten with a 0·25 mm thick niobium foil heated to 925 °C for 20 min under a pressure of 700 kgf/mm^2; titanium is first provided with a copper foil [51], or sometimes silver coated.

To increase the operating temperature whilst maintaining the strength, elements reducing the melting point are added to the filler metal. At the brazing temperature, these elements penetrate the base material, thus being extracted from the filler metal, so that its melting point increases. This technique gives the opportunity of soldering close to joints previously made, without melting them.

(iv) *Volatilization brazing*

For the same purpose as with diffusion alloying, melting point reducing elements are added. Brazing in a reducing gas is followed by heating in vacuum till below the solidus point. As a result, the volatile elements (P, Li, etc.) escape, so that the melting point rises.

(v) *Reactive brazing*

Here, too, the filler metal contains elements lowering its melting point. After brazing, they are eliminated by heating to above the original melting point, for one hour, say. This leads to the formation of stable compounds between the additives and the base or filler metals.

This technique looks promising for the superalloys, such as the heat-resistant metals, used in space travel. An application mentioned in [50] is brazing tungsten with Pt/3·5B. Continued heating gives rise to crystal growth and brittleness.

(vi) *Braze welding*

(Solder welding or weld soldering) is soldering (mainly of cast iron, but also of copper, which is then brazed with brass solder) of V-grooves filled by gravity only. The surface must be cleared of all graphite, it is pre-tinned and pre-heated to about 350 °C; the joint is tougher than it would be after welding [7,36,37,38].

REFERENCES

[1] B. Cambels *Metal cleaning methods*, Materials and Methods, **38**, No. 5, 1953.

[2] A. P. Schulze, *Electrostatic descaling*, Production Engineering and Management, **22**, No. 3, p. 67, 1948.

[3] W. R. Lewis *Einige Probleme beim Weichlöten*, Metall, II, Heft 5, p. 372, 1957.

[4] E. Lüder *Handbuch der Löttechnik*, Berlin 1952.

[5] *Automatic soldering machine*, Sheet Metal Industries, **31**, No. 325, 1954.

[6] T. E. Protz *How to dip-braze aluminium*, American Machinist, **99**, No. 2, 1955.

[7] N. F. Lashko, S. V. Lashko-Avakyan *Brazing and soldering of metals*, London 1961.

[8] A. J. Davis, A. J. Easton, J. Freezer *The analysis of soldering alloys by density*, Sheet Metal Industries, **32**, No. 335, 1955.

[9] W. E. Hoare *Hot tinning*, Tin Research Institute, 1948.

[10] *Proceedings of the second RETMA conference on reliability of electrical connections.* Sept. 11–12, 1956, RETMA Engineering Office, New York.

[11] E. S. Miller, A. A. Johns *Dip soldering printed circuits;* Symposium on solder, ASTM-STP No. 189/1956, p. 30–39, 115–126.

[12] C. J. Thwaites *Some experiments on the hot-tinning of small parts*, Metallurgia, Sept. 1961, p. 117.

[13] *Working instructions for hot-tinning cast iron*, Tin Research Institute, 1957, p. 381–385.

[14] V. Beatson, H. R. Brooker *Industrial brazing, torch brazing*, Welding and Metal Fabrication, **20**, No. 6, 1952.

[15] H. H. Grix, *Efficient torch soldering*, Schweissen und Schneiden, Aug. 1957, p. 381–385.

[16] *Equipment selection for soldering*, Proceedings of the second *RETMA conference on reliability of electrical connections*, 11–12 Sept. 1956, RETMA Engineering Office, New York.

[17] F. W. Curtis *High-frequency induction heating*, McGraw-Hill, New York, 389 pp.

[18] *Brazing Manual*, A.S.W., No. 1, New York 1955.

[19] V. Beatson, H. R. Brooker *Industrial brazing*, 1953.

[20] C. H. Yetman *Induction heating successfully solders aluminum*, Iron Age, **167**, No. 11, p. 108, March 1951.

[21] N. A. de Cecco, J. U. Perks *The brazing of aluminium*, Welding Journal, **32**, No. 11, p. 1031, 1953.

[22] Anon *The selection problem*, Materials and Methods, **35**, No. 3, p. 120, 1952.

[23] W. Reed, L. Edelson *Silver-brazing by carbon-arc process*, Metal Industry, **52**, No. 18, 1954.

[24] I. C. McGuire *This new method permits soldering difficult materials*, Materials and Methods, **44**, No. 1, 1956.

[25] Anon. *Method of soldering difficult materials*, Machinery, vol. **89**, July 6, 1956, p. 81.

[26] A. Kei *Über die Benetzungsfähigkeit von Löten*, Zeitschrift für Metallkunde, Heft 7, 1956.

[27] B. E. Noltingk, E. A. Neppiras *Cavitation produced by ultrasonics*, The Proceedings of the Physical Society (B), **63**, No. 9, p. 665, 1951.

[28] B. E. Noltingk, E. A. Neppiras *Ultrasonic soldering irons*, Nature (L), **166**, p. 615, 1950.

[29] A. E. Grawfords *Ultrasonic soldering of light metals*, Light Metals, **15**, No. 3, 1952.

[30] L. WALTER — *Ultrasonics, the answer to aluminium soldering*, Materials and Methods, **38**, p. 59, 1953.
[31] D. C. BURCH, R. L. SIMPKINS — *New tin-depositing flux for soldering aluminium*, Materials and Methods, **40**, No. 3, 1954.
[32] W. HOFMANN, H. J. HUSMANN, R. KOPPE — *Die Keilschweissung von Aluminium und Kupfer als Selbstlötung*, Zeitschrift für Metallkunde, **43**, Heft 10, 1952.
[33] A. KEIL — *Wie verbinden sich Metalle*, Zeitschrift für Schweisstechnik, No. 2, 1951.
[34] E. V. BEATSON, H. R. BROOKER — *Furnace brazing*, Welding and Metal Fabrication, **20**, No. 9, 1952.
[35] D. BROOKS — *Brazing titanium to titanium and to steels*, Metal Fabrication, **21**, No. 111, 1954.
[36] C. H. WANAMAKER — *How to bronze-weld*, Welding Journal, **29**, No. 3, p. 235–257, 1960.
[37] H. R. SCHMUCK — *Principles of bronze welding*, Welding Journal, **31**, No. 10, 954, 1952.
[38] *Werkstoff und Schweissung*, Band 2, Akademie Verlag, 1133, Berlin 1954.
[39] S. WATKINS — *Tube assemblies joined by vacuum induction brazing*, Metal Progress, April 1966, p. 73.
[40] B. J. COSTELLO — *Infrared soldering of printed circuits*, Design News, 27 Oct. 1965, p. 154.
[41] *Philips high frequency induction heating*, N.V. Philips' Gloeilampenfabrieken, Eindhoven 1961.
[42] *Cold tip puts solder where it counts*, Iron Age, p. 80, 5 Nov. 1964.
[43] H. H. MANKO — *Solders and soldering*, Mc-Graw-Hill Book Company.
[44] ANON. — *Soldering for microcircuits*, The Engineer, p. 168, 27 Jan. 1967.
[45] ANON. — *Brazing and brazing alloys.*
[46] ANON. — *Solder sleeves*, The Engineer, 20 Jan. 1967.
[47] F. Z. KEISTER — *Beware of solder slivers*, Electronic Packaging and Production, p. 162, Nov. 1966.
[48] H. B. G. CASIMIR, J. UBBINK — *The skin effect*, Philips Techn. Review, 28, 1967, No. 9, p. 271.
[49] R. CAILLER — *Le brasage au bain de sels de l'aluminium et de ses alliages*, lecture delivered for the Société des Ingenieurs Soudeurs, 17 Nov. 1966.
[50] L. SANDERSON — *A new brazing technique*, Tooling, July 1966, p. 39.
[51] K. R. PERUN — *Diffusion welding and brazing of Titanium 6A1-4V, Process Development*, Welding Journal Supplement, September 1967, p. 385–s.
[52] W. VANSCHEN — *Das Elektrische Stiftlöten*, Industrie-Anzeiger, 23 April 1968, p. 665 (73).
[53] W. B. ARCHEY — *Hot gas soldering for interconnection of integrated electronic packages*, IEEE, Dec. 1964, p. 1657–1660.
[54] D. E. SOLOMON — *Joining dissimilar metals by gas tungsten-arc braze-welding*, Welding Journal, March 1968, p. 181–191.
[55] ANON. — Inco Nickel, No. 20, Oct. 1967.
[56] V. B. BOIKO, YU. A. BAIGUDANOV — *Flash soldering by heating in glycerine*, Svar. Proiz., No. 8, 1967, p. 42.

[57] C. J. THWAITES *Some effects of abrasive cleaning on the solderability of printed circuits*, Metal Finishing Journal, Sept. 1968, p. 291.
[58] B. M. ALLEN *Soldering Handbook*, Multicore Solders Ltd. 1969.
[59]W. D. WILKINSON *High frequency induction brazing and soldering*, British Welding Journal, October 1965, p. 478–487.
[60] C. E. EADON-CLARKE *Modern production line methods for soldering and brazing* British Welding Journal, October 1965, p. 500.

3.8 Quality control of soldered connections

Good quality depends on the shape and dimensions of the soldering seam, the materials used and strict adherence to the directions given earlier. In every case, the function of a joint must be borne in mind.

The results can best be investigated on the joint itself, preferably under practical conditions. As this is often difficult, or even impossible, and it is desirable to make a correct forecast, many industries and official bodies have published guides and quality assessment standards for soldered joints and their behaviour.

Obviously, the investigation is usually carried out by experts in well-equipped laboratories. Testing methods are, for instance: **Visual assessment** of the product or a cross-sectional area of it (a smooth surface but rough at the top. When the joint is completely smooth, there is the risk of a "cold" joint; without holes, bridges, fractures and specks, fine grain; good connection along the whole periphery; the shape of the joined components should be visible through the solder; the seam should be concave; no insulating material in the solder; not too much bare wire-end; insulation not damaged by soldering heat*; sufficient wire-end around solder tags, etc.; a cylindrical hole should contain all the individual wires of a litzwire (type of twinned wire); wire-ends should be placed in the hole as deep as possible; the print panel should not be damaged; wire-ends bent flat along the panel should not protrude from that panel; plated-through holes, in which a wire-end is mounted, should be filled with solder, etc). **Tensile test** (short or long-lasting at a single or double overlapping seam or butt seam). **Bending test** around a mandrel. **Creep test. Other tests** are: Dynamic loads among which vibration, over and under pressure (soundness), melting test (of solder), wet chemical analysis (for instance, the silver nitrate test to prove the presence of chloride rests), spectrochemistry, X-ray fluorescence, radio activity, pigments and gas for investigation of porosity, ultrasonic and other means of non-destructive testing as ultraviolet light to demonstrate flux residue after cleaning of the joint, measuring the resistance of a residue of ionized flux dissolved in water, to prove the presence of flux, etc. The investigated quantities are:

(a) As regards the *solder alloys.*

Composition, contamination, melting point, specific mass, strength, coefficient of linear expansion and conductivity.

* When the insulating material produces corrosive products when heated (e.g. PVC, teflon), the distance from the beginning of the insulation to the joint should be 1·5 mm or 6 times the metal wire diameter.

(b) As regards *resin cored soldering wire* (those not mentioned under (c)).

Shape, diameter of wire, weight, flux content, homogeneous distribution of the flux in the wire and sputtering during soldering, chlorine content.

(c) As regards the *fluxes.*

Composition, contamination, content of solid matter, specific mass, melting point, viscosity, corrosivity, acidity, flow, dielectric constant, insulation resistance, specific resistance of the flux in water, resistivity against sagging in the solvent, colour, poisonousness and condition after soldering: slivering, pulverizing, hardness and stickiness.

(d) As regards the *solderability* of component parts.

The effect of the pre-treatment (for instance, metallic or organic coating), the percentage of coated area after dipping in flux and solder, the time required for complete adhesion of solder to an immersed sample (sheet or wire, wetting time test), the capillary rise in metal tubes, between two metal strips or twisted wires, the surface tension, the area covered by a standard quantity of flux and solder, the height of the flown drop or the boundary angle α, the minimum temperature at which a combination base metal-flux-solder wets within 0.2 sec [27], and the variation of the vertical force as a measure of wetting when soder tags, etc., are dipped vertically into the soldering bath [56,58]

(e) As regards the *soldered products.*

Adherence, strength, creep, plasticity, porosity, also of seams and grooves, degree of filling of grooves with flux and solder, maintenance of gap width, electrical properties, corrosion resistance, removability of flux remnants and the pre-treatment of the surface.

REFERENCES

(*a*) *Solder alloys*

[1] *Hartlote für Schwermetalle*, DIN 8513.
[2] *Silberlote zum Hartlöten von Edelmetallen*, DIN 1735.
[3] *Soft solders*, British Standards Specification, BS 219, 1959.
[4] *Hart- und Weichlote für Aluminium und Aluminiumlegierungen*, DIN 8512.
[5] *Weichlote für Schwermetalle*, DIN 1707, 4–1964.
[6] *Solder, lead alloy, tin-lead alloy and tin alloy*, Federal Specification QQ-S-571d, 1963, USA.
[7] *Tentative specification for solder metal*, ASTM B32-60aT, 1960.
[8] *Methods for chemical analysis of metals*, ASTM, 1960.
[9] *Ingot tin*, British Standards Specification BS 3252, 1960.
[10] *Soft solders for automobile use*, British Standards Specification, AU 90, 1965.
[11] *Methods for the sampling and analysis of tin and tin alloys*, Parts 1 to 13, 17 to 19 of British Standards Specification, BS 3338, 1961.

(*b*) *Resin cored soldering wire*

[12] *Resin-cored solder wire "activated" and "non-activated" (non-corrosive)*, British Standards Specification, BS 441, 1954.
[13] *Tentative specification for rosin flux cored solder*, ASTM.B 284-60T.
[14] *Special Technical Publications*, ASTM No. 189, 1956.
[15] *Report of the conference on reliability of electrical connections*, RETMA, pp. 46–54, 1954, Illinois, USA.
[16] *Soldering fluxing and fluxed solders*, UL 381-O-o (USA).
[17] *Resin flux cored solders, Japanese Industrial Standard*, JIS C2512–1965, Japanese Standards Association.
[18] *Weichlote für Flussmittelseelen auf Harzbasis*, DIN 8516.
[19] *Non-corrosive rosin-cored solder wire, containing rosin, mildly activated rosin or activated rosin. Activity of flux, corrosivity of residue*, QQ-S-571d, 1963, USA.

(*c*) *Fluxes*

[20] *Non-corrosive flux for soft soldering*, Aircraft Materials Specification DTD 599, 6-1953 (England).
[21] *Flux, soldering, liquid (rosin base)*, Military Specification Mil-F-14256c, 20.12.1963 (USA).
[22] P. M. Fisk, *Testing of solder fluxes*, Sheet Metal Industries, Sept. 1954, pp. 743–747.
[23] F. Hochberg, *The suitability of soldering fluxes for use in the assembly of military equipment*, Report of the conference on reliability of electrical connections, RETMA, 15/16 April 1954, pp. 46–54.
[24] *Flussmittel zum Löten metallischer Werkstoffe*, DIN 8511.
[25] *Composition of a salt flux solution*, DTD 81 (England).
[26] *Rosin*, DEF 72, (England).
[27] *Rosins, classified by source, colour and presence of additives to reduce stickiness*, LLL-R-626b, 1957 (USA).

(*d*) (*Solderability of*) *component parts, techniques*

[28] *Soldering of metallic materials—definitions*, DIN 8505, 1965.
[29] Draft Test T: *Solderability*, DIN 40046, 1967.
[30] Draft *Copper-clad laminate for printed circuit boards*, DIN 40802, 1965.
[31] *Artificial climates for testing purposes*, DIN 50015, 1959.
[32] Draft *Testing of electrical insulating materials*, DIN 53482, 1965.
[33] *Environmental test, Part 2T: Soldering*, British Standards Specification, BS 2011, 1966.
[34] *Copper-clad laminate for printed circuit boards*, British Standards Specification, BS 3888, 1965.
[35] *Electronic parts of assessed quality*, British Standards Specification, BS 9000, 1967.
[36] *Sampling procedures for electrical parts, British Standards*, BS 9001, 1967—Sampling procedures, British Standards DEF-131-A, 1964—Sampling procedures for inspection by attributes, USA-MIL-STD-105.
[37] *Environmental testing*, British Standard K 1007, 1963.
[38] Draft Environmental test T: *Solderability*, British Standard DEF 5011, 1967.
[39] *Printed wiring boards with plated-through holes*, British Standard DEF 5028, 1963.
[40] *General specification for soldering processes*, USA-MIL-S-6872A, 1954.

[41] *Method* 208 (*solderability*) *and Method* 210 (*resistance to heat*), USA-MIL-STD-202C, 1962.
[42] *High-reliability hand soldering*, USA-MIL-S-45743B, 1967.
[43] *Solderability of printed wiring boards*, USA Standard RS-319, 1965.
[44] *Meniscus test for printed wiring boards*, USA-MIL-P-55110.
[45] *Solderability of terminations*, Electronic Industries Association Standard, RS-178-A USA.
[46] *Pessel test*, IEEE, Transactions on product engineering and production, Jan. 1963, pp. 28–33.
[47] C. J. THWAITES, *A new solderability test apparatus*, Tin Research Institute, Publication No. 344.
[48] C. J. THWAITES, *Testing for solderability*, British Welding Journal, pp. 543–550, Nov. 1965.
[49] H. H. MANKO, *Solders and soldering*, McGraw-Hill Book Company.
[50] ANON., *Solderability of tin-coated copper*, Plating, p. 315, April 1965.
[51] J. A. TEN DUIS, *An apparatus for testing the solderability of wire*, Philips Techn. Review, 20, 1958/59, No. 6, p. 158.
[52] KAGEYAMA NOBUO, *On the spreading properties of solder on metal surfaces*, Review of the Electrical Communication Laboratory. vol. 13, No. 1–2, Jan.–Febr. 1965.
[53] L. Z. G. EARLE, *A quantitative study of soft soldering by means of the Kollagraph*, Journal Institute of Metals, pp. 45–73, 1945.
[54] W. B. HARDING, *Solderability testing*, Plating, p. 971–981, Oct. 1965.
[55] W. H. WADE, G. W. BROWN, *A system for numerical evaluation of the solderability of lead wires*, Plating, 53, p. 783, June 1966.
[56] J. A. TEN DUIS, E. VAN DER MEULEN, *Measurement of the solderability of components*, Philips Techn. Review, 28, 1967, No. 12, p. 362.
[57] L. E. HELWIG, P. R. CARTER, *Solder flow on galvanized surfaces*, Appendix, Metal Finishing, Febr. 1969, p. 63–68.
[57A] R. S. BUDRYS, R. M. BRICK, *Variables affecting the wetting of tinplate by Sn–Pb solders*, Metallurgical Transactions, vol. 2, January 1971, p. 103.

(*e*) *Soldered products*

[58] *Testing of printed circuits*, British Standards Specification, BS 4025, 1966.
[59] *Guide on the reliability of electronic equipment*, British Standards Specification, BS 4200, 1967.
[60] *Soldering and brazing inspection*, British Standards AP4089 D.405.
[61] *Prüfung von Hartlötverbindungen*, DIN 8525, 1965.
[62] *Prüfung von Weichlötverbindungen*, DIN 8526, 1967.
[63] *Criteria for inspection for highly reliable soldered connections in electronic and electrical applications*, USA Standard C 99.1-1966.
[64] *Lötspezifikationen entstanden im Projekt AZUR*, I. *Vorläufige Spezifikationen für die Herstellung zuverlässiger handgelöteter elektrischer Verbindungen*, *Satellitenprojekt* 625A1, Bölkow G.m.b.H., Ottobrun bei München, 1967.
[65] AWS and ASTM, *Brazing manual*, Reinhold Publishing Corporation, N.Y., 1955.
[66] H. H. MANKO, *Solders and soldering*, McGraw-Hill Book Company.
[67] R. M. MACINTOSH, *Technical aspects of soldering practices*, Welding Journal, **31**, No. 10, 1952, p. 881.
[68] W. H. ROMBACH, *Controlling quantity on soldered electrical connections*, Symposium on solder, ASTM-STP No. 189, p. 175–283, 1956.
[69] *Dip brazing aluminum assemblies*, Machine Design, 18 Aug. 1966, p. 158.

Chapter 4

Adhesive bonding

F. Abels

4.1 Introduction

When deciding on adhesive bonding for a joint, the designer will have to observe carefully all the requirements and rules for successfully applying this technique. Even at a very early stage, it is necessary to decide on the shape of the object or, at least, of the adjoining faces, to ensure maximum efficiency of the adhesive bonding as a joining technique. It is fairly obvious that close cooperation between designer, adhesive expert, machine manufacturer and workshop is of the utmost importance.

Furthermore, it should be fully realized that neatness and accuracy are indispensable ingredients for this technique: otherwise, the best of adhesives will give disappointing results. In some cases, the joint could even prove a complete failure.

As the formation of adhesive bonded joints is a physical and/or chemical process, influenced by temperature, time, pressure and sometimes humidity, it is necessary, for constant production quality, to control these factors and keep them absolutely constant in many cases[1].

For regular bonding, it is even advisable to select people for the job and train them, to some extent, or at least tell them something about the technica background.

Not all adhesive bonded joints are made as simply as sticking on a postage stamp.

For the characteristics and choice of adhesives see Volume 2 Chapter 8.

4.2 Application techniques and equipment

The method of applying the adhesive layer depends on the adhesive, the way in which the joint is made, and the shape and characteristics of the object.

The adhesive must often be adapted to the requirements dictated by the adhesive application equipment by changing the consistency and solvents. When applying adhesive that can be kept only for a limited period of time, such as the cold-hardening chemically reacting adhesives, this short period of usefulness should, of course, be taken into account when using the application equipment. Either very short replenishment times or so-termed metering and mixing equipment must be used [8].

The way in which the connection is made is often decisive for the type of adhesive to be used, and the application technique to be adopted. Where the object is concerned, a specific application technique can sometimes be taken into account in the design.

(a) *Brushes*

Although a primitive method, adhesive is often still applied with a brush (and in many cases from an open glue pot). The main advantage lies in the low investment made. The disadvantages are many, such as: usually too-rich an application, spilling of adhesive, drying-up of the store, which entails a considerable loss of adhesive, apart from a variation of the consistency due to evaporation, so that too-thick layers are applied. With care, some of these shortcomings can be overcome, such as supplying the adhesive through the stem to the inside of the brush, whether or not it is directly connected to the adhesive container (Fig. 4.1).

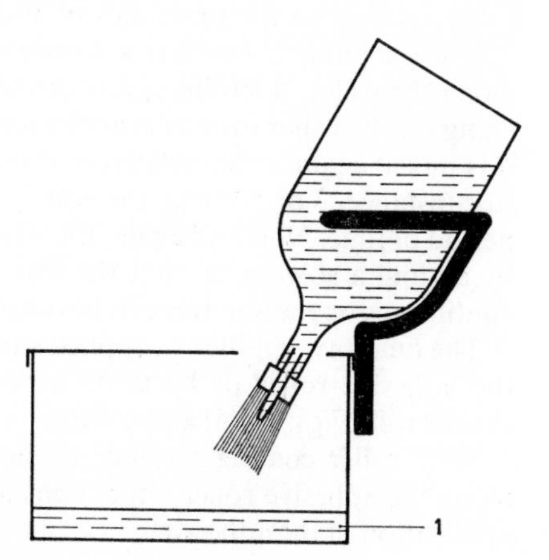

Fig. 4.1. Brush with adhesive holder in stand.
1 = solvent.

When not in use, the brush is placed in a holder containing a slight amount of solvent, to prevent the brush from drying out.

(b) *Roller spreading*

This is an application technique where the adhesive is applied to the surface via a roller. The simplest construction is the hand-roller with an adhesive vessel (Figs. 4.2(a) and 4.2(b)).

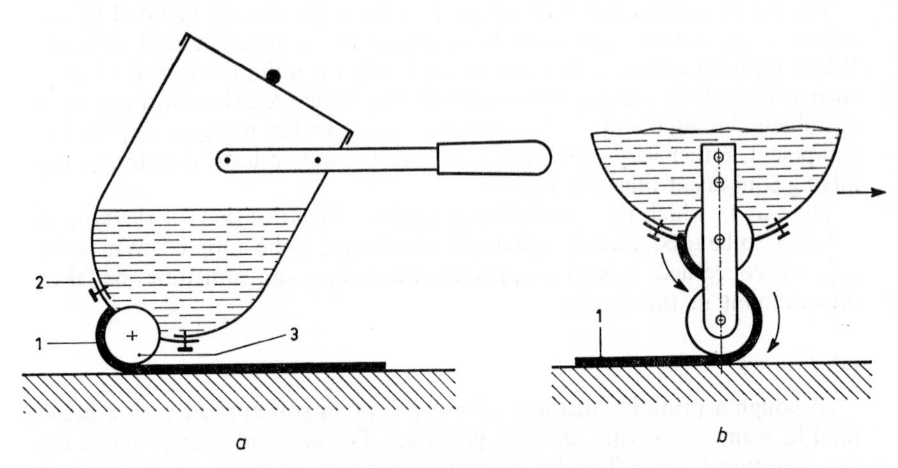

Fig. 4.2. *a.* Single adhesive roller with reservoir.
b. Adhesive roller with doctor roll and reservoir.
1 = adhesive. 2 = adjustable lips. 3 = knurled roller.

For a more mechanized use, for large or small areas, there are suitable rollers acting on the principles of Figs. 4.3(a) and 4.3(b).

Since a pump maintains a steady supply of adhesive from the reservoir along the roller, the roller always receives fresh adhesive. The adhesive thickening on the roller (due to evaporation) is forced back to the reservoir, where it is mixed again with a relatively thinner adhesive. By choosing a large store (for instance, 3 or 5 litres), the adhesive remains fairly constant over a longer period of time. Now and again, the viscosity of the glue itself must be corrected by adding a solvent, so that the loss due to evaporation is neutralized. For continuous operation, this can be done automatically by a drip installation.

The quantity of adhesive applied can be controlled by selecting the profile of the adhesive roller or by using a doctor knife, or by what is known as a doctor roll (Figs. 4.4(a) and 4.4(b)).

With roller coating systems, as shown in Fig. 4.3(b), the objects are led along the adhesive roller under constant pressure, and this ensures a uniform application of the adhesive.

Fig. 4.3.
a. Adhesive applicator with roller (transverse cross-section).
1 = pressure roller. 2 = adhesive roller.
3 = adhesive current. 4 = adhesive pump.
5 = motor.
b. Adhesive applicator with roller (longitudinal cross-section).
1 = driven roller. 2 = adhesive.
S = distance determining the thickness of the adhesive layer.

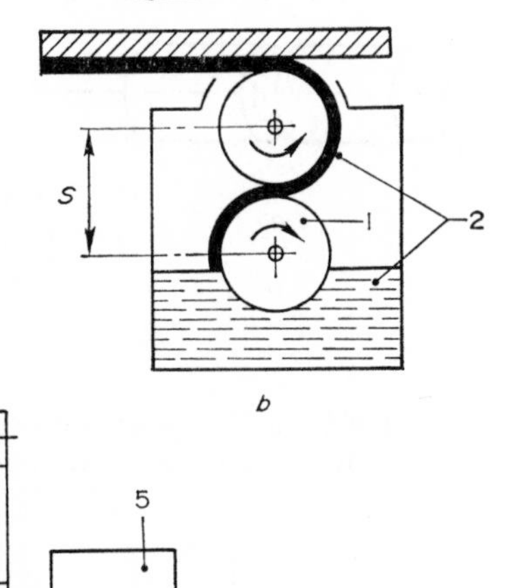

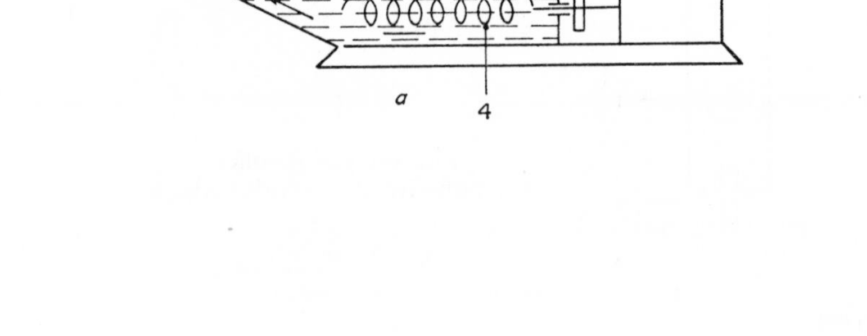

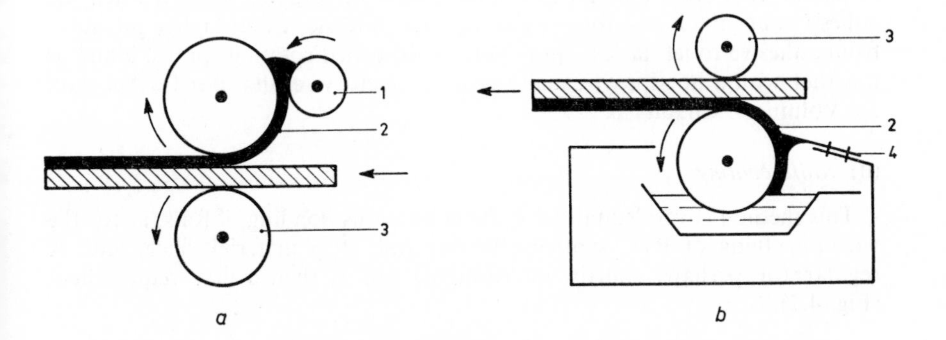

Fig. 4.4. *a.* Principle of adhesive roller-coating machine with doctor roll.
b. Principle of adhesive roller-coating machine with doctor knife.
1 = doctor roll. 2 = adhesive. 3 = transport roll. 4 = doctor knife.

The size and area of the adhesive roller can differ, depending on the quantities of adhesive consumed and the adhesive pattern (Figs. 4.5(a), (b) and (c)).

These rollers can also be used for manual operation, even including the simple "piping roller" (Fig. 4.6).

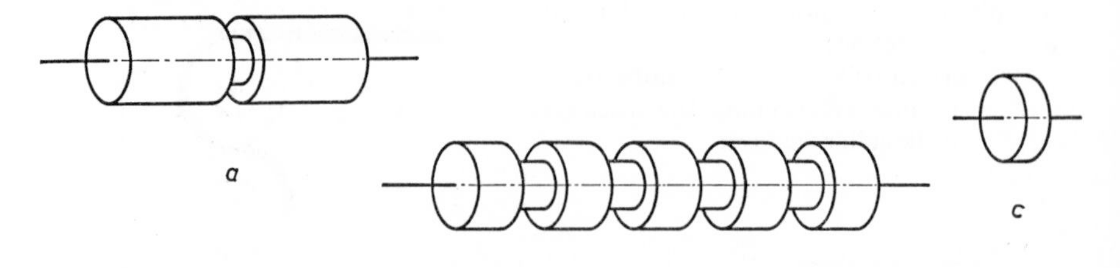

Fig. 4.5. Various rollers for two or more adhesive tracks of various widths. *a.* for 2 tracks. *b.* for various tracks. *c.* for very narrow simple tracks.

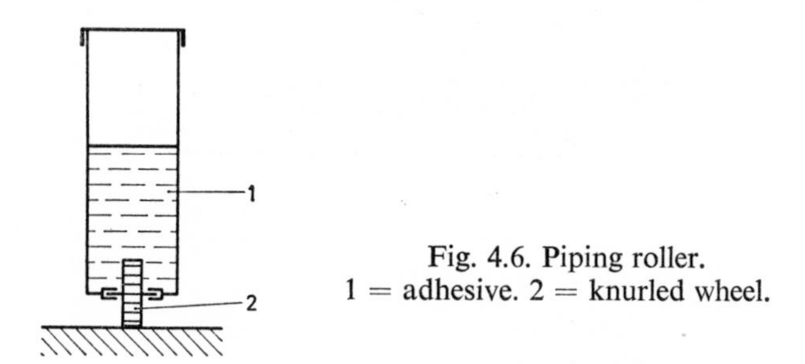

Fig. 4.6. Piping roller.
1 = adhesive. 2 = knurled wheel.

(c) *Spraying*

In principle, the same guns and methods can be used as for the application of paints. It is often difficult to obtain a suitable spraying pattern when the adhesives are thick and form webs. Special thinners, using higher pressures from adhesive to jet, airless spray guns or co-called rotating sprays, alone or together with a hot-spray, usually give satisfactory results. For further data see Volume 4, Chapter 6.

(d) *Knife-coating*

This method is very suitable for the continuous bonding of foils, as for the vacuum gluing of PVC synthetic leather foil, strip material, sheet, etc. A regular (or perhaps shortly intermittent) use is then a first requirement (Fig. 4.7).

(e) *Spray bottle*

Usually, the bottle consists of a somewhat flexible material, so that the adhesive can be squeezed manually through the "jet". All shapes of jet can be

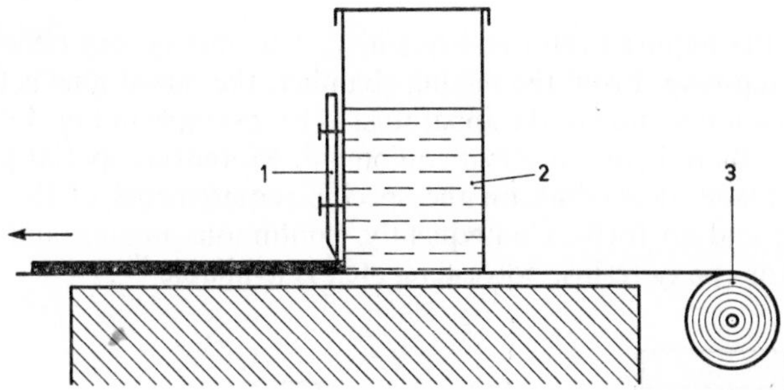

Fig. 4.7. Knife-coater.
1 = knife. 2 = adhesive. 3 = foil.

used, depending on the object to be bonded. In many cases, the jet is fitted with a trigger and, instead of the small squeeze bottle, a pressure feed container used. (Figs. 4.8(a) and (b)). Various spray jets with all kinds of openings are suitable.

(f) *Mixing and metering equipment*

These are specially used for chemically reacting adhesives with a short pot-life. There are various types, depending on the kind of adhesive and the requirements of the application method. Usually, they consist of reservoirs for the separate components, from where the components are fed to a mixing

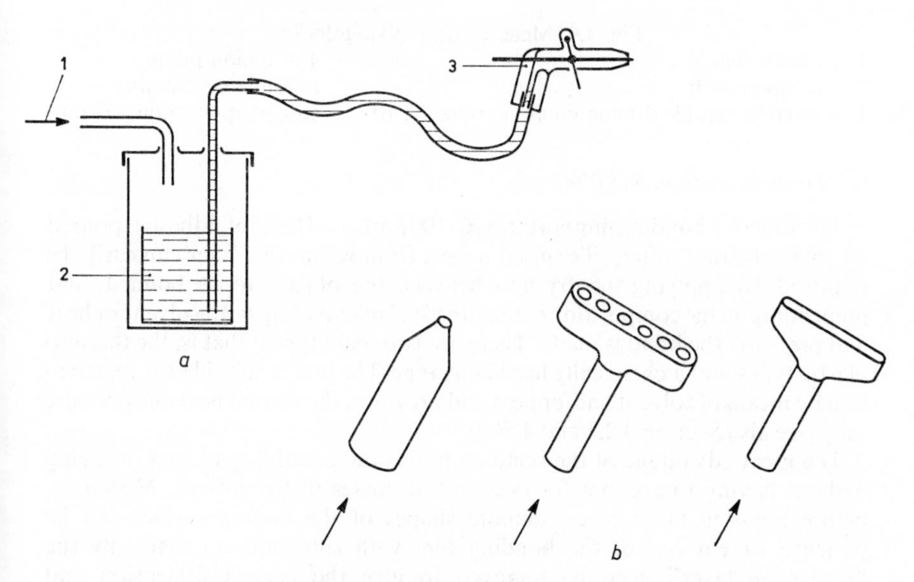

Fig. 4.8. *a*. Pressure feed container with gun.
b. Various types of jet.
1 = compressed air. 2 = adhesive. 3 = spray.

chamber in the required ratio and quantity. This mixing can either be continuous or step-wise. From the mixing chamber, the mixed glue is fed to the actual application section of the apparatus, as for example in Fig. 4.9.

In general, there is no universal equipment. Moreover, special provisions have to be made, depending on the specific requirements of the user, the glue, timing and so forth. Consequently, continuous mixing and metering equipment for epoxy resins, polyesters, etc., is available [8].

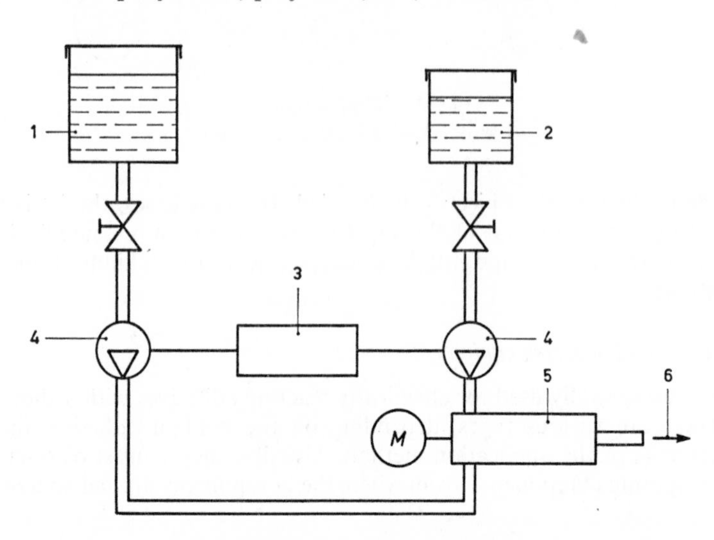

Fig. 4.9. Metering and mixing device.

1 = component A.
2 = component B.
3 = interchangeable driving unit (e.g. ratio 1 : 10).
4 = pinion pump.
5 = mixing chamber.
6 = to application section.

(g) *The adhesive bonding film*

The adhesive bonding film is a thin (20–100 μm) dried layer of adhesive poured on to a substrate (often siliconised paper) from which this layer can easily be removed. By applying the dry film between the objects to be bonded, and (depending on the composition of the adhesive) reactivating with solvent or heat and pressure, the bond is made. There are two main types: that is, the thermoplastic types and a chemically hardening type. The first is suitable for reactivation by means of solvent and/or heat and pressure: the second heat and pressure only (see also Section 4.2.1 and 4.5(c)).

The great advantage of the bonding film is the possibility of easy dosaging without having to account for evaporation losses of the solvent. Moreover, before bonding takes place, definite shapes of the mating surface can be punched or cut out of the bonding film with substrate, so that only the "protective layer" need be removed to give the correct dimension and quantity of adhesive available.

It is also possible to work from the roller, and to heat the object to some extent so that the film sticks slightly and the protective leaf can easily be

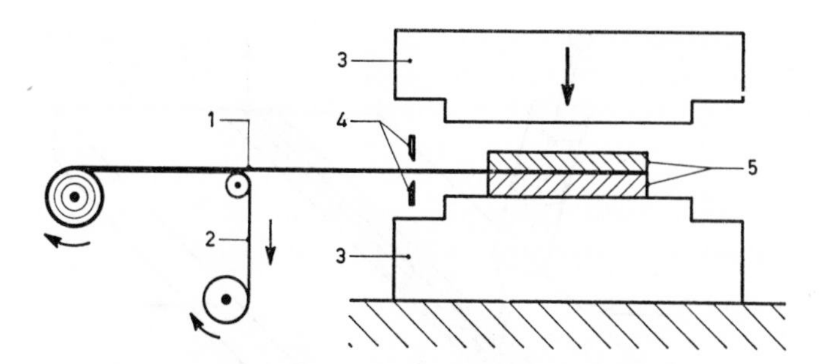

Fig. 4.10. Adhesive bonding film equipment.
1 = bonding film. 4 = cutting knife.
2 = carrier. 5 = parts to be glued.
3 = heated die.

removed and cut off. This is especially advantageous with thin adhesive bonding films, since they are difficult to handle without the substrate. (Fig. 4.10).

(h) *Silk screening*

This method can be used only to a certain extent, as quick-drying adhesives cannot be used for silk screening due to the rapid blocking up of the holes in the screen (see also Volume 4, Chapter 3). If, however, the adhesives do not contain solvents, or are composed of slowly-evaporating solvents, this is a good method of applying the adhesive according to a certain pattern in an accurate, reproducible way.

(j) *Bonding with solvent*

It is sometimes surprising how quickly some thermoplastics can be bonded firmly together with the proper solvents.

There are various methods of doing this:

(i) By placing the components accurately together and administering a drop of solvent to the seam with a syringe. The solvent is absorbed due to the capillary effect, often over greater distances, and thus makes a permanent connection. To prevent chemical attack from the overdose of solvent, "masking tape" should be used. It is also a good policy to chamfer the edges, so that the resulting groove provides guidance for the dosaging needle (Fig. 4.11).

(ii) By moistening the two, separate, surfaces with the solvent, and then pressing the components together immediately. After some time (depending on the solvents used and the kind of plastic), the pressure can be removed, as with method (i).

Moistening is often quite simple and easy to reproduce by means of a moistening device (Fig. 4.12).

The same procedures are used for the reactivation of previously applied and dried films of adhesive.

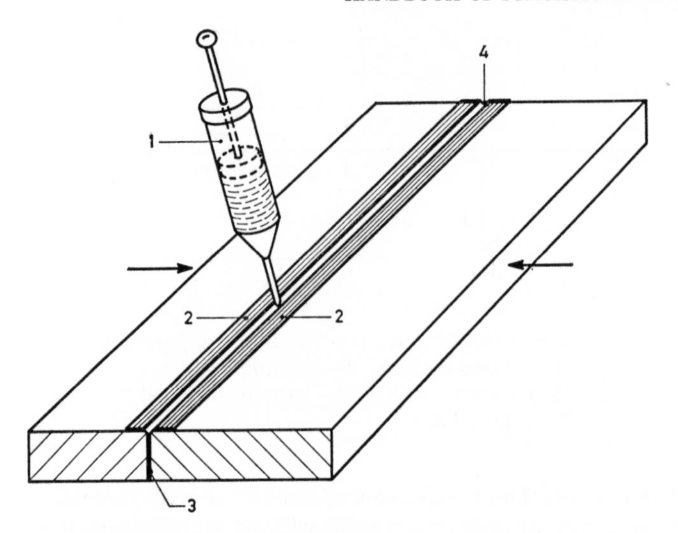

Fig. 4.11. Bonding with solvent-syringe.
1 = syringe with solvent. 2 = tape. 3 = bonding line. 4 = V-groove.

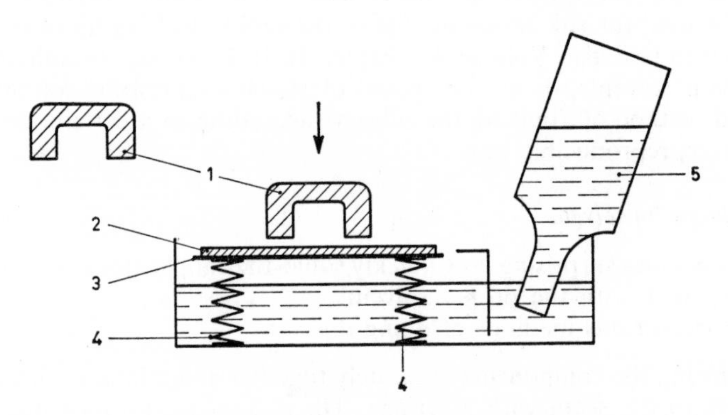

Fig. 4.12. Solvent-wetting equipment.
1 = parts to be glued. 2 = felt. 3 = sieve. 4 = springs. 5 = solvent.

(k) *Hot-melt applicators*

The application of the so-termed "hot melts" has recently become a centre of interest. The adhesives are supplied in a solid state and are melted down to a liquid that can be applied. The connection is formed by the adhesive cooling off in the bonding line. The technique itself is not entirely new, but thanks to good application equipment brought on the market in the U.S.A. and England, it has recently been gaining well-deserved interest. Extensive mechanization is another possibility, as has been proved in the packing industry, but manual methods give quick results too[4].

It is important to keep the reservoir of molten adhesive at the correct temperature and as small as possible, so as to prevent decomposition of the adhesive. A premelting chamber, operating at low temperature, is required to ensure a rapid supply. Preheating the components to be bonded together is sometimes desirable, to prevent too rapid cooling of the adhesive that has been applied [9].

Figs. 4.13(a) and (b) give examples of hot-melt applicators.

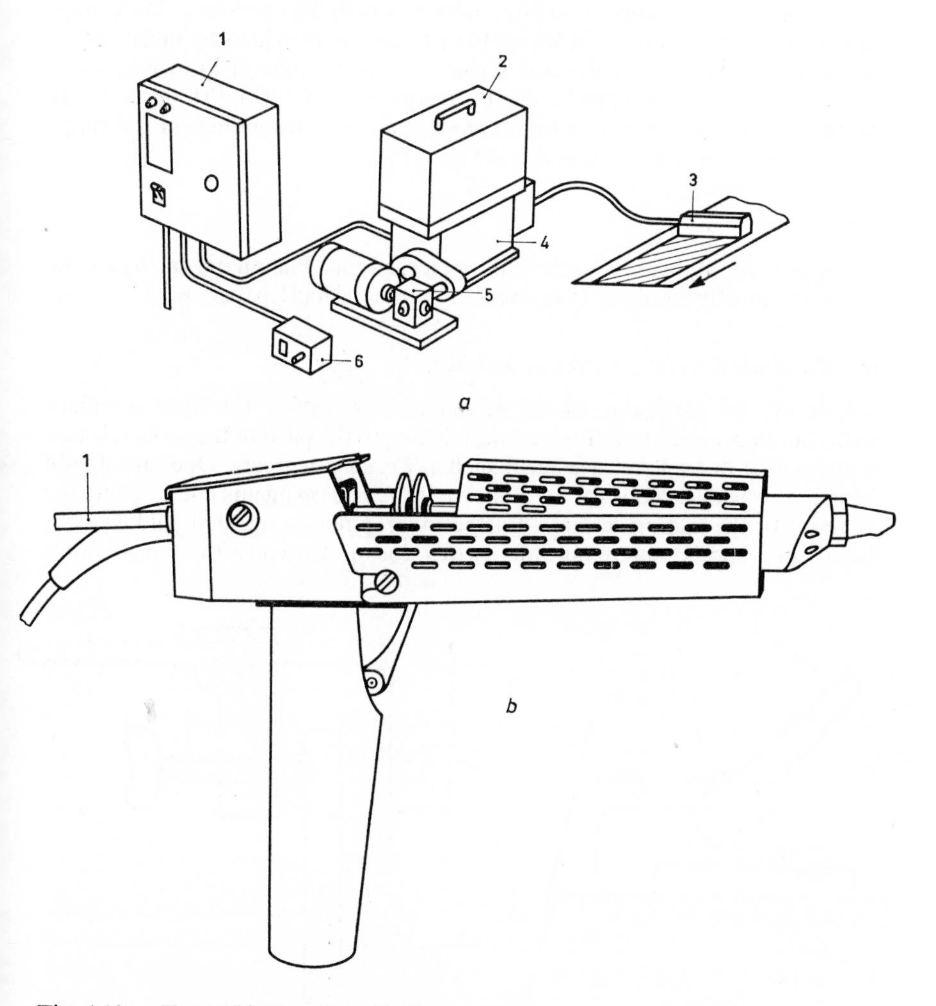

Fig. 4.13. *a*. Type of hot-melt applicator.
1 = switchboard. 3 = applicator. 5 = driving unit.
2 = adhesive reservoir. 4 = melting. 6 = control switch device.
b. Another type of hot-melt applicator.
1 = hot-melt rod.

4.3 Auxiliary tools

Practically all adhesive bonded joints require a certain pressure during the hardening process of the adhesive; firstly, to keep the components in place, and secondly to compensate for an often-occurring shrinkage when the adhesive hardens. For that purpose, there are clamps, jigs and presses.

(a) *Clamps and jigs*

Are employed in simple bonding under relatively low pressure. The clamps can be of the toggle type, in which the pressure is provided by means of an eccentric, or the commonly-used screw clamps. Springs of known pressure are more suitable for reproducible pressures (Figs. 4.14(a), (b), (c) and (d)). Moreover, certain parts require a compression jig, depending on the shape of the components glued together. (Fig. 4.15).

(b) *Presses*

Are used for higher pressures, and for larger areas. Pneumatic and hydraulic types are mostly used, as employed for pressing hard paper, plywood, etc.

(c) *Materials for clamps, presses and jigs*

Usually, no particular choice of material is needed for these auxiliary tools, but to prevent the adhesive from sticking to the parts of the tools, it is best to use a coat of some release agent such as f.e. silicon grease. Due care should be taken: it is disastrous for an adhesive if the release agents contaminate the adhesive or the mating surfaces. Furthermore, a release agent should be used, based on a special wax emulsion, as this does not involve the dangers of a

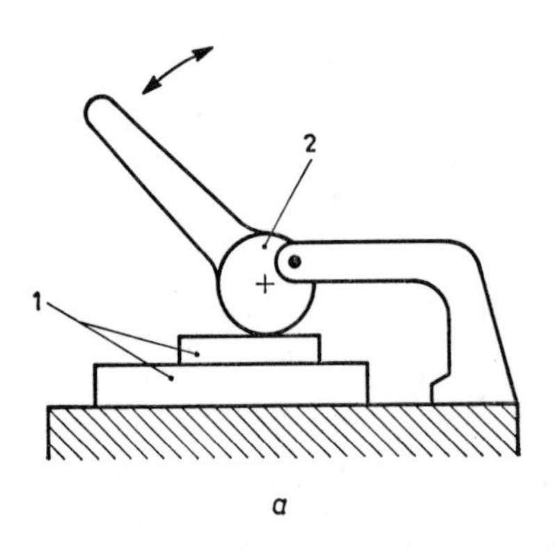

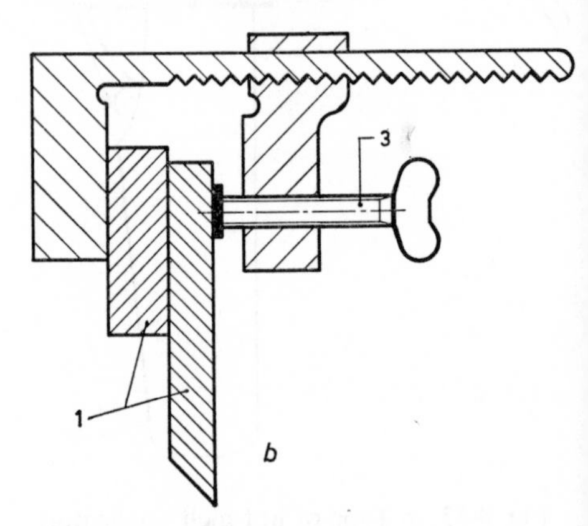

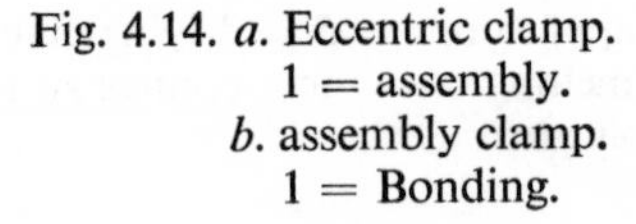

Fig. 4.14. *a.* Eccentric clamp.
1 = assembly. 2 = eccentric.
b. assembly clamp.
1 = Bonding. 3 = adjusting screw.

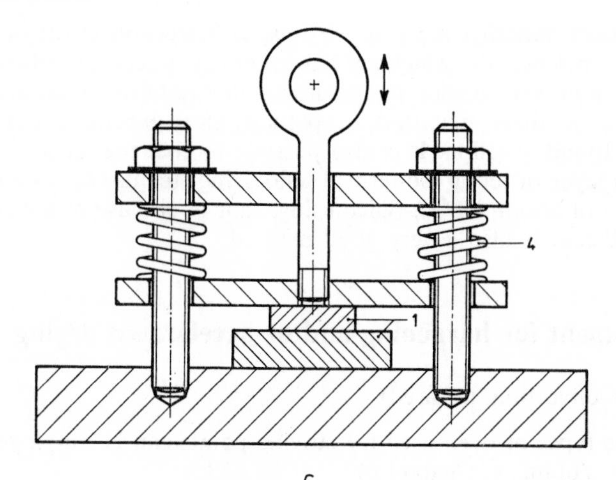

Fig. 4.14 *c.* Spring clamp.
1 = assembly. 4 = spring.

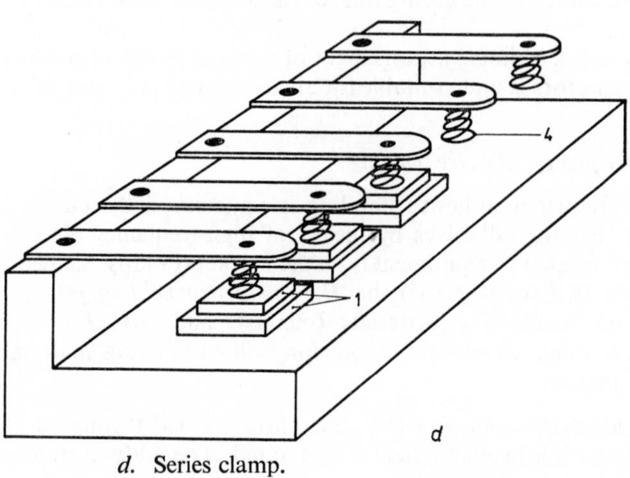

d. Series clamp.
1 = assembly. 4 = springs.

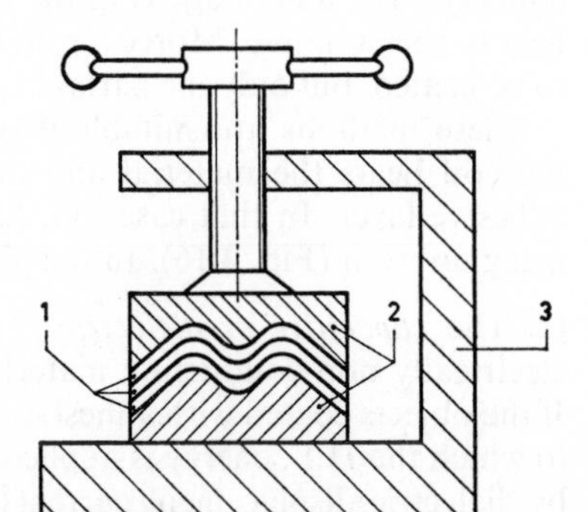

Fig. 4.15. Jig with clamp.
1 = assembly. 2 = jig. 3 = clamp.
1 = assembly. 2 = jig. 3 = clamp.

silicon product. Another measure is to use compression pieces of polyethylene, polypropylene, etc., which are known not to accept any adhesive. If for hardening, however, higher temperatures are required than the melting temperature of these materials, a perhaps-more-expensive but excellent solution is found in teflon. It is also possible to coat the metal components with a thin layer of teflon (so-termed teflonizing) for use up to a maximum temperature of about 250 °C. Siliconising, that is, the use of a coat of well-adhering silicon, is also a very good method.

4.4 Equipment for hardening and/or accelerated drying

(a) *Ovens* (convection, infrared)

The same types of oven used for stoving paints can be employed for this process (see Volume 4, Chapter 6).

The preheating times of the jigs and the objects to be bonded must be taken into account. The time necessary to bring the objects to the correct temperature must be added to the curing time of the adhesive prescribed by the manufacturer.

In many cases, a simple arrangement of infrared lamps or heating elements in special reflectors is quite suitable for an accelerated curing of adhesives.

(b) *High-frequency adhesive bonding*

By high-frequency adhesive bonding is understood the curing of usually chemically reacting adhesives by means of high-frequency electric currents. Special high-frequency equipment is on the market (Philips, Siemens, A.E.G.).

There are two methods: (i) the *H.F. inductive heating process*, which is applied with conductive materials (metals), and (ii) *H.F. dielectric or capacitive heating*, which is suitable for non-conducting materials such as wood and plastics.

(i) In the *inductive method*, a H.F. a.c. current is fed through a coil so that a H.F. alternating magnetic field is generated. The eddy currents occurring in the metal object inside the coil produce heat which can be accurately adjusted. The advantage is in the rapid heating of the material that transfers heat to the adhesive. Moreover, in most cases, the whole object does not need to be heated, but only the parts close to the bond line.

These methods are suitable if one or both components are of metal, for the coil heats the material and the heat is conducted by the metal to the adhesive layer. In that case too, far quicker curing can be achieved than by using an oven (Fig. 4.16). In the given example the curing time is 1 to 2 min.

(ii) The *capacitive or dielectric H.F. method* is suitable for the bonding of electrically non-conducting materials such as wood, synthetic materials, etc. If the objects coated with adhesive are placed between the plates of a capacitor to which the H.F. energy is applied, the materials and the adhesive are heated by dielectric displacement currents. The material with the highest tan δ (loss

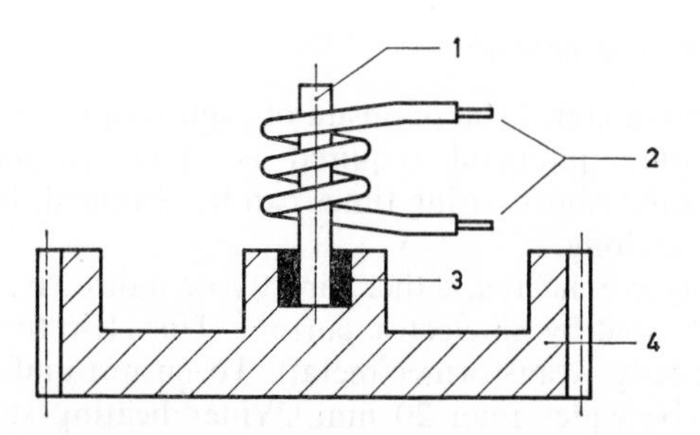

Fig. 4.16. Inductive H.F. adhesive bonding.
1 = metal rod. 4 = gear wheel.
2 = connector for H.F. device. 5 = isolating material.
3 = adhesive.

factor) will reach the highest temperature. In the adhesive bonding of synthetic materials (usually with a low tan δ) and not too humid wood ($<10\%$), the adhesive (provided it is of a composition suiting the purpose) will be quickest in reaching the highest temperature and consequently harden rapidly.

At the correct combination, the material to be bonded is hardly heated at all and most heat is developed *in* the bond line, that is to say, where the heat is actually required.

The method is also applied to chemically-reacting adhesives. This system has found wide applications in wood-working industries, where curing times of 1 to 3 min are now possible, instead of the usual hardening times of 15 to 30 min needed for conventional heating [5].

In principle, there are different electrode arrangements, such as given in Figs. 4.17(a), (b) and (c).

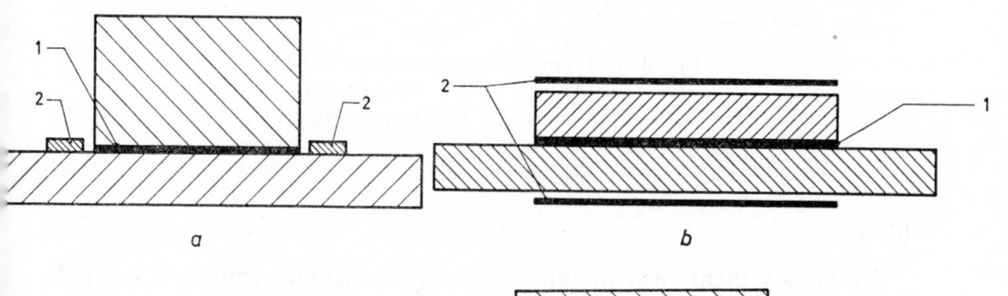

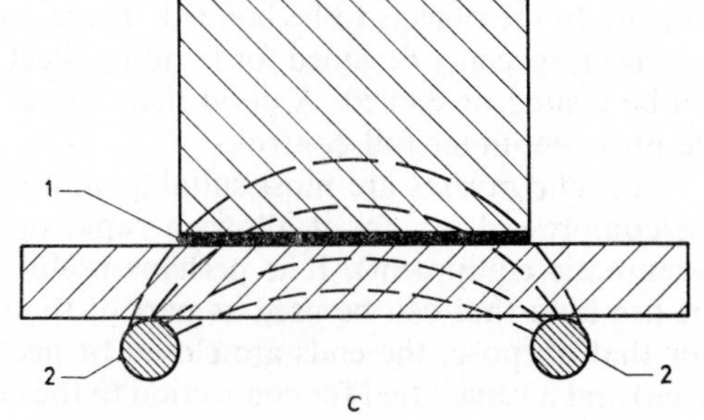

Fig. 4.17.
a. Capacitive H.F. adhesive heating.
b. Capacitive H.F. through and through heating.
c. Capacitive H.F. eddy field heating.
1 = adhesive layer.
2 = electrode for H.F. connections.

(c) *Low-voltage or strip-heating*

This method, often called the opposite of high-frequency heating, has the advantage that the equipment required is much cheaper. For certain applications, the same short curing times can be obtained, but most of them are 2 to 3 times as long.

The adhesive layer is heated, either directly or indirectly, by a metal strip which in turn is heated by an electric current. This strip for instance can be 0·1–0·5 mm steel strip (transformer metal). To ensure uniform heating, the strip should not be wider than 20 mm. Wider heating strips can best be subdivided into smaller ones of maximum 20 mm. The current source is a single-phase transformer, primary 220V, secondary 20–40V, preferably without step control or adjustable per 2V. Power is 4 kVA (Fig. 4.18). When using 0·2–0·4 W/cm^2 strip, for instance, the temperature of the glue on wood will vary from 60 to 75 °C. Certain limits are set, of course, by materials which cannot stand higher temperatures, same as metals where insulation must then be used [3].

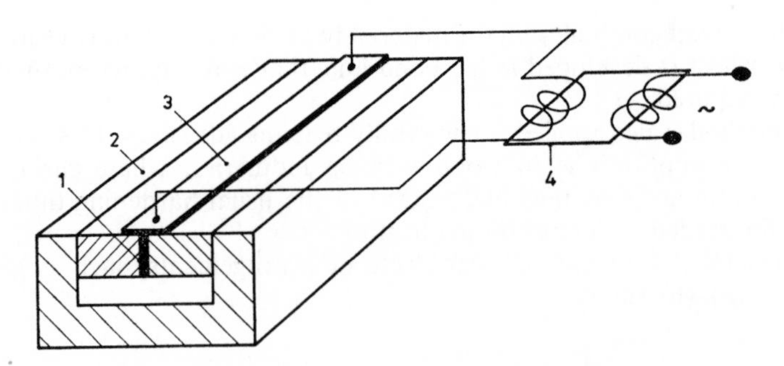

Fig. 4.18. Example of low-voltage heating.
1 = bond line. 3 = heating strip.
2 = clamp. 4 = transformer.

(d) *Presses*

As already mentioned in (b), presses are sometimes employed to apply pressure to the objects to be bonded. There are cold presses, and presses that have been specially designed for bonding sheet material, with press plates that can be heated or cooled. A good manometer is a first requirement, to keep the pressure under full control.

Hydraulic presses are most suitable, although it is often advantageous to use compressed air, for it allows a rapid opening and closing of the press (pneumatic equipment). It is perhaps useful in this connection to mention the fire hose that can be used as part of the pressing equipment (Fig. 4.19). For that purpose, the ends are closed (if necessary with some adhesive between) and a valve fitted for connection to the compressed air.

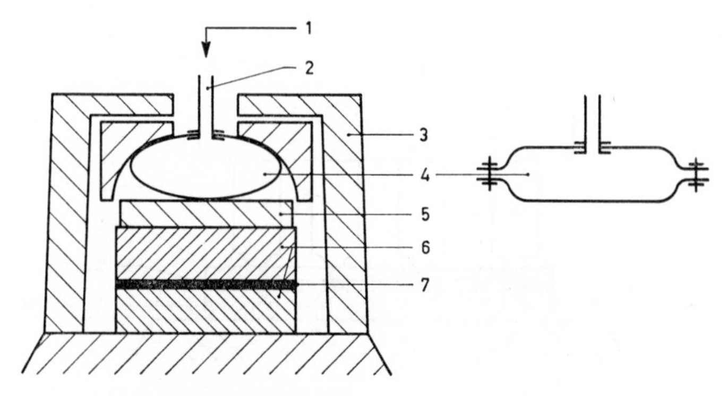

Fig. 4.19. Fire hose press.
1 = compressed air.
2 = valve
3 = yoke.
4 = fire hose with valve and closed ends.
5 = pressure part.
6 = assembly.
7 = bond line.

(e) *Ultraviolet*

For certain adhesives on a polyester base (and especially for those on an acrylic monomer base) curing by polymerization is obtained by ultraviolet light from special lamps (U.V. dark radiators). It is thus possible to make a very strong and invisible joint between clear acrylic sheets and between castings of clear acrylic. The resulting layer of adhesive becomes an acrylic polymer itself, and there are no objections to mechanical operations such as drilling, milling, etc. The technique is widely employed in workshop model design.

4.5 A few special techniques

(a) *Spot-wise accelerated curing process*

With cold hardening, chemically reacting adhesives, it is possible to accelerate the curing process by adding heat. This requires the usual jigs, presses and heating equipment. It has been found possible in some cases, by applying heat, to obtain such a strength at small spots (by rapid curing of the adhesive on these spots), that further curing of the rest of the adhesive can take place at room temperature, without any further pressing of the components. The strength of the joint directly after local curing is approximately 25% of the ultimate strength after complete curing [7]. The simplest method is to use

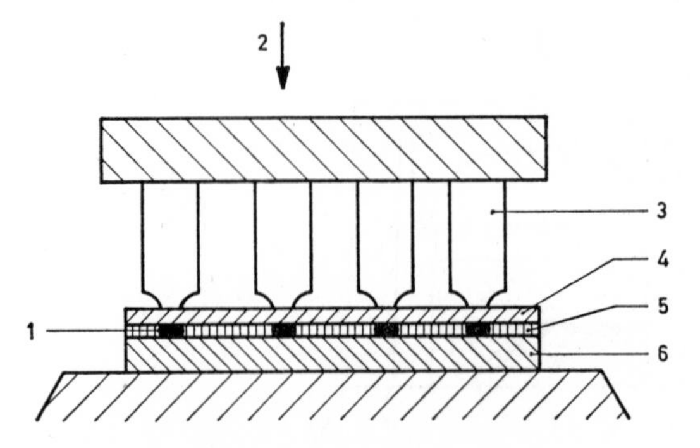

Fig. 4.20. Spot-wise curing.
1 = overheated hardened adhesive, by local heating.
2 = short pressure applied in the direction shown.
3 = heated dies.
4 = one of the components to be joined.
5 = adhesive layer.
6 = the other component.

one or more spaced soldering bits with adapted clips. There are many variations of the principle (Fig. 4.20).

(b) *Spot weld (or plunging) adhesive bonding combination*

As already known, spot weld or flanged joints have the disadvantage that the load on the joint is not distributed uniformly over the bond line. For certain applications, this disadvantage can be largely eliminated by applying extra adhesive between the joint. From the bonding-technical point of view, this means that the disadvantage of long clamping times, and the longer occupation of jigs and clamps, is done away with.[6]

(c) *Heat-seal technique*

In principle, this technique is based on reactivation by means of heat and pressure of the previously-dried thermoplastic layer of adhesive. There are, of course, various methods of obtaining heat and pressure (see Section 4.4 for heating equipment). Thermoplastic synthetic materials can often be connected according to this heat-seal technique, without the addition of adhesive, and the system is widely employed for sealing all kinds of plastic packages of polyethylene, PVC. etc. Various machines have been designed for this purpose, most of them based on high-frequency or strip heating for both continuous and non-continuous operation (Figs. 4.21(a) and (b)).

There is a simple foot-operated H.F. heat-seal apparatus for packing all kinds of small components in polyethylene bags, to protect them against corrosion and other chemical attack in store (see also Chapter 2).

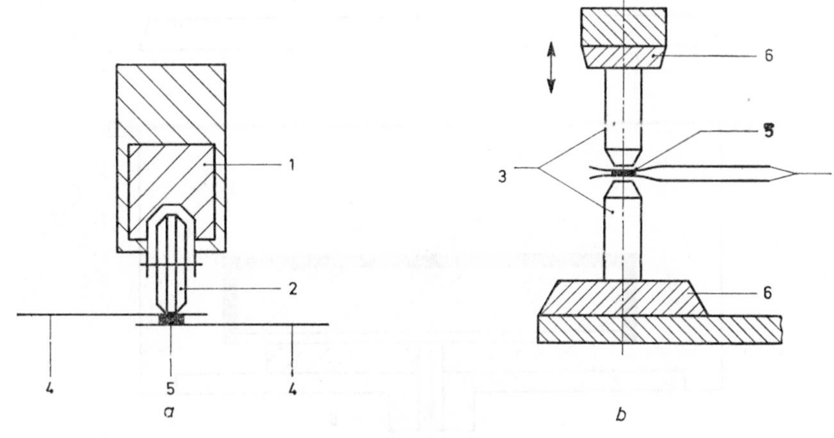

Fig. 4.21. *a.* Heat-seal apparatus with heated wheel.
b. Heat-seal apparatus with heated ruler.
1 = heating device. 3 = heated ruler. 5 = bond line.
2 = wheel. 4 = foil. 6 = insulation.

(d) *Vacuum forming technique in foil gluing*

Apart from coating cabinets, boxes, etc. with synthetic leather foils, this technique has been widely employed over the last few years for packing all kinds of tool components. The principle can best be explained by an example of how a portable radio cabinet is covered with PVC synthetic leather foil.

The adhesive used here is mainly neoprene based adhesive with a hardener. Both the cabinet and the foil are coated with adhesive. Adhesive is applied to the foil by rolling, squeegeeing or spraying, but the cabinet is just sprayed.

This done, the adhesive, depending on its composition, is given 20 to 40 min to dry. Then the cabinet is placed on a support in the vacuum heating apparatus and the foil inserted in the clamping frame. The foil is then heated by infrared radiators, as sometimes is the surface of the object to be coated.

After a few seconds heating, vacuum is applied, and at the same time the cabinet is lifted so that the now-quite-flexible foil is drawn completely round the cabinet. The heat seals the coats of adhesive together: after cooling, cutting and, if necessary, finishing the edges, the cabinet is fully covered (see Fig. 4.22). This method is economic and technically a most attractive form of coating or enveloping.

(e) *Locking techniques on screw threads*

The principal aim is, of course, to lock the thread with adhesives, paints, etc. Generally, adhesive or paint is applied in the screw thread, where it dries so that the connection cannot be loosened by vibration. The connections can be loosened mechanically or by means of a solvent or heat. There are special

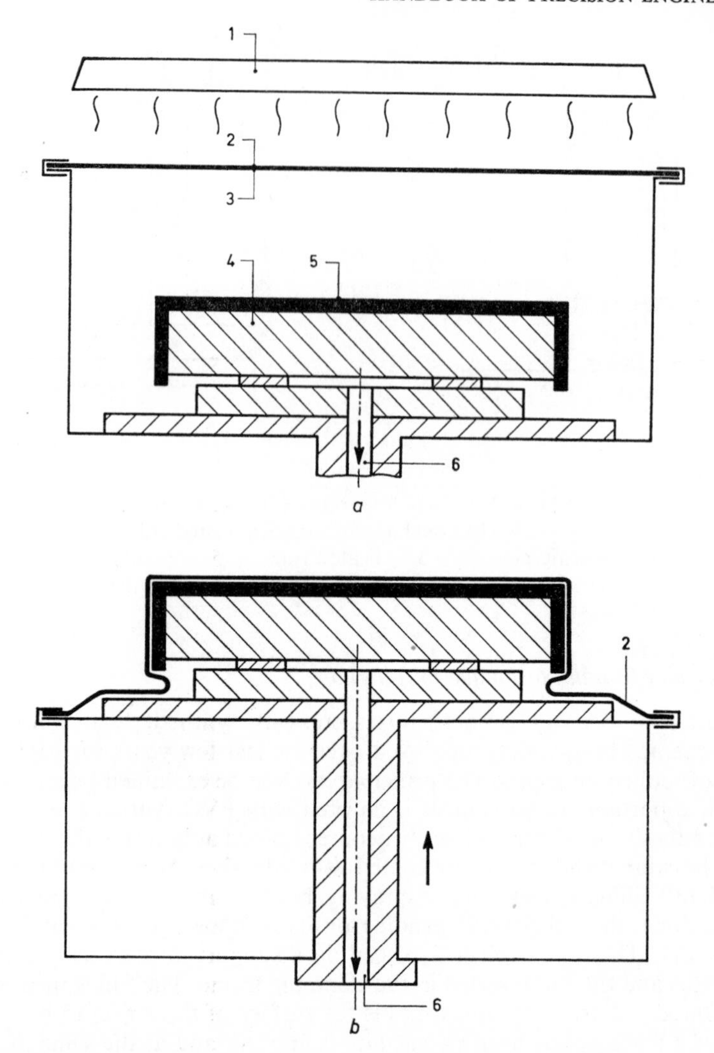

Fig. 4.22. Principle of adhesive bonding with the vacuum forming technique *a*. mounting the cabinet with adhesive. *b*. after vacuum sucking.
1 = infrared radiation. 3 = adhesive layer. 5 = object with adhesive layer.
2 = PVC foil. 4 = jig. 6 = connection for vacuum.

locking paints, but satisfactory results can often be obtained with rubber-based adhesive. Extra security is obtained by applying some extra adhesive or paint after the connection has been made (see Fig. 4.23).

Excellent securing is nowadays available under the name "Loctite" [2].

These are synthetic resins that harden catalytically when in contact with metals in the absence of air. There are various kinds for various applications,

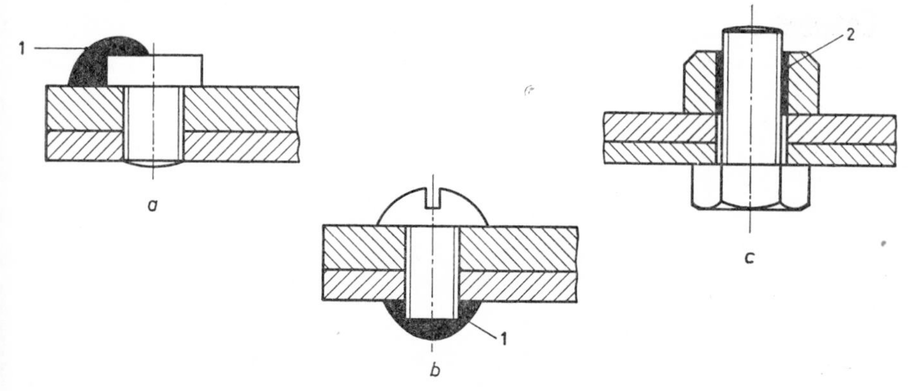

Fig. 4.23. Examples of joints secured with adhesive and paint.
1 = locking paint. 2 = locking adhesive.

which have proved their excellence in practice. Even in fixing bearings, spindles, etc. these products often give excellent results.

(f) *Ultrasonic gluing*

See Volume 3, Chapter 1, para 1.6.5 and Volume 4, Chapter 1 para 1.4.

REFERENCES

[1] I. SKEIST — *Handbook of adhesives*, Reinhold Publ. Co., London.
[2] — "*Loctite techniek*", VIBA, Den Haag.
[3] — *Stripheating*, Kunststoffe **43**, Jan. 1953, Heft 1, S. 37.
[4] — *Bostik-Thermogrip Hot-melt-apparatus*, Bostik Ltd., England.
[5] — *Radio-frequency heating*, CIBA Ltd, England, Technical Notes 213, 214.
[6] F. MITTROP — I. *Untersuchungen über die Kombination von Metallkleben und Punktschweißen;*
R. MITGAN — Essen, II. *Bindemittelfilme für die Metallverklebung und ihre Anwendung in der Praxis*, Mitteilung Forschungsgesellschaft Blechverarbeitung, 1965, No. 3.
[7] A. MATTING, K. UEMER, G. HENNING — *Punktweises Schnellaushärten von Metall-Klebverbindungen* Mitteilung Forschungsgesellschaft Blechverarbeitung, 1965, No. 8.
A. MATTING, G. HENNING — *Das Metallkleben in der Einzel u. Mengenfertigung*, Feinwerktechnik, Jhrg. **70**, 1966, Hft. 1.
Facia Panel manufacture, Aero Research Technical Notes, bulletin 77, May 1949.
[8] — *Semco sealant Gun-book*, Inglewood, California.
[9] M. BOL, VAN BOCHOVE — "*Hotmelt*" *impressies van de PMMI-show* **1965**, Was de Wit, Voorburg.
G. MENGES, J. EILERS — *Beitrag zum Kleben von Kunststoffen*, Plastverarbeiter, Jhrg. **18**, 1967, Hft. 3.

Chapter 5

Wire Winding

H. G. de Cock

5.1 Introduction

There are many different kinds of wire, differing both as regards material (metals, plastics, natural and synthetic yarns, rope) and as regards cross-section and finish. Wires are used for an enormous variety of products.

In all processes leading to wire products, winding takes up an important place. In many products, wire is found in the shape of a coil. But, before that point is reached, the wire is often wound up and off in previous operations such as drawing, annealing, lacquering, enamelling, twining, spraying, galvanizing, cleaning, etc.

During winding the wire is wound on a reel or bobbin. Perfect control of the wire tension, and thus other wire characteristics, is a first requirement[2]. Section 5.2 is devoted to wire tension control, as found in wire supply units (also termed de-reelers).

During the winding process, a rotating and a reciprocating movement can be distinguished. In general, the spindle on which the winding body is placed rotates and a wire guide or winding finger, putting the wire in the correct position, makes the reciprocating movement (Fig. 5.1). There are, of course, other combinations of these two movements. Both movements, depending on their application, give rise to their own particular problems.

Section 5.3 deals with winding in wire manufacture: controlling the winding speed to ensure a constant speed of the wire, forms the centre of interest.

Section 5.4 is devoted to the winding of wire products. The greatest variety is found in coils for electronic applications. A great deal of attention will be paid to the various coil shapes and the winding mechanisms adapted to them. For the winding of helical threads, reference is made to Volume 11, Chapter 4.

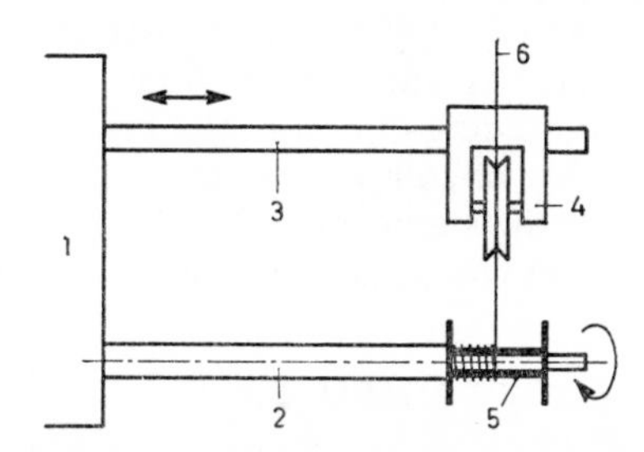

Fig. 5.1. Principle of a winding machine.
1 = winding machine. 3 = traversing spindle. 5 = coil.
2 = winding spindle. 4 = wire guide. 6 = wire.

REFERENCES

[1] WM. QUERFURTH *Coil Winding*, Geo Stevens, Mfg. Co. 1954, Chicago, Illinois, 192 p.

[2] F. O. GLANDER *Die mechanische Beanspruchung von Kupferdrähten bei der Weiterverarbeitung durch Biegen und Wickeln*, Zeitschrift Metallkunde, Bd. **47** (1956), Hft. 6, pp. 364 to 369.

5.2 Wire tension control

5.2.1 Subdivision of wire brakes

To preserve the characteristics of the wire as much as possible, it is important, both in winding store reels and for wire products, to keep the wire tension within certain limits. This is usually done by means of a wire braking system. There are many types, subdivided into two classes [1,3]:

- Braking the wire itself.
- Braking a wheel driven by the wire.

Furthermore, a distinction can be made depending on whether there is a compensating system or not. This is in fact a feedback of the wire tension to the braking device; as the wire tension increases, the braking force is reduced. Most braking systems can be used with and without compensation.

5.2.2 Requirements wire brakes must satisfy

A wire braking system must satisfy the following requirements:
(a) The *wire tension* must be kept constant within certain limits. It depends on the kind of wire and its application, what the level of this tension will be and what tolerances are given. Table 5.1 gives permissible tensile loads for copper winding wire of various diameters, for instance. It is, however, desirable to stay considerably below these values, since, in passing guide wheels and wire guides, the wire is subjected to an extra bending force. (The latter will be greater, the smaller the radius of curvature, so small radii must be avoided.) Especially should anything that interferes with wire tension be borne in mind. Accelerations and decelerations in wire consumption, as periodically occur in the winding of non-round coils, for instance [2], require the lowest possible number of components moving with the wire. However, if there is interference at the supply side (for example, due to heavy points in the supply unit), a good compensation system is a first requirement.

TABLE 5.1

Permissible tensile load in copper wire

Nominal wire diameter (microns)	30	50	70	100	140	200
Yield strength (mN)	200	500	1120	2250	4400	9000
Permissible tensile load (mN)	100	250	430	780	1430	2920

(b) The *speed sensitivity*. Apart from the periodical speed variations in winding non-circular coils, an important point is the gradual speed increase occurring when the machine is started up, or during the time that the bobbin reel driven at a constant speed becomes full. Also, under these conditions, the wire tension must be kept within the set limits.

(c) The *tension traject* for which a braking system can be used.

(d) The *minimum tension* that can be maintained at low speeds.

(e) *Accessibility*, as regards insertion of the wire.

(f) *Susceptibility against soiling*.

(g) Development of heat in the wire.

Since the friction coefficient plays an important part in most braking systems, the characteristics vary considerably for various kinds of wire. A special problem is presented by stranded wire (copper litze, twined rayon); the braking action must not untwine the wire.

5.2.3 Kinds of wire braking system

(a) *Braking takes place on the wire itself.* A general advantage of this category of braking systems is that no parts are moving with the wire, so the wire need not transfer acceleration forces.

On the other hand, there are the drawbacks that any wire insulation may suffer from the braking action, for such a braking system is usually rather susceptible to soiling and stranded wire runs the risk of being detwined. The following versions are found:

(i) *A piece of felt* through (or round) which the wire is led (Fig. 5.2). The possibilities of setting a certain wire tension are limited. In many cases the piece of felt is used to produce a certain pre-tension for other braking systems.

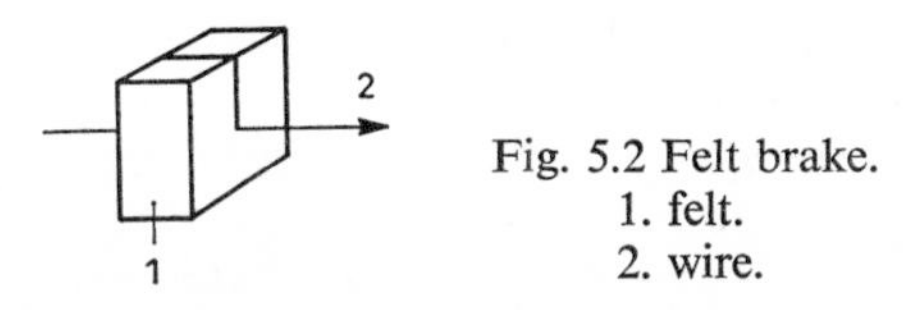

Fig. 5.2 Felt brake.
1. felt.
2. wire.

(ii) Various kinds of *surrounding brake.*

Fig. 5.3 shows a brake with an interrupted surround and a compensation device. The pins around which the wire is led are alternately positioned on two levers. As the wire tension increases, the levers open, so that the angle over which the wire embraces the pins becomes smaller, with the result that the contribution of the braking force to the wire tension decreases. The wire tension can be adjusted with an adjustable spring. A drawback of brakes of this kind, however, is that the pins around which the wire is led are usually of small diameter. This produces considerable additional bending stresses in the wire.

(iii) *Cup and shoe brakes.* The wire is led between two planes that are forced together. The wire tension is adjusted by varying the pressure between the

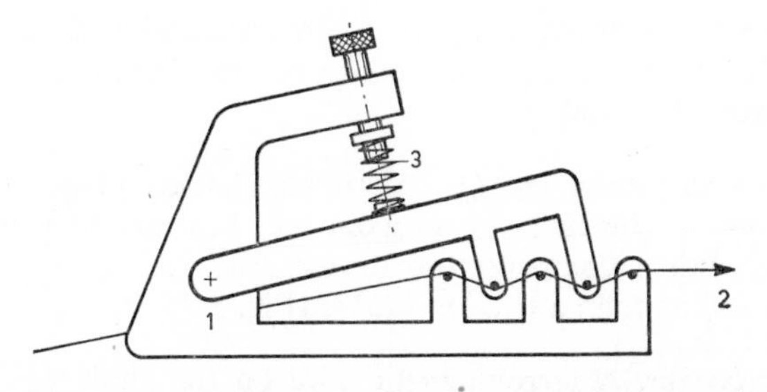

Fig. 5.3. Wire brake with interrupted surround.
1. pivot. 2. wire. 3. adjustable spring.

two planes. See the cup brake of Fig. 5.4. It can also be done by a number of adjustable spring-loaded pawls resting on the wire (Fig. 5.5).

Braking forces exercised directly on the wire are usually employed in cases where high (and strongly variable) wire speeds combine with low tensions.

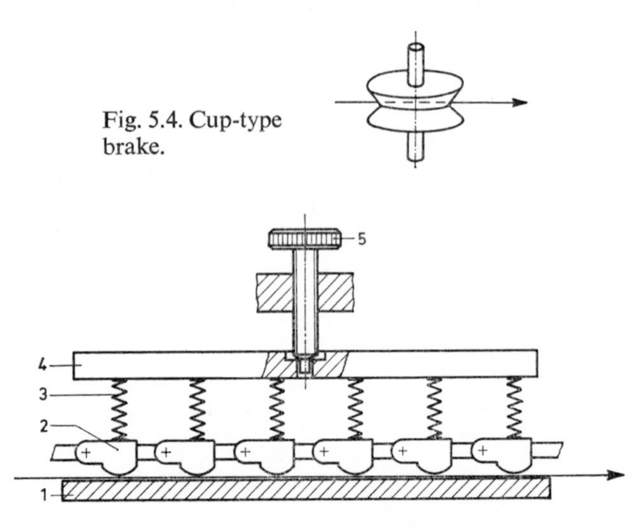

Fig. 5.4. Cup-type brake.

Fig. 5.5. Spring-loaded shoe brake.
1. wire guide plate.
2. pressure pawls.
3. springs.
4. support strip.
5. adjusting screw.

(b) *A wheel driven by the wire is braked.* Of course, this brakewheel is kept as light as possible. The wire does not slide over the wheel, but runs along with it. Consequently, the insulation will not suffer; also the risk of the wire detwining is much less than it would be if the braking force were applied directly to the wire.

Here, the friction problem is shifted to the braking area of the wheel. Depending on the design, the reproducibility of the set tension and the susceptibility to soiling may vary considerably. Various designs are:

(i) *Friction brakes.* A lateral plane or a circumferential plane of the wheel is braked with a brake lining. This can be very well combined with a compensation device. (Fig. 5.6).

(ii) *Braking on distortion energy.* A rubber lateral plane of the wheel functions as one of the races of an axial ball bearing. The pressure on it can be varied with a compensation device. Such a type of brake is less susceptible to soiling than a friction brake. (Fig. 5.7).

(iii) *Magnetic brakes.* A ferromagnetic core on the guide wheel is braked by stationary magnets. By varying the distance between the core and the magnets, the wire tension can be adjusted. (Fig. 5.8).

(iv) *Eddy current brake* (Fig. 5.9). By means of induction currents, a variable magnetic field has a braking effect on a rotor of conductive material attached to the guide wheel. Very soon the result will be that the wire has to drive a relatively heavier rotor and this entails several difficulties. But the eddy current brake is very suitable for a feedback between wire tension and braking.

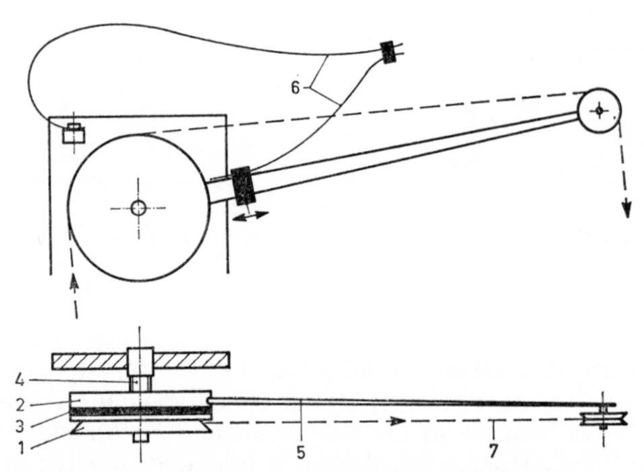

Fig. 5.6. Friction brake with compensation device.
1 = guide wheel. 3 = brake lining. 5 = lever. 7 = wire.
2 = brake disc. 4 = threaded spindle. 6 = springs.

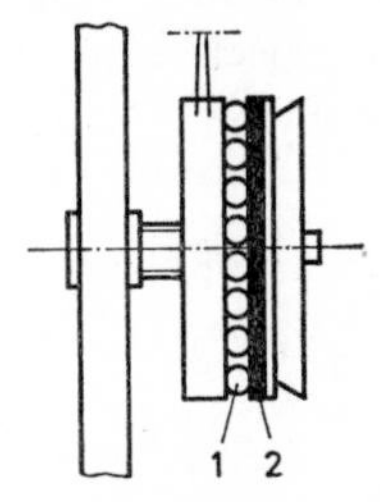

Fig. 5.7. Brake based on distortion power.
1 = thrust bearing. 2 = rubber lining.

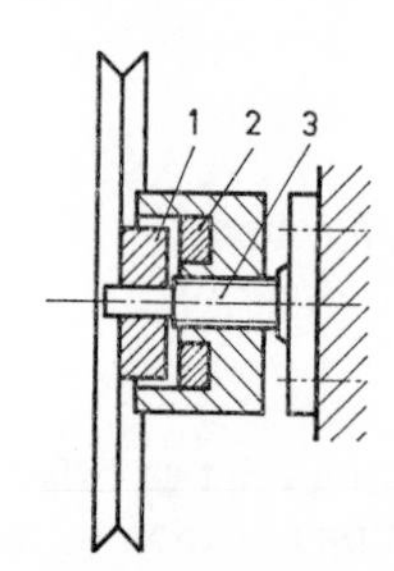

Fig. 5.8 Magnetic brake.
1 = ferromagnetic core.
2 = permanent magnets.
3 = threaded spindle.

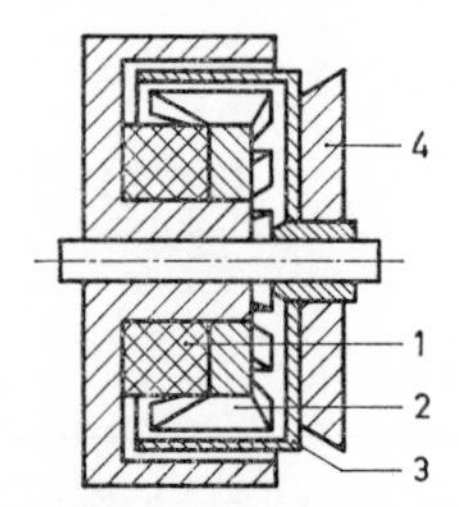

Fig. 5.9. Eddy current brake.
1 = controlled coil.
2 = stator coils.
3 = braked rotor (copper).
4 = guide wheel.

Types (iii) and (iv) are, of course, little susceptible to soiling.

With all braking systems where the wire (whether or not slipping) is led over a curved surface, the wire tension increases over the enclosed angle α from F_1 to F_2. That is, $F_2/F_1 = e^{\mu\alpha}$ (Fig. 5.10).

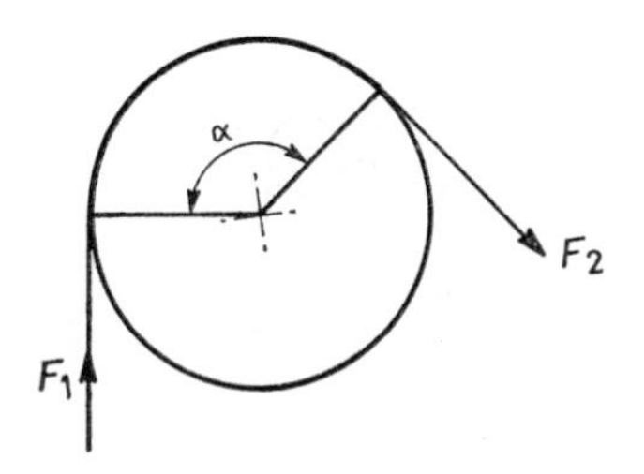

Fig. 5.10. Tension increase in wire led over a curved surface.

$$F_2/F_1 = e^{\mu\alpha}$$

Apparently, this increase is independent of the curvature. The friction coefficient, however, has a great effect; it is usually about 0·2. For an enclosed angle of 360° (2π radians), F_2 can become about 3 F_1. An initial tension F_1 is always required. Usually this tension is produced by a felt or a similar item and cannot become too high. In order to obtain a sufficiently high final tension F_2, it is often necessary to wind the wire several turns around the same wheel, or round a series of wheels or guide pins.

There are special instruments for measuring the wire tension. Usually, the sensing element of these instruments consists of three wheels along which the wire is led (Fig. 5.11). The central wheel is mounted on a lever, to enable

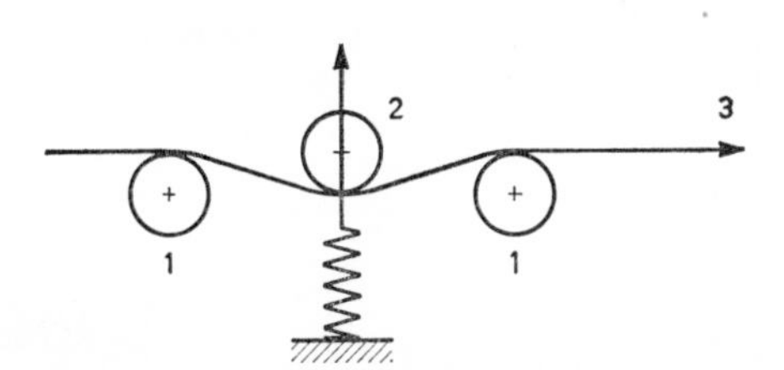

Fig. 5.11. Principle of the wire tension gauge.
1 = guide wheels. 2 = measuring wheel. 3 = wire.

the wheel to move perpendicular to the line connecting the two other wheels. The wire tension is decisive for the respective positions of these wheels. The transfer of this displacement and the indication can be mechanical. This wire tension indicator can only follow slow tension fluctuations.

For dynamic measurements, use is made of electronic transmission (inductive or capacitive) and indications. In the latter case the moving parts must, of course, be extremely light.

REFERENCES

[1] J. A. KALKMAN — *Over de eigenschappen van garenremmen*, Rayon Revue, May 1953, pp. 54 to 80.

[2] R. ELDER — *Drahtspannungen beim Wickeln von Vierkantspulen*, Elektrotechnik und Maschinenbau, Jg. **70**, Hft. 15/16.

[3] J. HENNO, R. GUIGAL, A. LYONNET — *Etude des tendeurs utilisés dans l'industrie textile*, Rayonne et fibres synthetiques 1965, Nos. 9 and 10.

5.3 Winding in wire manufacture

5.3.1 Requirements for winding-up

The process requires a constant wire speed, and the fully wound reel, bobbin or drum must satisfy certain demands. The most important are the following:

- It must be easy to wind off the wire again. The greater the speed at which this is done, the more attention must be paid to winding up.
- Where metal wire is concerned, it should not be overstretched or deformed in any other way, in view of the required specific ohmic resistance and/or flexibility. This implies that, in winding-up the wire, the tension must be kept as low and constant as possible.
- Any insulation layers should not be damaged.

5.3.2 Shapes of reels and bobbins

Reels, bobbins or drums (Fig. 5.12) are used for winding-up the wire. Wire is wound on reels in closely-wound layers between flanges. Wire is wound on bobbins with such changing motion that a self-supporting body results. Drums contain coils of wire.

Since reels and bobbins are usually driven in the winding-up process, they pull at the wire, and thus add to the tension and deformation in the wire. As a result, there is compression in fully wound reels. This compression increases from the outside to the inside, so the diameter ratio between the full and empty reel should not be too large.

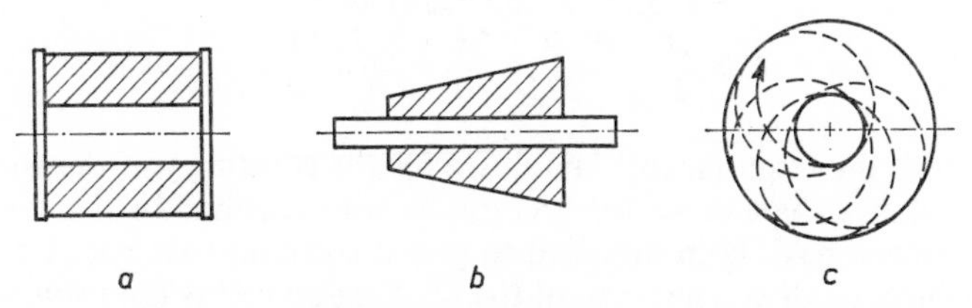

Fig. 5.12. Coil forms.
a. reel. *b.* bobbin. *c.* drum.

Because of their flanges, reels are only suitable for limited winding-off speeds; in tangential winding the relatively great moment of inertia plays a part. Furthermore, the wire gets jammed now and again between windings of underlying layers, especially near the flanges.

Bobbins, on which the wire is cross-wound, allow a greater winding-off speed. The driving mechanism for the cross-winding of bobbins is, however, more complex. Yarns are often wound in this way. With drums, the wire is placed in an absolutely relaxed condition and with a constant winding radius. This can be of advantage when great speeds are combined with low permissible tensions. The winding-off process can develop very rapidly. The wire is, indeed, torqued one complete turn per loop, but this can be compensated for in winding-off.

5.3.3 *Winding-up*

When winding-up reels and bobbins, there are two movements: (a) rotary, and (b) reciprocating. Although the reverse also occurs, the bobbin usually rotates while the wire guide performs the reciprocating movement.

(a) *Rotary movements*

For the drive of the winding-up reel, the starting point should be the constant wire speed (v) the process requires, and the constant wire tension (F) aimed for in winding-up. In principle, the product $F \times v$, the winding-up power, should be a constant: the winding-up characteristic, the product between the driving couple (M) and number of revolutions (n) of the winding-up reel is therefore a hyperbola (Fig. 5.13).

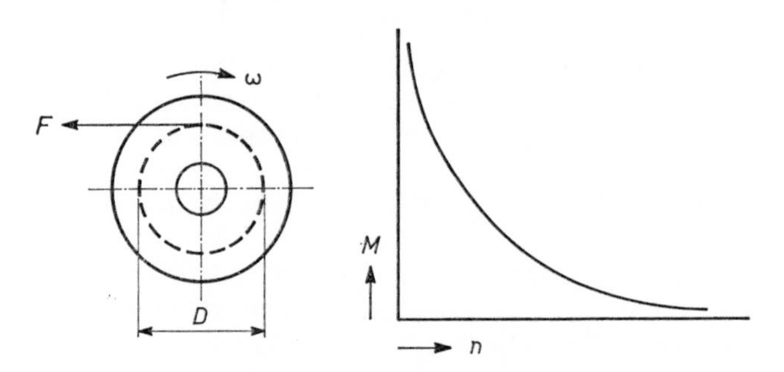

Fig. 5.13. Winding characteristic.

$$F.v = F.\omega.\frac{D}{2} = \frac{F.D}{2}.n.\frac{2\pi}{60} = M.n.\frac{2\pi}{60} = \text{constant}$$

In general, the requirements laid down for the constancy of the wire speed are more severe than those for a constant wire tension. In the first place, control systems have been designed to give a constant wire speed by means of a pull-through disc. The drive of the winding-up reel is then so controlled that the wire tension remains constant throughout the winding-up process. This is known as *tension control*, although it will be clear that this control has no effect on what takes place before the wire reaches the pull-through disc; nor has it any effect on the tensions introduced into the wire before the process. This principle is most commonly used, and has many variations. As regards control technique, *speed control* goes one step further. In this case, the wire speed is determined by the winding-up reel. The draw-through disc is no longer used. This, of course, sets more stringent requirements for the drive control. It is endeavoured, by means of braking systems or auxiliary drives, to keep the tension in the wire as constant as possible before and after the process.

A few examples of *tension control* will now be given:

(i) In the simplest form, the reel spindle is driven by a *clutch*. A brake on the reel spindle, operated by the wire via a lever system, matches the speed of the reel to the wire tension (Fig. 5.14).

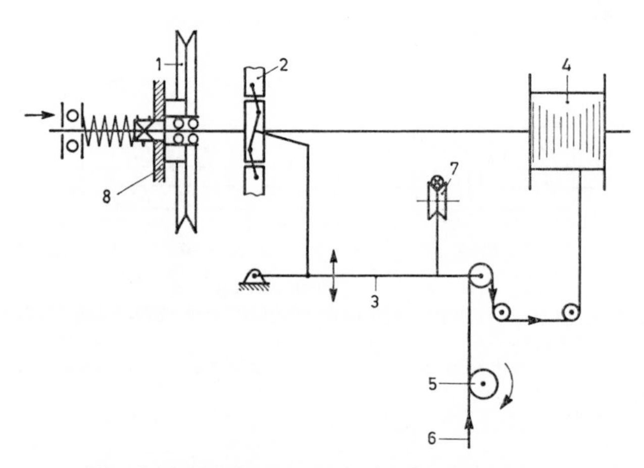

Fig. 5.14. Tension control with friction coupling
1 = friction drive. 3 = lever system. 5 = pull-through disc. 7 = weight.
2 = brake. 4 = reel. 6 = wire. 8 = friction-lining.

(ii) The reel spindle can also be driven by a *variable speed drive*, whose transmission ratio is controlled by the wire itself. A refined version of this uses an eddy current coupling. With a guide wheel on a lever controlling a differential transformer, the wire tension can be translated into an electric signal which, after amplification, can be used to control the eddy current coupling (Fig. 5.15).

There are several other tension control methods.

(b) *Reciprocating movements* (changing movement, transverse movement)

In winding reels, the most important point is accurate adjustment of the reciprocating movement with respect to the reel width, in order to avoid the wire being drawn between the flange and the wound part of the bobbin. The pitch is slightly more than the wire diameter.

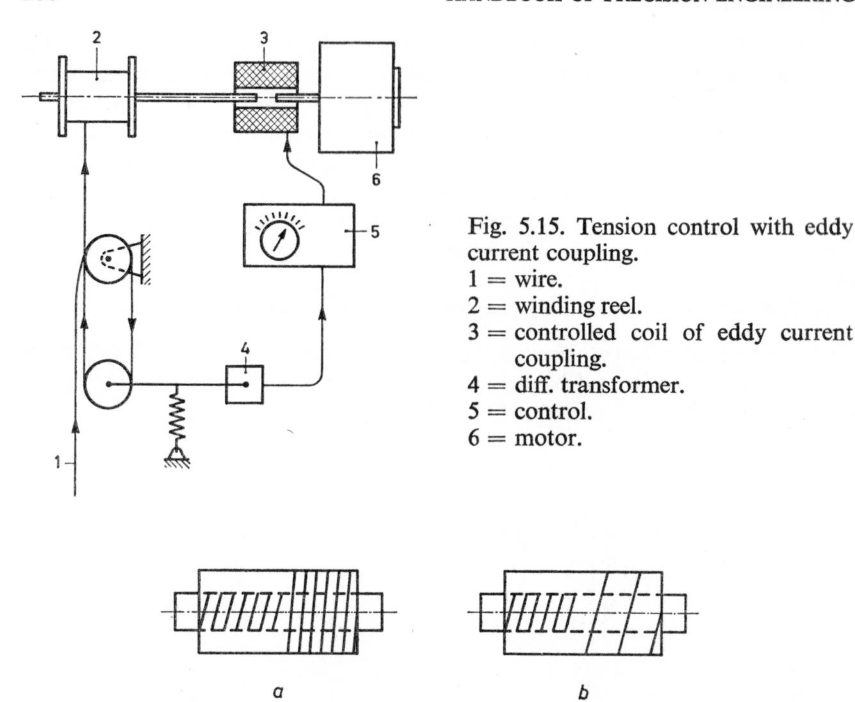

Fig. 5.15. Tension control with eddy current coupling.
1 = wire.
2 = winding reel.
3 = controlled coil of eddy current coupling.
4 = diff. transformer.
5 = control.
6 = motor.

Fig. 5.16. Wire configuration on bobbin:
a. traversing motion coupled with bobbin drive; pitch constant, pitch angle decreases towards the outside.
b. independent traversing motion; pitch angle constant, pitch increases towards the outside.

It is best to give the reciprocating movement its own drive with constant speed, so that the pitch increases as the reel is getting fuller. Coupling the reciprocating movement to the reel drive, giving a constant pitch for all layers, would result in a less compact body. (Fig. 5.16).

When winding bobbins without flanges, the result must be a self-supporting unit with a uniform wire distribution. The whole thing becomes self-supporting by applying a cross-winding, that is, a winding with a large pitch (Fig. 5.17), which prevents windings sagging at the edges of the bobbin. Especially in the textile industry, is use often made of cross-wound store bobbins, both cylindrical and conical.

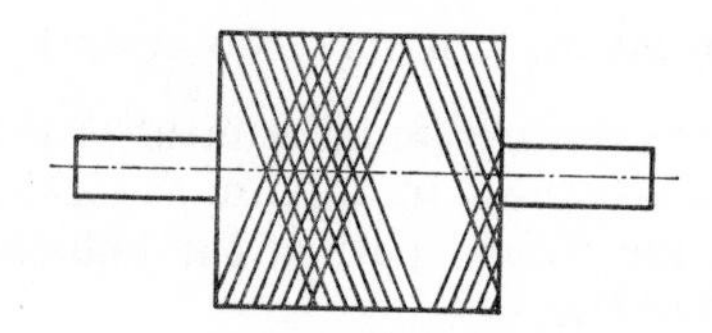

Fig. 5.17. Appearance of a cross-winding.

The following are the reasons:

(i) The necessity for winding-off yarn very rapidly for high production speeds. Cross-wound bobbins can rapidly be unwound "overhead".

(ii) A cheap, light packing is sufficient. One type of bobbin is suitable for coils of various dimensions.

(iii) In certain cases, the yarns can be dyed and bleached on the bobbin, especially if the bobbin is perforated.

A disadvantage of cross-winding can be that weak yarns suffer by the rapid reciprocating movement of the wire guide.

A distinction can be drawn between *scrambled cross winding* and *precision cross winding*. This is a similar distinction to that just mentioned in winding reels regarding the coupling of rotary and reciprocating movement.

In the first case (Fig. 5.18), the reciprocating movement (of the wire guide) and rotation (of the coil) are driven independently from each other. The reciprocating movement maintains the same frequency, while the rotational speed of the coil adapts itself to a constant rotary speed, that is, it decreases slowly as the coil becomes thicker. The result is that the angle of the pitch at which the wire is applied to the coil remains constant throughout the whole winding. The pitch increases, however, as the coil becomes thicker. The small difference in pitch between successive layers leads to irregularities in the manner in which the wire settles.

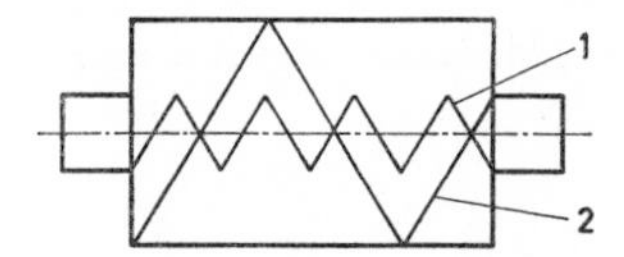

Fig. 5.18. Scrambled cross-winding. 1 = first layer. 2 = last layer.

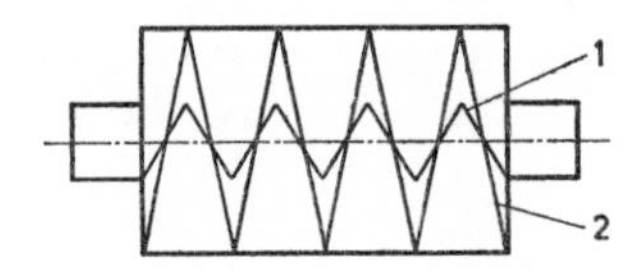

Fig. 5.19. Precision cross-winding. 1 = first layer. 2 = last layer.

In the precision cross-winding [1,2], (Fig. 5.19), there is a fixed relation between the rotary and the reciprocating movement (the transmission ratio) so the problem does not occur. The pitch remains the same from inside to outside, but the angle of pitch decreases. The correct value of this transmission ratio is fixed in two steps. The first step follows from the angle of pitch at which the first layer of the winding is applied to the bobbin. The angle of pitch at which the wire still assumes the prescribed pattern is determined by the diameter of the bobbin, the friction coefficient between wire and bobbin, the rigidity of the wire and the distance of the wire guide to the winding point (where the wire comes into contact with the bobbin).

Angles vary from 3–40° in the winding of yarns. The required angle of pitch, bobbin diameter and coil width give the first approximation of the transmission ratio. The exact value of the angle of pitch follows from a calculation which is aimed at distributing the reversing points uniformly over the circumference. The changing ratio, that is, the ratio between the rotary and reciprocating motion, is then given too.

5.3.4 Faults

It will be clear that *imperfections* in the winding process can later on give rise to the following:

(a) Wire fracture caused by the wire jamming between windings of the underlying layers.

(b) Stiff wire, due to the high winding tension.

(c) Irregular wire thickness due to tension variations in the winding process.

(d) Curling of the wire; caused by the wire running along guide wheels or rods with too small a radius; a combination of tensile and bending forces then results in permanent distortion.

(e) Sharp bending of the wire caused by high compression in the points where the wires cross into tightly wound or over-full bobbins.

(f) Bad insulation quality due to damage of the lacquer or enamel coat caused by excessive deformation.

REFERENCES

[1] R. WASSMANN *Moderne Präzisions-Kreuzspulmaschinen für Chemiefäden feiner Titer*, Chemiefasern in der Spinnerei, Hft 1, 2, 3, 4, 1961.

[2] GEORG SAHM *Die Präzisions-Kreuzspulung*, Melliand Textilberichte, 1965/5 pp. 465 to 469.

5.4 Winding wire products

Whereas in the foregoing, the wire production process was the important item, here it is the finished product at which winding must be aimed.

From the supply reel to the finished product, the wire runs along the dispenser and the wire guide. The wire dispenser usually consists of a dereeling device, a wire braking system and if necessary a "buffer", a small stock of wire able to cope with sudden large demands (Fig. 5.20). The wire guide presents the wire at the correct place of the coil. Finally, the winding mechanism determines the relation between the movement of the winding spindle and wire guide and consequently the configuration of the coil.

5.4.1 Wire dispenser

(a) The supply reel can be wound off tangentially or axially (overhead). During overhead *de-reeling*, the supply reel stands still while the wire is spooled-off via a flyer situated in the centre of the reel (Fig. 5.20). Each turn round the reel causes a twist in the wire but this will not usually be a drawback. This is the simplest method of avoiding the great disadvantage of

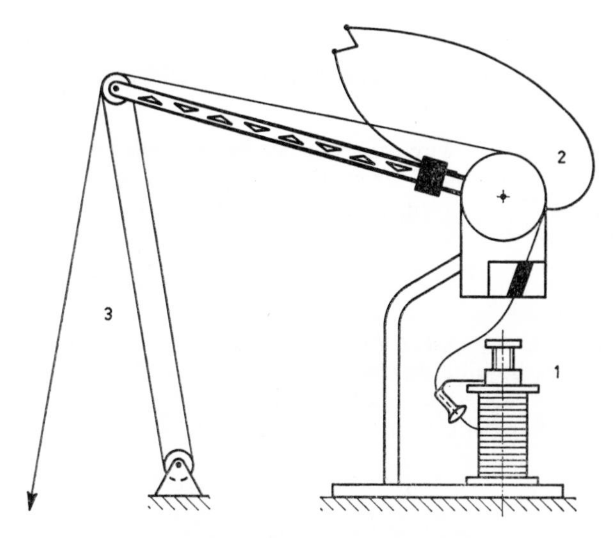

Fig. 5.20. Survey of wire dispenser.
1 = overhead flying off. 2 = brake system. 3 = wire buffer.

tangential de-reeling. For, in that case, the wire will usually have to drive the reel itself and must therefore transfer acceleration forces. It is, of course, also possible to overcome this drawback by fitting a special drive for the supply reel, controlled by the wire tension, for instance, by means of an eddy current coupling, as described in Section 5.3.3(a). This is, however, far more complicated and more expensive than overhead de-reeling and only justified in cases where very special requirements are to be met, as for example, at very great accelerations and for very thin wire.

(b) For *wire braking*, see Section 5.2.

(c) The last component in the wire dispenser is the "buffer", the stock of wire that varies between certain limits, while maintaining the required wire tension. The necessity for this buffer increases, the greater the variations in wire consumption and/or the masses to be accelerated by the wire. In general, this wire buffer consists of a number of passages of the wire along a set of stationary wheels, and a set of wheels fitted on the lever controlling the wire braking mechanism (Fig. 5.20).

A particular case in which such wire stock can also prove of value is where wire must be repeatedly withdrawn from the winding process. For instance, when making coils with taps.

The three elements of wire dispension discussed here, the de-reeling mechanism, the tension control and buffer, can be united in one mechanism if the supply reel is driven by a system controlled by the wire tension, and can be used both for reeling and de-reeling. There are various applications, also in the field of coil winding.

5.4.2 Wire guide

Apart from the movement made by the wire guide, the following points are important:

(a) It is the last point the wire passes before being applied to the coil. The distance the wire guide must be from the coil depends on the stiffness of the wire and the winding pattern. Fig. 5.21 shows a wire guide for the winding of single-layer coils.

(b) The wire can be led over a wheel (Fig. 5.22), round a pin or through a hole (Fig. 5.23).

In special cases, that is, for the complicated coils used in deflection units for picture-tubes, the winding mandrel assists in guiding the wire. Most coils, however, are not wound on mandrels but on coil-formers or directly on parts of the mechanical circuit.

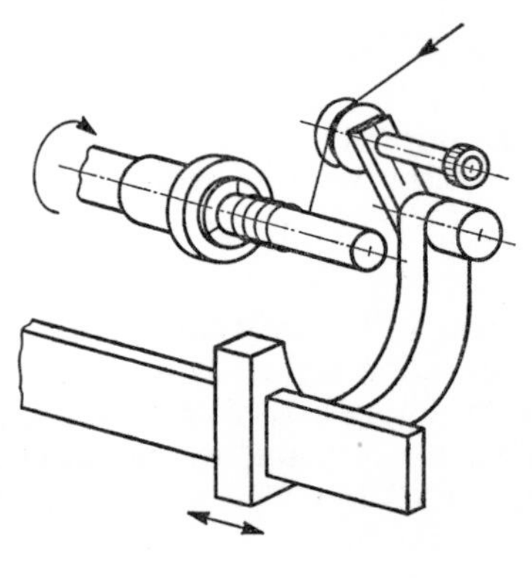

Fig. 5.21. Wire guide for single-layer coils.

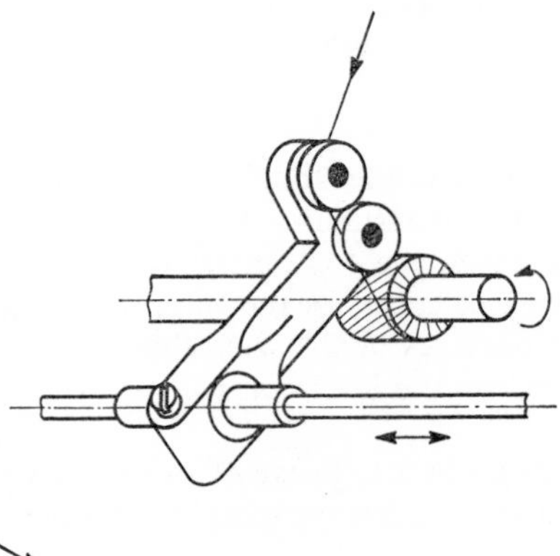

Fig. 5.22. Wire guide for cross-windings: wire guided round wheel.

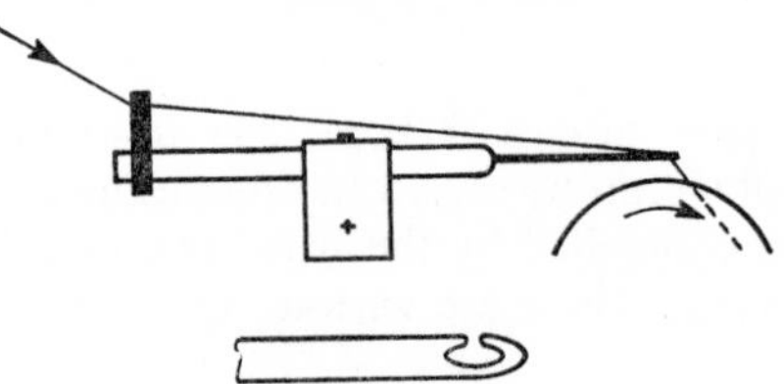

Fig. 5.23. Wire guide for cross-windings: wire led through hole.

5.4.3 Winding mechanism

The actual winding mechanism consists of a winding spindle and a wire guide, the motions of which are intercoupled. There is a rotating motion and a reciprocating motion, which can be assigned to both elements in four different ways:

The winding spindle rotates: the wire guide moves to-and-fro. This is the most common, since the construction is quite simple (Fig. 5.24(a)).

The winding spindle is stationary: the wire guide rotates and reciprocates. This construction will be found on rotary indexing machines on which a number of winding mandrels pass the same winding position in a sequential order. Driving the winding mandrel would then be too complicated. Moreover, the wires need not be fixed directly after winding; the coils can form a temporary link; the wire leads can then be fixed on a next position (Fig. 5.24(b)).

The winding spindle rotates and reciprocates: the wire guide is stationary. This combination seldom occurs; and then only in places where secondary operations do not allow of a moving wire guide (Fig. 5.24(c)).

The winding spindle reciprocates: the wire guide rotates. This is especially useful if only one layer is applied, so that, after winding, the coils can be led away in a continuous movement (Fig. 5.24(d)).

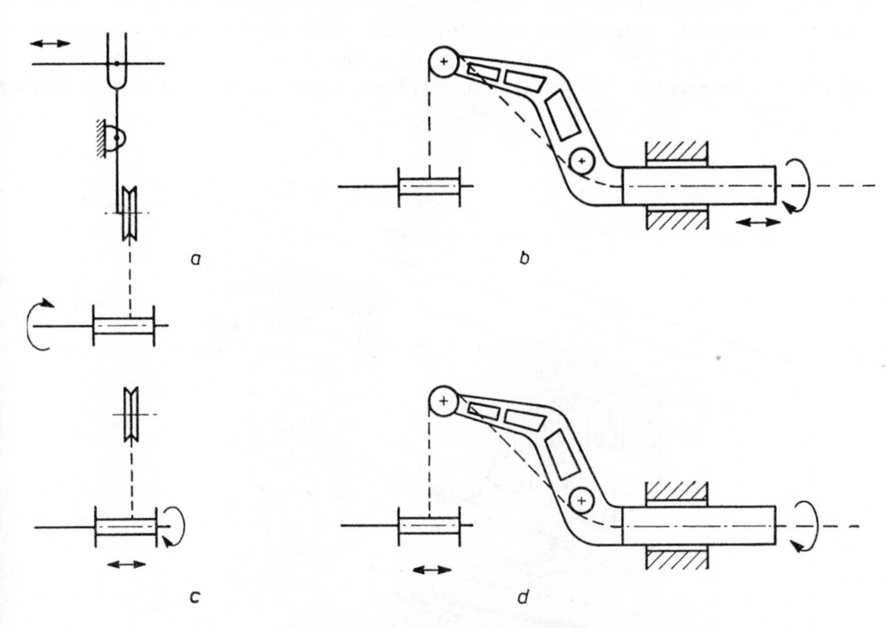

Fig. 5.24. Winding principles.
a. winding spindle rotates: wire guide moves to-and-fro.
b. winding spindle stationary: wire guide rotates and moves to-and-fro.
c. winding spindle rotates and goes to-and-fro: wire guide stationary.
d. winding spindle moves to-and-fro: wire guide rotates.

So that the windings are distributed uniformly over the whole width of the bobbin, the reciprocating motion must have a constant speed in either direction. Theoretically, this leads to an infinite acceleration at the reverse points. It is important that the construction should be light, rigid and without backlash at high winding speeds, especially on cross-winding machines, where the relative frequency of the reciprocating movement is much higher than that of layer winding machines, and, furthermore, high standards have to be met as regards the accuracy of the winding pattern.

In general, the rotary motion is the primary; in principle, the mechanism changing the rotary motion into a reciprocating motion has three functions:

(1) The actual conversion of the rotary into the reciprocating motion.

(2) The transmission ratio, which ensures that each layer gets the correct number of turns.

(3) Setting the correct pitch and winding width.

In various mechanisms, these functions are combined in various ways. Furthermore, it depends on the coil design what requirements of accuracy and reproducibility these functions have to satisfy (see Section 5.4.4).

Examples of applied mechanisms:

(i) *Rack and pinion* (Fig. 5.25). Two gearwheels rotating in opposite directions can be alternately coupled to a toothed rack. The wire guide is connected direct (or via a linkage) in this rack. The coupling is governed by stops which set the winding width. The number of turns per layer can be varied by means

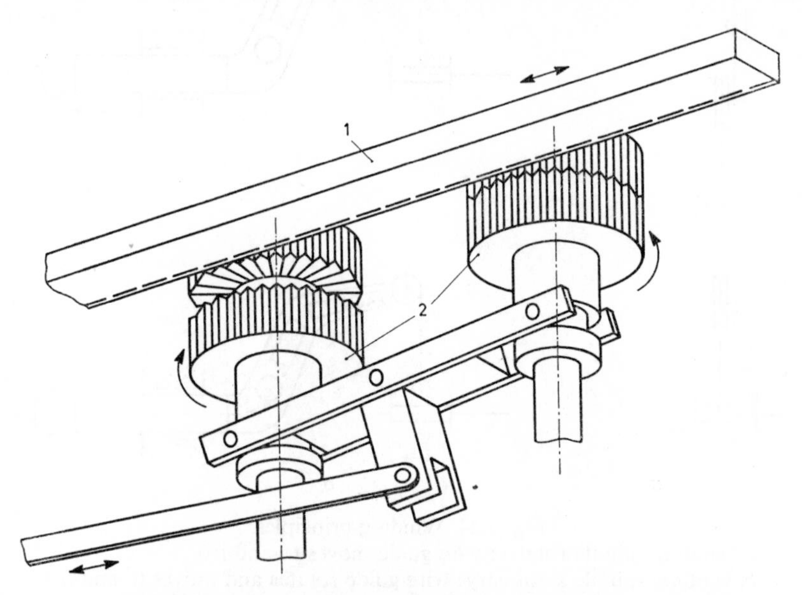

Fig. 5.25. Traversing movement obtained with a rack and pinion. 1 = toothed rack. 2 = gears rotating in opposite directions.

of a gear transmission or variator. The reverse point is not accurate, and the mechanism is not suitable for rapid transverse movements.

(ii) *A screw-spindle rotating alternately in opposite directions* (Fig. 5.26). A nut moving along this spindle takes along the wire conductor. The direction of rotation of the spindle can be reversed by coupling the spindle alternately with gears rotating in opposite directions. The couplings can be operated by stops or microswitches.

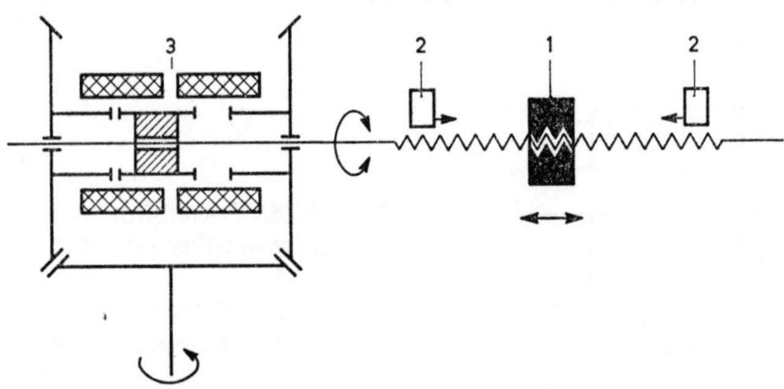

Fig. 5.26. Traversing movement obtained with one screw spindle, guide nut and reverse coupling.
1 = traversing nut. 2. micro switches. 3 = electromagnetic coupling.

(iii) A very simple mechanism is formed by a reversing ratchet wheel. Using a linkage system and a spring-loaded pawl, a continuously rotating eccentric causes a ratchet wheel to rotate stepwise. Adjustable stops on the ratchet wheel force the pawl into the opposite position, and so the direction of motion is reversed (Fig. 5.27). Because of the stepping motion, the mechanism is suitable only for slow winding.

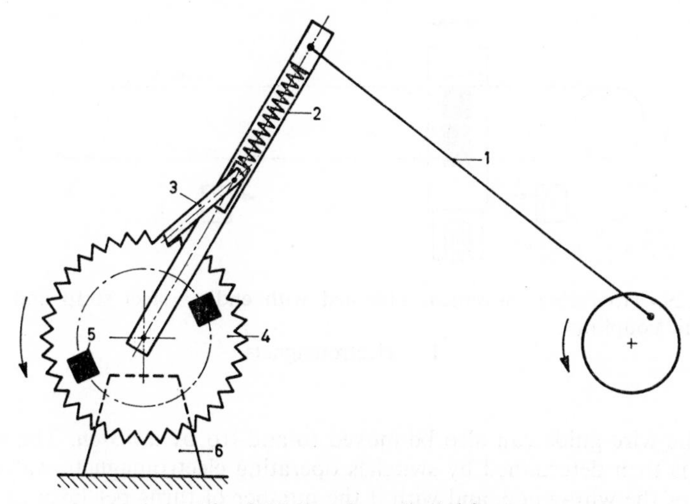

Fig. 5.27. Ratchet wheel drive.
1 = connecting rod. 3 = pawl. 5 = adjustable stops.
2 = spring. 4 = ratchet. 6 = frame.

(iv) A roller resting on a rotating spindle, forming a certain angle with it, describes a helix round this spindle (Fig. 5.28). The holder carrying this roller can slide, and supplies the reciprocating movement of the wire guide. The holder of the roller contains a mechanism which is operated by stops, so that the angle of the roller is reversed at the return points. The angle of the roller can be varied, thus providing satisfactory adjustment for the pitch. However, the points of return cannot be fixed exactly and for high transversal speed, this mechanism is not suitable.

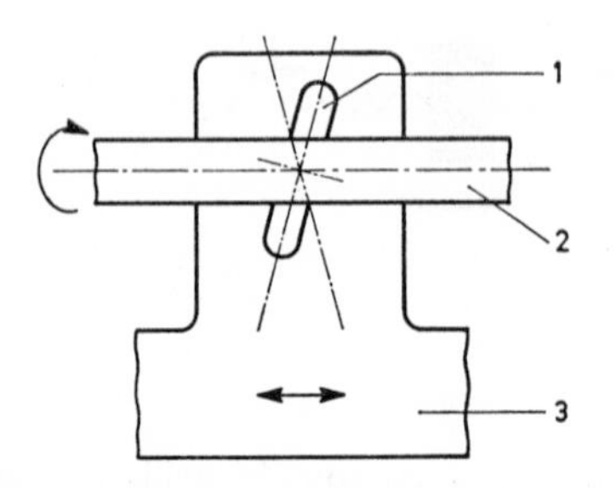

Fig. 5.28. Traversing movement obtained from roller on rotating spindle.
1 = roller.
2 = drive shaft.
3 = traversing block.

(v) *An endless steel strap* runs over two rollers at a constant speed (Fig. 5.29). A slide connected to the wire guide carries two electromagnets, alternately coupled to one of the parts of the steel strap. The magnets are governed by switches operated by the slide. A certain stroke corresponds with a certain number of turns per layer; the correct winding width can, for instance, be adjusted with a linkage mechanism.

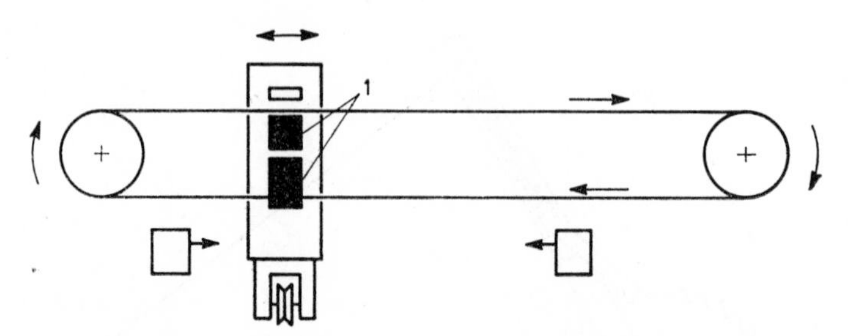

Fig. 5.29. Traversing movement obtained with endless steel strap and electromagnetic couplings.
1 = electromagnets.

(vi) The wire guide can also be moved to-and-fro by a *piston*. The winding width is then determined by switches operating electromagnetic valves. The speed of the wire-guide, and with it the number of turns per layer as well as the pitch, are controlled by driving the main shaft by an oil pump with an adjustable capacity (Fig. 5.30).

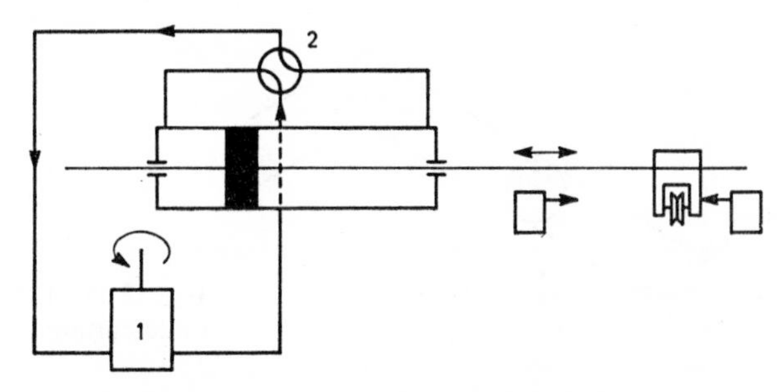

Fig. 5.30. Principle of hydraulic winding machine.
1 = oil pump. 2 = change-over valve.

The mechanism can easily be adjusted, but the point of return is not exact.

So far, the discussion has been concerned with a relatively-large transmission ratio between rotating and reciprocating movement. Some examples are given below of mechanisms with transmission ratios considerably smaller, and therefore suitable for rapid reciprocating movements.

(vii) *Cylinder cam* (Fig. 5.31)

A cylinder contains a groove with a right-hand pitch over half the circumference of the cylinder, while the other half of the cylinder has a groove with a left-hand pitch. Both groove halves can be joined, but is difficult since the cylinder consists of two separate halves. Another possibility is to put the two halves axially apart (Fig. 5.32). This solution requires more space than the first, and, in addition, two followers are needed.

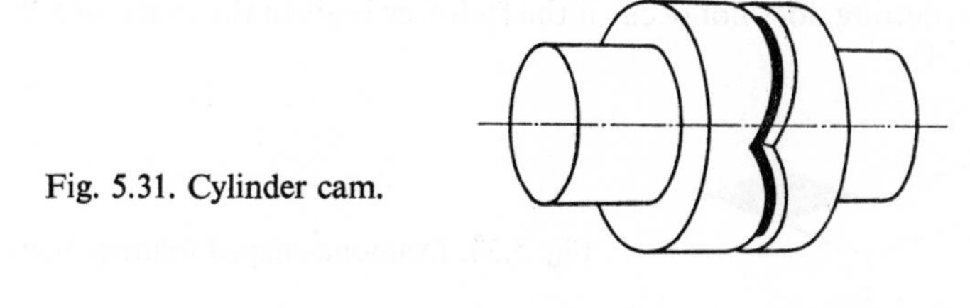

Fig. 5.31. Cylinder cam.

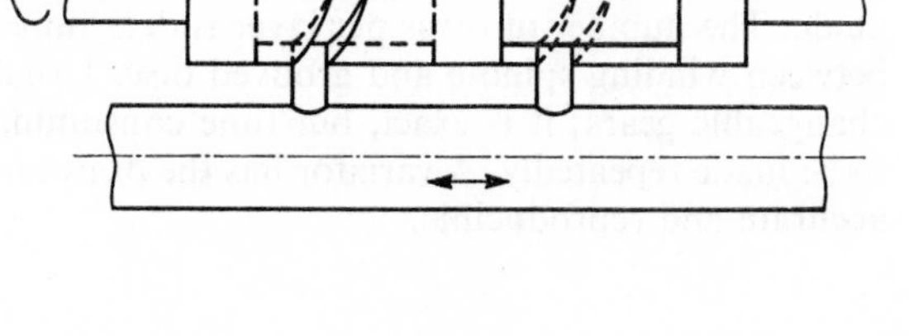

Fig. 5.32. Two half-cylinder cams.

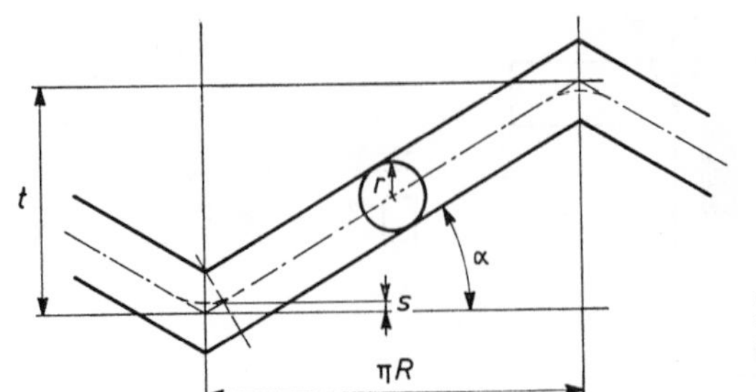

Fig. 5.33. Principle of undercutting.

Usually the follower running in the groove is a roller or a pin. At the point of return, the grooves are undercut (Fig. 5.33). It is thus possible for the roller to follow the top of the curve. The centre of the roller then lags a distance s behind the prescribed track, the play—

$$s = \frac{r}{\cos\alpha} - r = r\left(\frac{1}{\cos\alpha} - 1\right)$$

Since $\tan\alpha = \frac{t}{\pi R}$ and $\frac{1}{\cos\alpha} = \sqrt{\tan^2\alpha + 1}$

it follows that:

$$s = r\left(\sqrt{\tan^2\alpha + 1} - 1\right) = r\left(\sqrt{\frac{t^2}{\pi^2 R^2} + 1} - 1\right)$$

Hence, play is minimized by:

Choosing a small r for the follower.

Choosing a small stroke t for the grooved disc.

Using a disc of large diameter ($2R$).

Undercutting does not occur if the follower is given the shape of a diamond (Fig. 5.34).

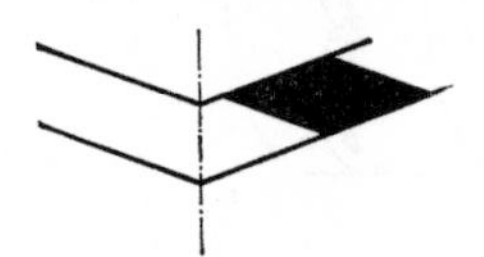

Fig. 5.34. Diamond-shaped follower body.

Since the grooved disc has a fixed stroke, the winding width must be varied by using a lever with an adjustable pivot between follower and wire guide. The number of turns per layer is determined by the transmission ratio between winding spindle and grooved disc. Usually, this is done with interchangeable gears; it is exact, but time consuming when readjustments have to be made repeatedly. A variator has the drawback that the setting is seldom accurate and reproducible.

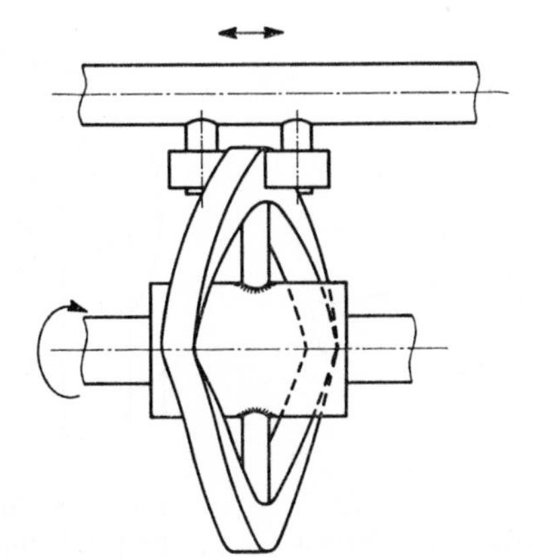

Fig. 5.35. Traversing cam.

(viii) *Traversing cam* (Fig. 5.35)

The traversing cam is, in fact, an external cylinder cam. The same problems of undercutting are encountered. For the winding width and the transmission ratio, the same methods are employed as for the grooved disc. Two rollers are required, which, apart from the high requirements the pitch must meet, also sets demands for the thickness tolerance of the disc. In this respect, the cupped disc (Fig. 5.36) which needs to be machined only on one side, is

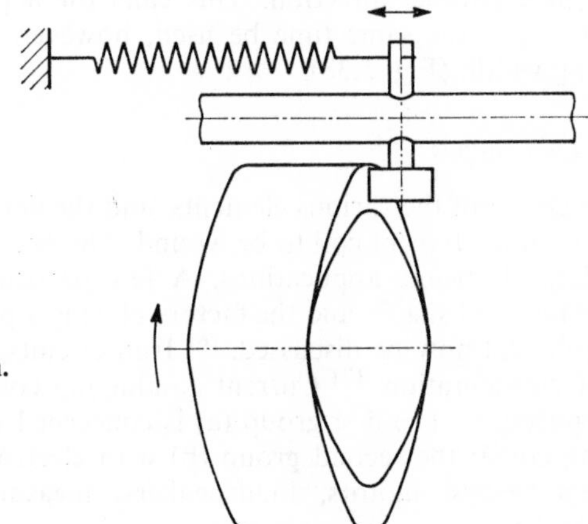

Fig. 5.36. Cup-shaped cam.

simpler. But the heavy spring, required to keep the cam roller constantly pressed against the profile, has drawbacks as regards heavy running and wear.

(ix) The *heart-shaped cam* (Fig. 5.37), used for reciprocating movement, is not followed in the axial direction as is the case with the grooved cam and traversing cam, but in the radial direction.

The heart-shaped cam is a cam plate with a profile consisting of two symmetrical spiral parts. The distance between the diametrically-positioned

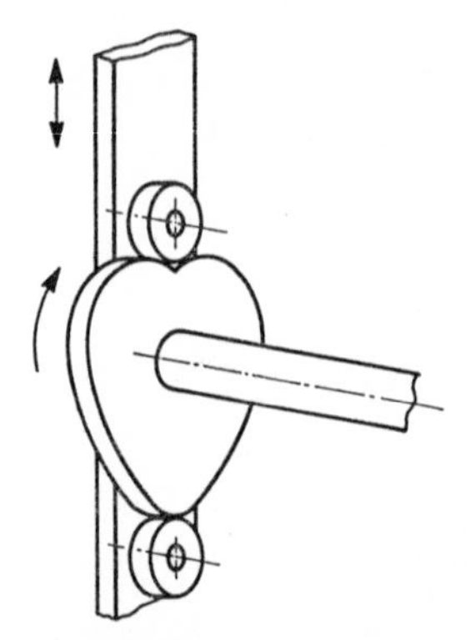

Fig. 5.37. Heart-shaped cam.

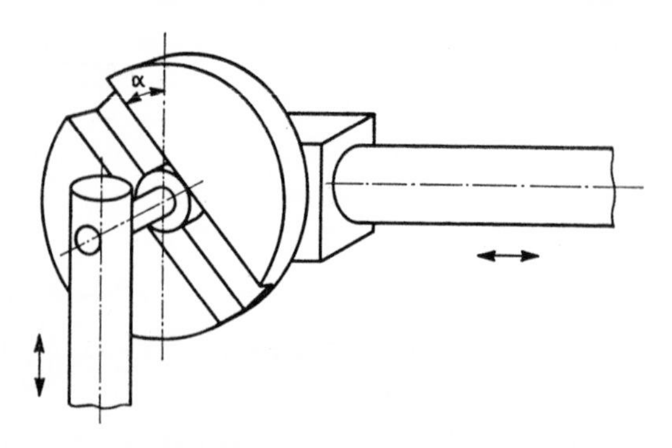

Fig. 5.38. Sliding block to adjust the stroke and perpendicular change of direction.
a = adjustable angle

points is constant. This makes it possible to use a dog (slide) with two cam rollers and to realize bilateral locking. This method, too, involves the problem of undercutting. For the winding of single-layer coils, a single linear profile is sufficient; the return stroke can, for example, vary sinusoidally or even be accelerated.

A disadvantage of such radial profiles is that the reciprocating movement takes place perpendicular to the rotating shafts to which the wire guide must move in a parallel direction. This calls for a perpendicular transmission, which can at the same time be used, however, to vary the stroke for the winding width. (Fig. 5.38).

5.4.4 *Coil shapes*

The choice of the various elements, and the design of the winding machine depend on the type of coil to be wound. The greatest variety in coil shapes is found in electronic applications. A few particulars, therefore, concerning the various coil shapes and the factors playing a part in the choice of winding principle will now be discussed. [6] Iron circuits, types of wire, etc., are left out of consideration. [7] Current conducting coils can be designed for two main purposes. The first group (a) is concerned mainly with self-inductance (tuning coils): the second group (b) with electromagnetic field coils (transformers, relays, motors, loudspeakers, measuring equipment, deflection coils and so on.

(a) *Self-inductance* (*tuning coils*)

In designing a coil primarily to obtain a certain self-inductance, the appearance and the winding configuration play an important part. For a rather long coil, the self-inductance follows rather exactly from the number of windings. For a shorter coil the self-inductance also depends on the length-to-diameter ratio and on the ratio between winding height and diameter. Another important aspect is the characteristic capacitance of the coil. To keep the losses in the coil within certain limits, and to guarantee its utility to the highest possible frequency, the characteristic capacitance must be at a minimum.

This means that the voltage differences occurring between adjacent turns must be kept as low as possible. For this, the single-layer coil is ideal, especially if the turns are applied with a certain spacing between them. (Fig. 5.39). The best method is to keep the pitch equal to 1·6× the wire diameter, and the length of the coil at least equal to its diameter. For low self-inductances, this is the indicated design. Because of the large number of turns required, the single-layer coil is not practical for high self-inductances The alternative, of using a winding with a number of layers, leads to impermissibly high values of the characteristic capacitance of the coil. This difficulty has been avoided by using the so-called "bank winding". (Fig. 5.40).

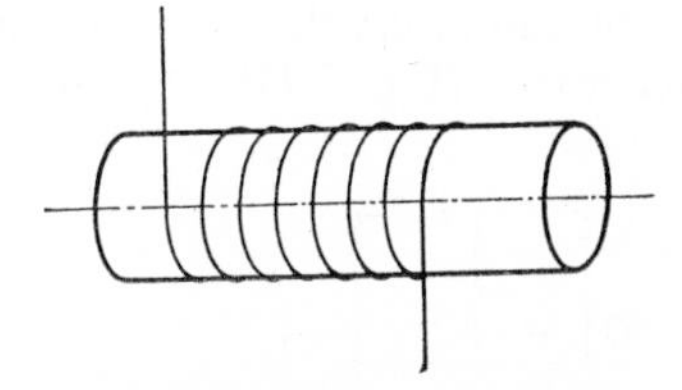

Fig. 5.39. Single-layer winding.

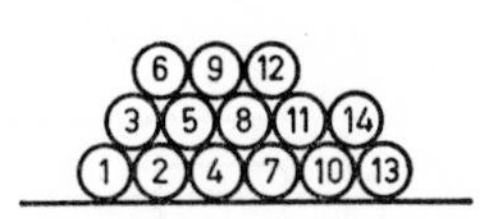

Fig. 5.40. Space (bank) winding.

The turns are distributed over a number of layers, in such a way that the turns with a large voltage difference between them are not close together. This winding process cannot take place mechanically, but fortunately there are two other methods. One is universal winding (Fig. 5.41), and the other progressive winding (Fig. 5.42), both of which can be mechanized.

Fig. 5.41. Universal winding.

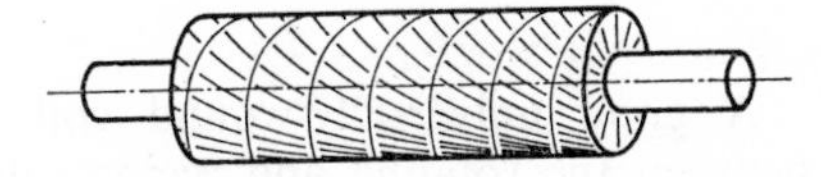

Fig. 5.42. Progressive winding.

(i) *Universal winding*

In universal winding the turns are wound at a large pitch. The result is a kind of zig-zag pattern with many crossings, which ensures that the turns of successive layers are positioned accurately side-by-side. As the coil is gradually wound from the inside to the outside over its full width, the turns with the greatest voltage differences between them are farther apart than is the case with an ordinary cylindrical layer winding.

(ii) *Progressive winding*

Further spreading is possible by using progressive winding, a combination of universal winding and layer winding, in which the wire guide making the cross-stroke moves gradually along the coil at the same time. The ends of the coil are not spaced radially, as is the case with ordinary universally-wound windings, but axially, and therefore further apart, as a rule. In both cases the result is a mechanically sturdy and in all respects reproducible winding. The characteristics of the progressive winding are practically equal to those of the bank winding, but it can be made on a machine. The maximum number of layers is five, for too large a number of turns would lead to a greater length. In that case, the ordinary universal winding is indicated. The shape of a universally-wound coil can vary from long and low, to short and high (Fig. 5.43). This has its consequences, both on the shape of the field and the characteristic capacitance of the coil. The characteristic capacitance decreases as the coil becomes shorter and higher. If the characteristic capacitance is to be reduced even further, the coil can be divided into sections wound on the same former a certain distance apart. (Fig. 5.44).

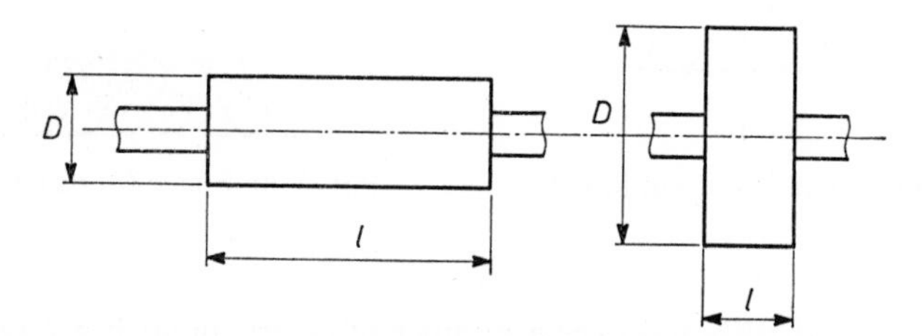

Fig. 5.43. Length-diameter proportions of universally-wound coils.
l = length. D = diameter.

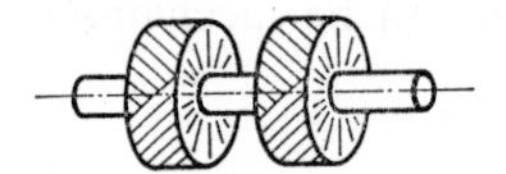

Fig. 5.44. Two sections on one bobbin.

A good universally-wound coil requires a special transmission ratio between the rotating and reciprocating movement [1,2,3,4,5]. A scrambled cross-winding, as used for the winding of yarn (Section 5.3.3), is not suitable. There is, however, a certain resemblance with the precision cross-winding mentioned there. In this case, too, the transmission ratio is determined in two steps. The overall value follows from the required pitch angle of the wire on the bobbin, which must be between 4° and 15°, the most favourable value being 12°. The ultimate transmission ratio is not chosen, however, to get a uniform distribution of the reversing points over the circumference, but so that the windings of successive layers are close together. The result is a beautifully compact coil. The transmission ratio then deviates only little from a whole number or a simple fraction. Depending on whether the wires of successive layers are positioned slightly before or behind the previous layers,

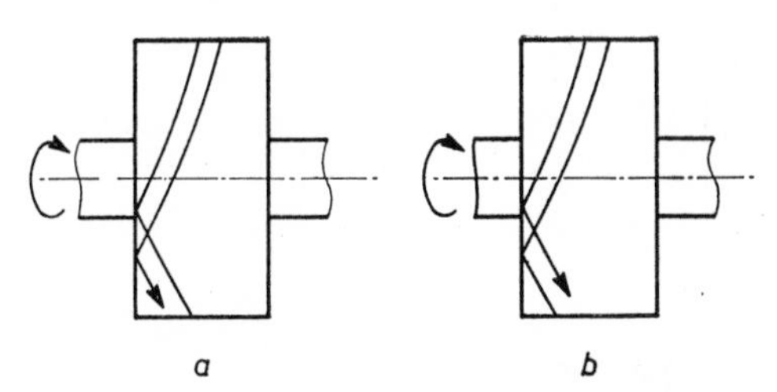

Fig. 5.45. *a*. Progressive cross-winding.
b. Retrogressive cross-winding.

gives progressive or retrogressive universal windings (Fig. 5.45). A good approximation to the transmission ratio (i) between the number of revolutions of the winding spindle and the number of reciprocating movements of the transversing motion is:

$$i = \frac{2}{n}\left(1 \pm \frac{1{\cdot}25 \times \mathrm{d}}{q \times b}\right)$$

where

n = number of single strokes per revolution;

d = wire diameter (including any insulation);

q = number of single strokes per total winding cycle. After one winding cycle, the wire, apart from the winding pitch, has returned to the same position;

b = coil width.

See the example in Fig. 5.46.

This transmission ratio can also be found graphically (Fig. 5.47), as follows:

From the origin draw a line under a certain angle whose complement is the required pitch angle. For instance, 78° = (90–12)°. Along the abscissae

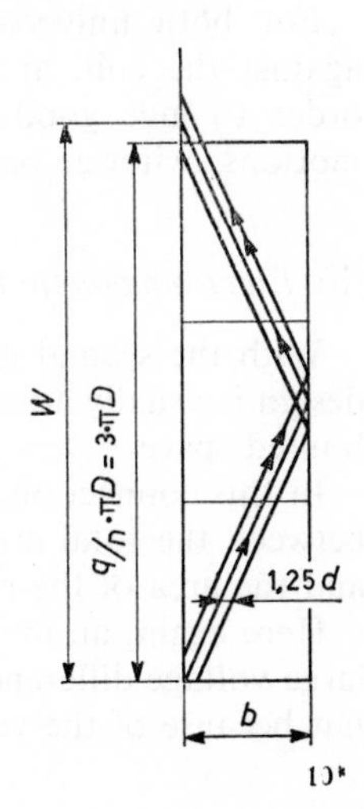

Fig. 5.46. Calculation of the transmission ratio of the universally-wound coil.
W = winding cycle. $n = 2/3$. $q = 2$. $q/n = 3$.

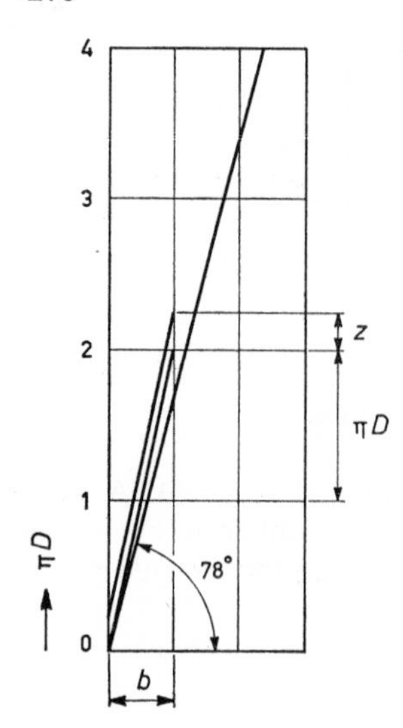

Fig. 5.47. Graphical determination of the transmission ratio of the universally-wound coil.

$$i = 2\left(\frac{2 \pm z/\pi D}{b}\right)$$

plot the winding width, and along the ordinate the periphery of the coil former is plotted a few times on the same scale.

The first approximation of the transmission ratio follows from the point located on the pitch line closest to the origin, for which a number of reciprocating movements corresponds about with a total number of coil peripheries. The pitch angle is first corrected so that the pitch line is made to run through this point. Then a second pitch line is drawn parallel to the first pitch line at a distance slightly greater than the diameter of the wire to be wound. This is winding step z; the line connecting the origin with the point located at the distance z above the point indicating the overall transmission ratio is then the definite pitch line. The ultimate transmission ratio can then be read from the figure.

For both universal and progressive winding, the winding finger rests against the coil, and must satisfy special constructional requirements, in order to give good guidance to the wire during the rapid reciprocating motions, without breaking the wire or damaging the coil.

(b) *Electromagnetic field coils*

With the second group of coils, where the field is the main interest, the design is usually aimed at accommodating as many windings as possible in a limited space.

In this connection, the term "*filling factor*" is used, to represent the ratio between the total cross-section of the windings (including their insulation) and the area of the rectangle enclosing these windings. (Fig. 5.48).

Here again, an important point is the proximity of the windings with the large voltage difference between them, not because of their own capacitance, but because of the voltage differences the insulation can take.

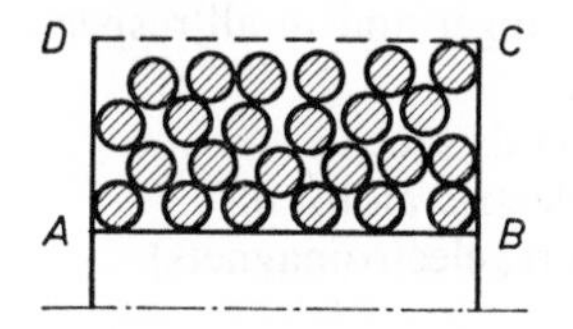

Fig. 5.48. Filling factor $= \dfrac{X.\pi/4\, d^2}{AB \times BC}$

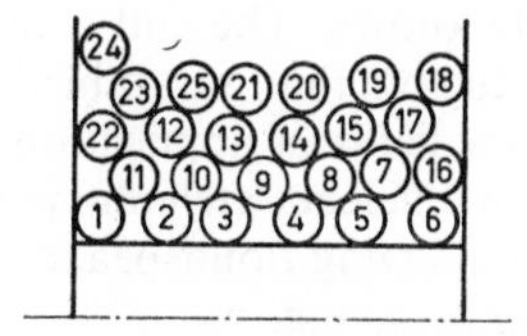

Fig. 5.49. Scrambled winding.

In order to keep the winding process simple and within the available winding speeds, layer winding is usually preferred.

There are various types:

(i) *Scrambled winding* (Fig. 5.49)

With a small filling factor; it is not known exactly where the turns will land. This method is only suitable where the voltage is not too high.

(ii) *Semi-scrambled wound* (Fig. 5.50)

Endeavours to keep all turns in their own respective layers. The danger of breakdown is less; the filling factor is higher.

(iii) *With intermediate layers* (Fig. 5.51)

One or several layers are alternated with a layer of insulating material High voltages can then be used. The filling factor is, of course, very low.

(iv) *Orthocyclic* (Fig. 5.52)

The maximum filling factor for round wire (91 %) is obtained by positioning the wires in the grooves of the preceding layer. The position of the turns is

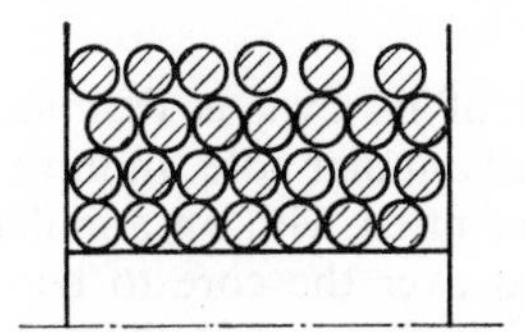

Fig. 5.50. Semi-scrambled winding.

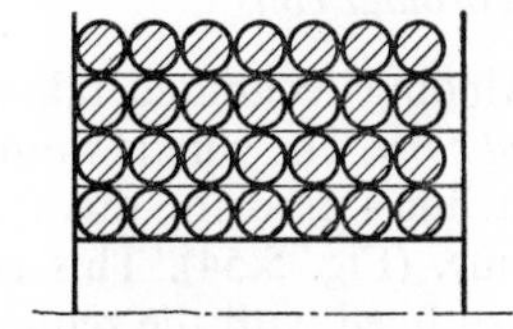

Fig. 5.51. Winding with intermediate layers.

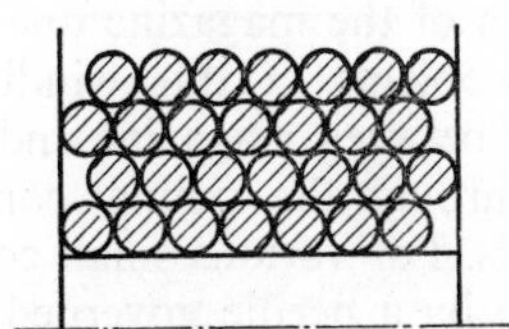

Fig. 5.52. Orthocyclic winding.

exactly known. The coil is more compact, stronger and in all respects more reproducible than scrambled wound coils [8].

A few fields of application for orthocyclic coils are:
High efficiency (telephone relays, battery electric motors).
Space-saving (loudspeaker coil, small motors, electromagnets).
Weight saving.
To obtain an accurate field (focusing coils in TV sets).
Mechanically strong coils.
Reproducible coils (bridge circuit in measuring equipment).

(v) *Pilgrim walk winding* (Fig. 5.53)

This is a layer winding for which the stroke of the wire guide, which is less than the length of the coil, is gradually displaced along the coil. Section I is first wound with an increasing stroke, then section II with a constant stroke gradually moving forward. Finally, section III is wound with a gradually decreasing stroke. One might say that the pilgrim walk winding is a laterally shifted layer winding, just as the progressive winding is a laterally displaced universal winding. The result is that turns with too great a voltage difference are not brought into close contact. The pilgrim walk winding is sometimes used for transformers.

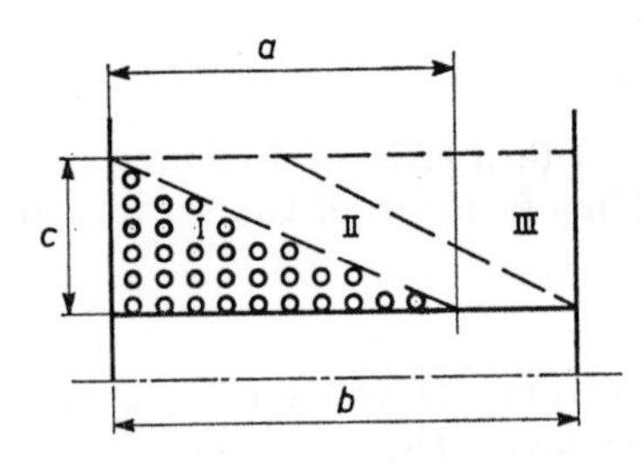

Fig. 5.53. Pilgrim walk winding.
a. max stroke.
b. coil width.
c. coil thickness.

5.4.5 *Toroidal coils*

A winding technique taking up a position of its own is that used for *toroidal coils.* The wire is wound around a closed annular core. In most cases, the machines used for this type of winding are fitted with an annular wire magazine. (Fig. 5.54). This magazine is closed over the core to be wound and then filled with the required amount of wire. The wire is then attached to the annular core and wound out of the magazine. The annular core rotates in its own plane, thus determining the pitch of winding. The wire magazine rotates perpendicularly to this plane through the annular core, and for each revolution of the magazine one turn is completed.

The problems of this winding method are concentrated on the wire guidance between magazine and coil. Since the wire magazine must be able to pass through the annular core, applications are limited to comparatively large coils. For various small coils, attempts have been made to replace the magazine by a needle governed by a magnet.

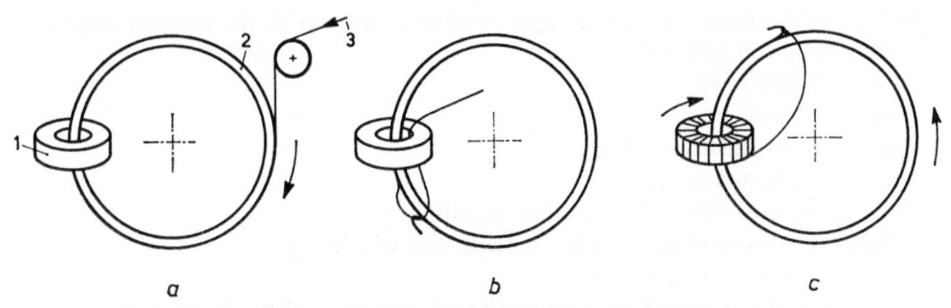

Fig. 5.54. Principle of toroidal winding.
a. filling the annular magazine. *b.* attaching wire to core. *c.* winding.
1. coil core. 2. annular magazine. 3. wire.

REFERENCES

[1] L. M. HERSHEY, *The design of the universal winding*, Proc. I.R.E., August 1941, pp. 442 to 446.

[2] F. E. PLANER *Toroidal Winding Machines*, Journal of Scientific Instruments, December 1943, pp. 185 to 188.

[3] A. W. SIMON *On the winding of the universal coil*, Proc. I.R.E., January 1945, pp. 35 to 37.

[4] MYRON KANTOR *Theory and design of progressive and ordinary universal windings*, Proc. I.R.E., December 1947, pp. 1563 to 1570.

[5] A. W. SIMON *Correspondence on the article of Myron Kantor*, Proc. I.R.E., September 1949, pp. 1029 and 1030.

[6] RENS and RENS *Handboek der radiotechniek*, deel 3, hoofdstuk VI, par. F: *Spoelen voor hoge frequenties*, H. M. VISCH, A.E. Kluwer, Deventer 1951.

[7] KURT RABE *Die konstruktive Ausbildung von Gerätespulen*, Feinwerktechnik, Jg. **60**, Hft. 6, 1956.

[8] W. L. L. LENDERS *The orthocyclic method of coil winding*, Philips Technical Review, vol. 23, 1961/62, No. 12, pp. 365–404.

[9] G. MÖLLENSTEDT, E. KRIMMEL *Herstellung von stromleitenden Wendeln mit extrem kleinen Durchmessern*, Zeitschrift für angewandte Physik, XVI Band, Hft. 2, 1963, pp. 121 to 124.

5.5 Coil winding in manufacture

So far, attention has been directed to the technique of winding. This is only one of the many operations in the manufacture of a coil, but an operation well-suited to mechanization.

The relationship between winding and the remaining manual operations is important in coil manufacture.

These operations include:

Placing the coil former or winding jig on the machine.

Attaching the beginning of the wire.

Placing the wire guide in its initial position.

Placing the turn counter in its initial position.

Starting the machine.

Stopping the machine in the right position, as soon as the correct number of turns has been applied.
Finishing the wire.
Making taps.
Applying insulating layers.
Finishing the whole coil.
Taking the coil from the winding machine.
Baking out the coil and taking it from the winding jig.
Mounting the coil in a frame.
Connecting the ends of the coil to the contact pins of the bobbin or frame.

The above summary is not complete, and not all the operations have to be carried out for all coils: they vary from coil to coil.

Apart from these operations required for each coil, others have to be carried out periodically, such as inserting wire, readjusting the machine for another type of coil, etc., which cannot be discussed in detail here.

It is most important, in setting up coil production, to make the best use of the labour and machinery available. It is then important to decide what operations must be carried out by the machine, and what operations remain for the operator [1].

In earlier days, coils were even wound manually, that is, the winding machines were operated manually, as is still done for some simple coils. However, winding can easily be done on machines, and a winding machine is usually a universal type, suitable for various types of coil. In most cases, therefore, a winding machine will prove profitable.

It is a different matter with secondary operations, however, which often cannot be carried out by machines and differ from coil to coil, so it is impossible to design a universal machine for them.[3,4,5,6].

A distinction can be drawn between the operations forming part of the actual winding cycle, and for which the winding machine must "wait", and the operations directly connected with winding, that must be carried out separately. Attaching the starting wire, for instance, belongs to the winding cycle, while connecting the ends of the coil to the terminals on the winding bobbin stands alone. These last operations are the ones the operator can carry out during the winding time of the machine. In order to obtain maximum tuning between the two, the winding machine can be of multiple design [2]. Especially expensive winding machines should preferably be made to operate in multiple, even if it means that one man can no longer cope with the secondary operations. In that case, more labour will have to be employed. This may entail difficulties if large quantities of (expensive) tools have to be distributed.

Manual operations forming an integrating part of the machine cycle, such as inserting the bobbin, attaching the beginning wire, etc., have the drawback of making the man machine-dependent and the machine man-dependent. Therefore, many mechanization systems have been aimed at building these operations into the machine. For example, the insertion of insulation layers, making taps, and finishing wires. Again, a distinction can be drawn between turret-type machines, on which the above-mentioned operations are carried out in successive positions, and machines on which all operations are carried out in sequence on one position only. The cycle time of the latter is, of course, longer and the various mechanisms are turned to less efficient use, but a multi-position machine is far more vulnerable.

This explanation will give an idea of the problems involved in selecting a system of coil winding.

REFERENCES

[1] W. AUMANN *Verfahrenstechnische Betrachtungen über das vollautomatische Wickeln von Spulen*, „Draht", Jg. **5**, Hft. 9, 1954.

[2] P. AUMANN *Möglichkeiten und Grenzen des Vielfach-Wickelns*, Elektrotechnik, Jg. **39**, Hft. 17, 1957.

[3] W. OSTERLAND *Die Herstellung der Gerätespule*, „Draht", Jg. **8**, Hft. 8, 1957.

[4] ANON. *H.F. Coil Winding Machine*, Electronic Engineering, July 1961, p. 441 (1 pg.).

[5] R. A. LANCASTER *Winding Machine (which wraps automatically leads around terminal posts)*, Western Electric, Indianapolis, 1963.
1965), Drahtwelt Düsseldorf, **51**, (1965), No, 6. pp. 303–304.

[6] W. D. SCHUPPE *Drahtverarbeitung in der Elecktrotechnik* (Hannover Messe 1965), Drahtwelt Düsseldorf, **51** (1965), No. 6, pp. 363–304.

[7] W. OSTERLAND *Der Einfluß der Leiter-Eigenschaften auf Wickelfaktor und Wickelarbeit*, Feinwerktechnik, Okt. 1967, **71** Jg., No. 10.

Chapter 6

Applying Texts

J. O. M. van Langen

6.1 Introduction

Products and components are usually provided with symbols, texts, text plates or name plates. Texts may be necessary to indicate the function of a product. The aim can also be to display the name of the manufacturer, to supply technical data or directions for use.

For the application of texts, etc., there are the following methods:

- The text is moulded together with a product or component.
- The direct application of a text on an otherwise ready product or component.
- By using a label, text plate, name plate, word or emblem which is attached to a product or component.

The choice of method and the existing techniques for a certain application depend on a number of factors, such as:

(a) The appearance of a text. A word mark or a name plate on a permanent article will usually contribute to its appearance. Appearance, however, is not bound to certain techniques; with most of the techniques a certain attractiveness can be obtained by good industrial design, proper tools and a good finish. In all other cases, an attempt must be made to match the text aesthetically to the object.

(b) The dimension of a text or symbol and the required precision and detail sharpness. Some techniques are more suitable for accurate work than others. For very small texts, it is sometimes necessary to avoid all dust, so special installations are required.

(c) The required duration of a text, that is, its resistance to mechanical damage, adherence to a certain background, resistance against chemical attack and high temperatures, behaviour under certain climatic conditions, and so on.

(d) The cost price of a text. This must be in reasonable proportion to the price of a product or component. The cost price is usually not determined by the material cost, but by the required tools and other means.

(e) The type and form of a product and the type of material on which the text must be written. Some techniques are only suitable for certain groups of materials; not all techniques are suitable for curved surfaces.

For labels or loose name-plates, text-plates and emblems, it is usually best to obtain them from suppliers specializing in these fields.

It is not intended here to give a detailed discussion of the various techniques, but merely a summary for use as a guide towards a correct choice.

For technical information concerning the techniques mentioned, for instance, the mechanical and plastic-forming techniques, see the relevant chapters in Volumes 3 and 11 of this Handbook.

6.2 Diecasting, casting, pressing, etc.

(a) During the forming of a product a recessed or embossed text can be obtained by applying the text in negative in the matrix or mould used. This method is employed in pressing glass, casting metal, diecasting light metals and thermoplastics, and pressing thermo-hardening synthetic materials.

In the case of recessed texts, there is furthermore the possibility of filling up the letters with some other material, like paint. The paint is then applied by spraying or brushing, after which the surplus paint is removed simply by wiping it off.

In objects made of synthetic materials, sometimes a local, slight recess is pressed or cast in, and an accurately-fitting separate text or name plate is afterwards glued into the recess.

(b) Word marks, and also loose letters and symbols, are often made by diecasting thermoplastics or light metals, each item being provided with a pin to attach it to a product. The objects can be finished by lacquering or metallizing.

6.3 Forming techniques

(a) *Warm or cold pressing*

An example is the pressing of a name-plate out of a strip of metal (usually brass) between two matrix halves. See also Volume 11 of this Handbook, Chapter 4.

(b) *Embossing*

This is done with special dies and is a method of giving letters printed on *paper* some relief, as on letter headings. The method is used too for metal foil or thin sheet metal for producing number plates, dials for clocks or sealing caps for bottles.

A particular application method of this technique is the use of *letter pliers* for making letters or figures in adhesive tape (with protective foil). The tape consists of a transparent synthetic material with a pigmented layer. Due to the deformation of the tape, the transparent synthetic material shows an opaque white, so the result is a white text on a coloured background. The method is suitable for texts on cables, pipelines, cabinets or meters and tools for personal use.

(c) *Vacuum forming*

This technique is used for thermoplastic synthetic materials. The matrix into which the foil or sheet of synthetic material is sucked at the softening temperature of the material contains the required text. An example is the text on a vacuum formed synthetic beaker. See Volume 3, Chapter 1.

(d) *Impressing*

The text is impressed with a die into cardboard, wood, thermoplastic or metals. The possibility of curling edges makes this method less attractive. A simple version of the method is the application of *letter punches*, for instance, to punch a serial number in a product or into a standard text plate. The imprinted texts can then be filled with paint.

6.4 Cutting and machining techniques

(a) *Punching*

Consecutive word marks or loose letters can be stamped out of sheet metal, like brass or aluminium, followed by painting or metallizing.

Sometimes, the lettering is also punched out of the plating of larger machines.

(b) *Engraving*

This is a method used for glass, metals and synthetic materials. The tool can be a scissel of hard material or a rotating cutter. Apart from hand-work use is also made of engraving machines (e.g. copying cutters), equipped with a pantograph. A template is traced with a pin and the engraving tool (cutter) does the actual engraving. The engraved pattern can be filled with paint (see Volume 3, Chapter 2).

(i) *Synthetic materials*

Name of text plates can be made, from a sheet of synthetic material consisting of two differently coloured layers. The upper layer is removed (milled or cut) until the lower layer becomes visible.

(ii) *Metals*

Instead of milling, a text can be burned-in electrically with an arc, a less-beautiful method used for the marking of tools.

Here, too, the 2-layer method is employed, the upper layer being removed. One example is to apply symbols by cutting through the silver layer on brass (on watch dials).

Another technique of embellishment is known as *guilloching*, the engraving of entwined curls and lines with a diamond on jewelry and clock dials. The method is used too for engraving the plates employed for printing bank notes, etc.

Accurate scale divisions, are made with *cutting tools* (see Volume 11, Chapter 2).

(iii) *Glass*

For finer scratches, diamond tools are used. Texts or symbols can also be applied with a rapidly-rotating copper wheel (on a horizontal spindle), onto which is dripped amaril powder, suspended in oil.

6.5 Etching techniques

(See Volume 4, Chapter 3).

Etching is used for glass and metals.

(a) *Glass*

The text is supplied with *etching inks*, by means of stamping. After the etching agent has taken effect, and the etching ink has dried, the remnants are brushed away.

Deeper etching is used for the application of scale divisions and other symbols on glass equipment. The glass is coated with a layer of wax, which is removed from the places where the etching agent is to be brought in contact with the glass. The etched symbols can later be made more visible by filling them with paint.

(b) *Metals*

The etching of metals can be used to make name plates and text plates, scale divisions on foot rules, and for the application of texts in matrices. The parts not to be etched are coated with a protective layer, resistant to the etching agent.

A choice can be made between *matt etching* and *deep etching*. After deep etching, the etched parts can be filled up with paint or enamel.

Word marks and symbols are made by etching through metal foil, such as brass or aluminium. One side of the foil is coated with a carrier: the other is given a local acid-resistant coating. Etching takes place through the foil on to the carrier. The emblem can then be stripped off the carrier. It is a method of making products that cannot be made by punching them out of foil. Word marks made in this way are used on wood (they are stuck-on before the wood is varnished). The foil thickness varies from 40 to 300 μm.

6.6 Metallisation

For this technique, see Volume 4, Chapter 2.

The vapour depositing of metal, usually aluminium, is sometimes applied for the manufacture of emblems or word marks. The base can be a die-moulded product of a transparent thermoplastic. This is coated locally on one side with a pigmented layer. The same side is metallized with aluminium and subsequently coated with a protective lacquer. The metallic text is then visible through the synthetic material against a coloured background (or vice versa). See Fig. 6.1.

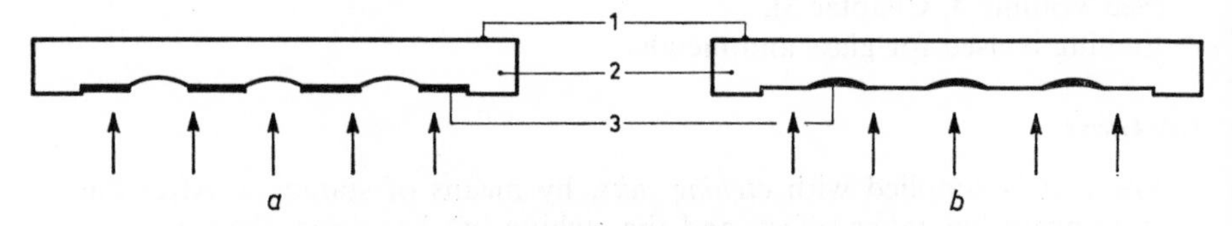

Fig. 6.1. Possible applications of metallizing for emblems.
a. Painted flat areas of the die-cast workpiece.
b. Recessed parts of the casting filled with paint.
1 = front side of the emblem. 2 = transparent thermoplastic. 3 = pigmented layer.

6.7 Transfer techniques

(a) *Thermo impression*

This is a technique where use is made of a carrier of paper or synthetic foil covered with a layer of pigment or metal. The tool is a polished metal die, on which is the pattern to be produced. The foil is applied between the object and the die, with the pigmented side facing the object. As soon as the hot die comes down, the pigment or metal bonds to the object and comes off the carrier.

The method is used for paper, cardboard (book covers), felt, leather, thermoplastics and some thermo-hardeners. The impressed depth in the material usually varies from 0·1 to 0·3 mm (see Fig. 6.2).

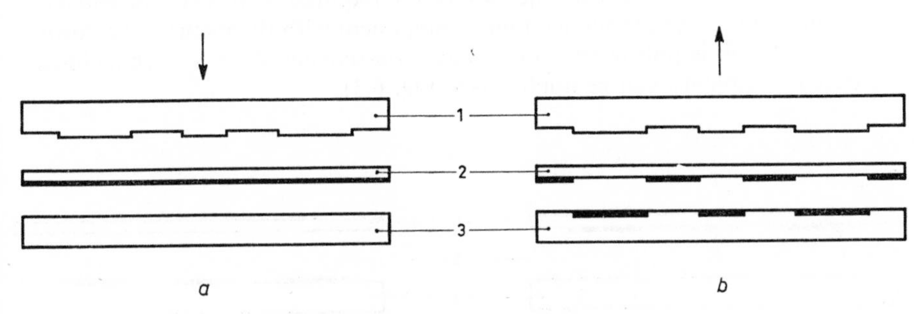

Fig. 6.2. Warm impression of a text.
a. Before pressing. *b*. After pressing.
1 = die. 2 = carrier with layer of pigment. 3 = object.

The job is done with a simple hand-operated press, but there are also automatic pneumatic presses.

Sheets of transparent thermoplastics, such as acrylates, for instance, are printed from the rear, to protect the text from mechanical damage. The text can be given an extra protection by lacquering it afterwards.

(b) *Transfers*

Transfers consist of a paper carrier or foil of synthetic material on which the text or a figure is printed in reverse by a screening process or other printing technique.

The printing is not bonded to the carrier. The image on the transfer is stuck on the object, after which the carrier foil is removed.

There are various types. With one of them, the transfers must be stuck on a layer of paint which is not completely dry. After thorough drying of the paint or lacquer, the carrier of the transfer is removed. With other versions, the transfers are first soaked in water and then stuck on the object. After drying the carrier can be removed. Transfers can also be applied to cylindrical planes. Applications are: for emblems on bicycles, texts on pipelines and so on. The system is also used in the ceramics industry. The pigments are then enamel-burned into the ceramic. The carrier foil is a material that burns without leaving remnants and need not be removed afterwards.

A variation of the transfer method is the *glide-off transfer*. This consists of a carrier, a glue film, a positive image and a cover layer. After soaking in water, the carrier comes off and the image with the cover layer can be slid on to the object. When completely dry, the cover layer can be removed.

Transfers based on all possible principles are obtainable from specialist firms.

(c) *Etched transfers*

The through-etching technique of metal foil mentioned in Section 6.5(b) can also be used for the manufacture of transfers. Etching then takes place in such a way that a mirror-reflection of the required word mark or emblem remains behind. After sticking it on a component with the picture-side down, the carrier foil is pulled off. In this way, non-sequential texts, such as loose letters or emblems, can be applied (see Fig. 6.3).

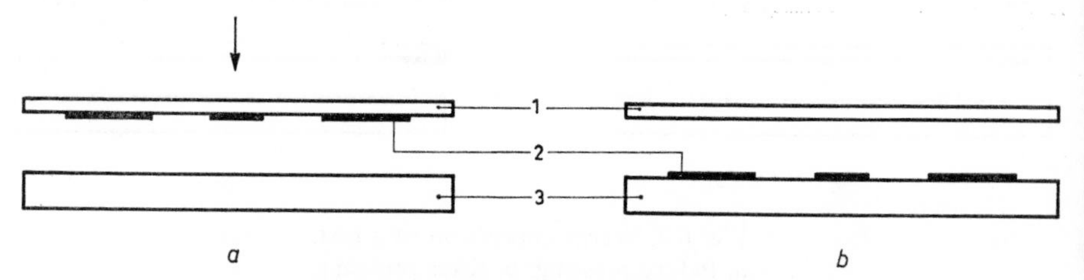

Fig. 6.3. Attaching a through-etched pattern to an object.
a. Situation before application.
b. Situation after applying the text and stripping off the carrier.
1 = carrier. 2 = etched pattern. 3 = object with glue film.

6.8 Silk screening

In silk screening, a squeegee is used to force paint on to an object through a tense gauze that is locally screened off, as shown diagrammatically in Fig. 6.4. See also Volume 4, Chapter 3, Section 3.3.1(b).

Silk screening can be used on practically all surfaces, such as glass, metal, wood, paper, cardboard, lacquer and synthetic materials, and is not limited to the printing of flat objects. Cylindrical (and, in general, rolling surfaces) can be printed with a flat screen. Highly complicated silk screening machines have been developed for such applications. In one method, the squeegee is stationary and the screen moves while the object rotates under the screen. It is possible, too, to roll the object under a stationary screen, the moving squeegee being all-the-time immediately above the place where the object is

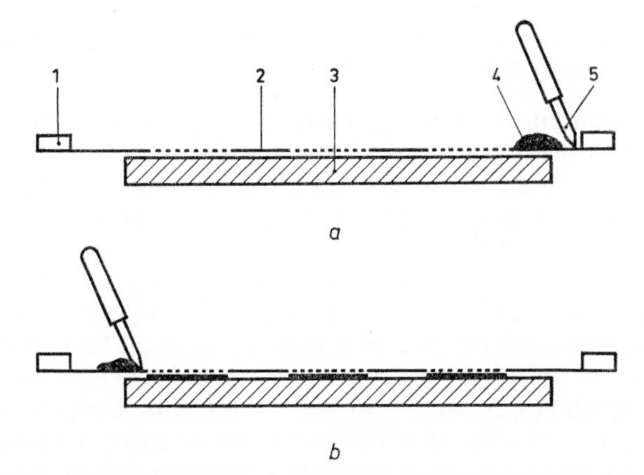

Fig. 6.4. Diagrammatic representation of silk screen printing.
a. Before printing. *b*. After printing.
1 = frame. 2 = screen. 3 = object. 4 = paint. 5 = squeegee.

contacted. These techniques are used for printing bottles. The silk screen paint can be an enamel, afterwards burned into the glass.

Because of the flexibility of the screens, slightly convex surfaces can be printed with a flat screen, with some loss in the sharpness of the image. The number of possibilities is expanded by choosing a somewhat flexible frame structure. A further advantage of silk screening is, that a thick layer of paint can be applied in one printing operation (in any colour), even on surfaces that are not very smooth, like an irregular paint surface.

Another advantage of silk screening, as compared with offset printing, is that the equipment, in its simplest form, is quite cheap. Due to wear of the stencil, the number of prints is limited to about 10 000 items.

Silk screens, provided with the required stencil, as well as silk screening paints, are available on the market.

The picture definition depends on the gauge of the silk and the quality of the stencil. Normal quality screens can be used to print lines varying in thickness from 0·2 to 0·3 mm.

The silk screen method can be seen as a refinement of the open template. Open templates are used for printing packing chests (by brush or spray gun), for the sand-blasting of glass and for manually applying letters and figures. Templates or jigs are obtainable in all sizes.

6.9 Other printing techniques

Among these techniques are relief printing, intaglio printing, surface or litho printing and offset printing. The principle of these graphical techniques will be discussed in brief and a few examples given of the printing of a text on products, components, labels, etc. Details of the special possibilities of such graphical techniques, as well as the reproduction techniques, are beyond the scope of this Handbook.

(a) In relief printing the elevated parts of the print form are coated with ink, after which the ink is transferred to the object to be printed. The print form consists of an etched copper plate (see Volume 4, Chapter 3). Relief printing is employed in book printing, where the print form consists of cast letters.

Apart from the graphical technique, and especially for smaller texts, the term stamping has found its way. Hand stamps, with the printing parts made of rubber or metal, get the ink from a stamping pad. An application is for stamping figures and other symbols on meter scales. There are many types of stamping machines, specially designed for the type of object to be printed. If a flexible stamp is selected, a flat stamp can be used to apply printing to slightly curved objects, at some expense in the sharpness of the print. There are also curved stamps, for use on concave or convex objects.

(b) For intaglio printing, use is made of a copper print form, the recesses of which are filled with printing ink which is then transferred to the object to be printed. Intaglio printing is suitable for materials that can absorb paint or ink (paper). A particular version is indirect intaglio, where the ink is first transferred from the printing form to a flexible gelatine cushion, after which the gelatine transfers the ink to the object to be printed. This method is suitable for very small and very accurate texts, as, for example, watch dials.

(c) With surface or litho printing the printing and non-printing parts of the form are practically in the same plane. The form is a plate of zinc or aluminium, kept moist and inked by a roller. The print form is so prepared that the printing parts repel moisture and accept the ink, while the non-printing parts accept moisture and repel the ink.

Surface printing is nearly always an indirect method, that is, the printing ink is transferred from the cliché to the printed object by rubber cylinders. The current name for this technique is offset printing. We also speak of "wet offset" in contrast to "dry offset", in which the printing parts of the cliché are slightly higher than the rest. Dry offset can be seen as an intermediate form between normal offset and indirect relief printing.

Offset printing is widely employed; a few examples are the printing of meter scales, tubes, bottles of synthetic material, tins and cans, anodized aluminium (text plates), labels, self-sticking synthetic foil (with carrier foil), and so on.

6.10 Anodised aluminium

This material is treated separately, as it is particularly suitable for text plates, etc., and offers many more excellent possibilities (see also Volume 4, Chapter 3).

When aluminium is anodized, which takes place in an acid bath, a porous layer of oxide is formed on the surface. This oxide layer can be given any required colour (as gold or bronze) by means of organic pigments (in a watery solution.) If necessary, the pigments can easily be removed with a simple chemical treatment.

Finally, the layer is sealed by giving it another chemical treatment, that is, the porous layer of oxide is sealed off, to make it rather resistant to corrosion and mechanical damage, whilst at the same time the pigmentation gains in permanence.

To produce texts, etc., a surface must be locally shielded-off with a special coat that resists the agents used. For coating methods, see Volume 4, Chapter 3.

It is then possible to use white anodized aluminium, and colour it locally. Or, use completely pigmented material and remove the pigment in certain places. After removal of the protective coating, sealing takes place.

Multi-colour texts can be produced by repeating the processing of local coating, pigmenting, etc.

The next possibility is to matt the aluminium with a soft etching agent (ammonium bifluoride) to a certain degree. It is then possible to produce a glossy text on a matt background, or a matt text on a bright background. Matting can also be done mechanically, for instance, by giving the material a length texture by brushing. Finally, beautiful effects can be obtained by deep etching, which results in name plates with coloured embossed texts.

Anodizing and pigmenting aluminium are relatively subtle processes that cannot be used for all Al-alloys. Firms specializing in this field have sample books at their disposal, to give a good indication of the various possibilities.

Those wishing to make their own text plates simply can obtain specially pre-treated aluminium. One example is fully-pigmented anodized aluminium sheet, coated with a light-sensitive film. After exposure via a photo negative or positive, the non-exposed areas can be washed away and de-coloured.

Another example is anodized aluminium plate, with the layer of oxide impregnated with a light-sensitive silver halide, and treated in the same way as traditional photographic material. By local exposure, development and fixation, a black text on a metal-coloured background is obtained, or vice versa. These text plates can be fitted mechanically or be glued. For gluing purposes the market offers double-sided sticky foil, with a protective foil on each side.

The application of anodized aluminium is not restricted to making loose name and text plates. The material is also employed for front panels of measuring equipment (showing the necessary texts, scale divisions, etc.) thus forming part of the structure.

REFERENCES

The periodicals mentioned below sometimes contain articles, or give excerpts from other periodicals, on the application of texts, etc.:

Product Finishing
Metal Finishing
Design
Industrial Design
Finishing News—Metals and Plastics
Industrial Finishing
Materials in Design Engineering
Machine Design

Index